**PHYSICS
FOR CAREER
EDUCATION**

SECOND EDITION

# PHYSICS FOR CAREER EDUCATION

**DALE EWEN**

*Parkland Community College*

**RONALD J. NELSON**

**NEILL SCHURTER**

PRENTICE-HALL, INC. *Englewood Cliffs, New Jersey 07632*

*Library of Congress Cataloging in Publication Data*

EWEN, DALE (date)
  Physics for career education.

Previous ed.: Dale Ewen . . . [et al.]. 1974.
  Includes index.
  1. Physics.  I. Nelson, Ronald J.  II. Schurter, Neill.  III. Title.
QC23.E9 1981      530      81–10581
ISBN  0–13–672329–2      AACR2

Printed in the United States of America

10  9  8  7  6  5

Editorial/production supervision
and interior design by Ellen W. Caughey
Drawings by George Morris
Cover design by Jorge Hernandez
Art production by Diane Sturm
Manufacturing buyer: Gordon Osbourne


ISBN  0-13-672329-2


Prentice-Hall International, Inc., *London*
Prentice-Hall of Australia Pty. Limited, *Sydney*
Prentice-Hall of Canada, Ltd., *Toronto*
Prentice-Hall of India Private Limited, *New Delhi*
Prentice-Hall of Japan, Inc., *Tokyo*
Prentice-Hall of Southeast Asia Pte. Ltd., *Singapore*
Whitehall Books Limited, *Wellington, New Zealand*

# Contents

## CHAPTER 3    *Motion*    37

## CHAPTER 4    *Forces in One Dimension*    53

## CHAPTER 5    *Vectors and Trigonometry*    74

## CHAPTER 6    *Concurrent Forces*    95

## CHAPTER 7    *Work and Energy*    104

*Appendix B*
*Scientific Hand Calculators—*
*Brief Instructions on Use    358*

*Appendix C*
*Tables    370*

# Preface

*Physics for Career Education* has been written as a preparation for students considering a vocational-technical career. It is designed to emphasize physical concepts as applied to the industrial-technical fields and to use these applications to improve the physics and mathematics competence of the student.

This text is written at a language level and at a mathematics level that is cognizant of and beneficial to *most* students in vo-tech programs that do not require a high level of mathematics rigor and sophistication. The authors have assumed that the student has successfully completed one year of high-school algebra or its equivalent. Simple equations and formulas are reviewed and any mathematics beyond this level is developed in the text. The manner in which the mathematics is used in the text displays the need for mathematics in technology. For the better prepared student, the mathematics sections may be omitted with no loss in continuity.

Sections are short and each deals with only one concept. The need for the investigation of a physical principle is developed before undertaking its study, and many diagrams are used to aid students in visualizing the concept. A large number of examples are included. The problems at the end of each section allow students to check their mastery of one concept before moving on to another.

This text is designed to be used in a vocational-technical program in a community college, a technical institute, or a high-school class for students who plan to pursue a vocational-technical career. The topics were originally chosen with the assistance of technicians and management of several industries and teaching consultants in various vo-tech areas. Suggestions from users of the first edition were used extensively in this second edition.

One of the most admired features of the first edition was the successful problem-solving method—both its development and its consistent use throughout the text. For each set of problems, where it can be appropriately used, students are reminded to use the method by the appearance of the figure at the left. This method is easily remembered and provides a valuable skill that can be used and applied daily in other technical and science courses, in industry, and in many other areas.

The text is divided into five major areas: mechanics, matter and heat,

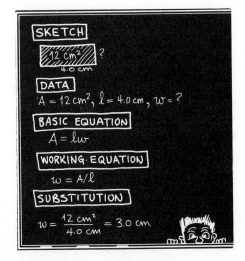

wave motion and sound, electricity and magnetism, and light. The emphasis on mechanics is a result of our belief that it is basic to all technical programs.

The chapter on measurement introduces students to basic units and some mathematics skills. Measurements are presented as approximate numbers and are used consistently throughout the text. Chapter 2 introduces students to a problem-solving method that is used in the rest of the text. The need for vectors is developed in the chapter on motion, where graphical addition of vectors is used. One-dimensional dynamics is then discussed and the need for trigonometry in more complex problems is developed. The chapter on trigonometry includes right-triangle trigonometry and the component method for addition of vectors. A more thorough treatment of dynamics and other standard topics follows. Chapters on simple machines and gear systems are also included.

The section on matter includes a discussion of the three states of matter, density, fluids, and pressure, and Pascal's principle. The section on heat includes material on temperature, specific heat, thermal expansion, gas laws, and change of state.

The section on wave motion and sound deals with basic wave characteristics, interference and diffraction, the nature and speed of sound, the Doppler effect, and resonance.

The section on electricity and magnetism begins with a brief discussion of static electricity followed by an extensive treatment of dc circuits, including sources, Ohm's law, and series and parallel circuits. The chapters on magnetism, generators, and motors are largely descriptive in nature, but allow for a more in-depth study if desired. Ac circuits and transformers are treated extensively.

The chapter on light briefly discusses the wave and particle nature of light, but deals primarily with illumination. The chapters on reflection and refraction develop the images formed by mirrors and lenses.

A laboratory manual is available that is designed to be used with this text. The experiments have been classroom-tested in community colleges and high schools.

The authors thank the Research and Development Unit, Vocational and Technical Education Division, Illinois Board of Vocational Education and Rehabilitation, and Parkland College for their support in the initial version of this text prior to the first edition and to the many pilot Illinois high schools and community colleges for using and testing the initial materials. We also recognize the special efforts of Clifton H. Matz, Gayle W. Wright, Sidney E. Barnes, William O. Smith, Walter Miller, Paul N. Thompson, and John Costello, who served as consultants on the initial version.

We give special recognition and thanks and extend our best wishes to our friend, colleague, and coauthor of the first edition, LaVerne McFadden, for his loyal support and work.

The authors especially thank the many teachers who have used the first edition and who, in offering suggestions, gave freely of their time. If anyone wishes to correspond with us regarding suggestions, criticisms, questions, or errors, please feel free to contact Dale Ewen directly at Parkland College, 2400 W. Bradley, Champaign, Illinois 61820, or through Prentice-Hall.

Finally, we are especially grateful to our families for their patience, understanding, and encouragement, and to Joyce Ewen for her excellent proofing assistance.

DALE EWEN
RONALD J. NELSON
NEILL SCHURTER

# CHAPTER *1*

# *Measurement and the Metric System*

**1-1 STANDARDS OF MEASURE** It should be obvious that when two people work together on the same job that they both use the same standards of measure. If not, the result can be disastrous.

 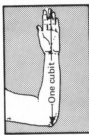  

Throughout history, there have been many standards by which measurements have been made:

1. *Chain:* a measuring instrument of 100 links used in surveying. One chain has a length of 66 feet.
2. *Rod:* a length determined by having 16 men put one foot behind the foot of another man in a straight line. The rod is now standardized as 16½ feet.

|←————————————— 1 rod —————————————→|
Distance divided by 16 equals one foot.

1

Yard

3 barleycorns

1 inch

3. *Yard:* the distance from the tip of the king's nose to the fingertips of his outstretched hand.

4. *Foot:* the distance in item 2 divided by 16 was a legal foot. It was also common to use the length of one's own foot as the unit foot.

5. *Inch:* the length of three barley corns, round and dry, taken from the center of the ear, and laid end to end.

Even today, the U.S. gallon and the British Imperial gallon are not the same. The U.S. units of measure are a hodgepodge of makeshift units of Anglo-Saxon, Roman, and French-Norman weights and measures.

After the standards based on parts of the human body and on other gimmicks, basic standards were agreed to by world governments. They also agreed to construct and distribute accurate standard copies of all the standard units.

During the 1790s, a decimal system based on our number system was being developed in France. Its acceptance was gained mostly because it was easy to use and easy to remember. Many nations began adopting it as their official system of measurement. By 1900, most of Europe and South America was metric. In 1866, metric measurements for official use were legalized in the United States. In 1893, the Secretary of the Treasury, by administrative order, declared the new metric standards to be the nation's "fundamental standards" of mass and length. Thus, indirectly, the United States *officially* became a metric nation. Even today, the English units are officially defined in terms of the standard metric units.

Throughout U.S. history, several attempts were made to convert the nation to the metric system. In 1951, Japan announced it was going metric. In 1965, Great Britain decided to go metric. In 1968, an exhaustive three-year metric study began. The study made three fundamental recommendations:

1. "Increased use of the metric system is in the best interest of the United States.

2. The nation should change to the metric system through a coordinated national program.

3. The transition period should be ten years, at the end of which the nation would be *predominantly* metric."

In the 1970s, the United States found itself to be the only nonmetric industrialized country left in the world. But the government actually did little to implement the system. Industry and big business, however, found that their foreign markets were drying up because they wanted metric products. Then in 1975, President Gerald R. Ford signed the "Metric Conversion Act of 1975," which "declared that the policy of the United States shall be to coordinate and plan the increasing use of the metric system in the United States and to establish a United States Metric Board to coordinate the voluntary conversion to the metric system."

So sometime in your lifetime, you should see the benefits gained from having the whole world accepting and using the same standards of measure.

**1-2
THE MODERN
METRIC SYSTEM**

The modern metric system is identified in all languages by the abbreviation "SI" (for Système International d'Unités—the international system of units of measurement written in French). The SI metric system has seven basic units:

| Basic unit | SI abbreviation | Used for measuring |
|---|---|---|
| metre* | m | length |
| kilogram | kg | mass |
| second | s | time |
| ampere | A | electric current |
| kelvin | K | temperature |
| candela | cd | light intensity |
| mole | mol | molecular substance |

All other SI units are called derived units; that is, they can be defined in terms of these seven basic units. (See page 4.) For example, the newton (N) is defined as 1 kg m/s$^2$ (kilogram metre per second per second). Other commonly used derived SI units are:

| Derived unit | SI abbreviation | Used for measuring |
|---|---|---|
| litre* | L | volume |
| cubic metre | m$^3$ | volume |
| square metre | m$^2$ | area |
| newton | N | force |
| metre per second | m/s | speed |
| joule | J | energy |
| watt | W | power |

**PREFIXES FOR SI UNITS**

| Multiple or submultiple[a] decimal form | Power of 10 | Prefix | Prefix symbol | Pronun- ciation | Meaning |
|---|---|---|---|---|---|
| 1,000,000,000,000 | 10$^{12}$ | tera | T | tĕr'ă | one trillion times |
| 1,000,000,000 | 10$^9$ | giga | G | jĭg'ă | one billion times |
| 1,000,000 | 10$^6$ | mega | M | mĕg'ă | one million times |
| 1,000 | 10$^3$ | kilo[b] | k | kĭl'ō | one thousand times |
| 100 | 10$^2$ | hecto | h | hĕk'tō | one hundred times |
| 10 | 10$^1$ | deka | da | dĕk'ă | ten times |
| 0.1 | 10$^{-1}$ | deci | d | dĕs'ĭ | one tenth of |
| 0.01 | 10$^{-2}$ | centi[b] | c | sĕnt'ĭ | one hundredth of |
| 0.001 | 10$^{-3}$ | milli[b] | m | mĭl'ĭ | one thousandth of |
| 0.000001 | 10$^{-6}$ | micro | μ | mī'krō | one millionth of |
| 0.000000001 | 10$^{-9}$ | nano | n | năn'ō | one billionth of |
| 0.000000000001 | 10$^{-12}$ | pico | p | pē'kō | one trillionth of |

[a] Factor by which the unit is multiplied.
[b] Most commonly used prefixes.

* At present, there is some difference of opinion in the United States on the spelling of metre and litre. We have chosen the "re" spellings for two reasons. First, this is the internationally accepted spelling for all English-speaking countries. Second, the word "meter" already has many different meanings—parking meter, electric meter, odometer, and so on. Many feel that the metric unit of length should be distinctive and readily recognizable—thus the spelling "metre" and "litre."

*The chart below shows graphically how the 17 SI derived units with special names are derived in a coherent manner from the base and supplementary units. It was provided by the National Bureau of Standards.*

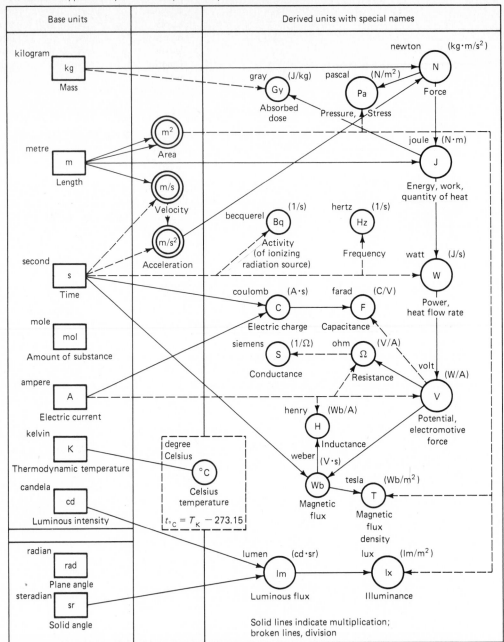

Because the metric system is a decimal or base 10 system, it is very similar to our decimal number system and our decimal money system. It is an easy system to use because calculations are based on the number 10 and its multiples. Special prefixes are used to name these multiples and submultiples, which may be used with most all SI units. Because the same prefixes are used repeatedly, memorization of many conversions has been significantly reduced. The table on page 3 shows these prefixes and the corresponding symbols.

**1-3
SCIENTIFIC NOTATION**

Scientists and technicians often use very large or very small numbers which cannot be conveniently written as fractions or decimal fractions. For example, the thickness of an oil film on water is about 0.0000001 m. A more useful method of expressing such very small (or very large) numbers is known as *scientific notation* or *exponential notation*. Expressed this way, the thickness of the film is $1 \times 10^{-7}$ or $10^{-7}$. For example:

$$0.1 = 1 \times 10^{-1} \text{ or } 10^{-1}$$

$$10000 = 1 \times 10^{4} \text{ or } 10^{4}$$

$$0.001 = 1 \times 10^{-3} \text{ or } 10^{-3}$$

---

*To write a number in scientific notation, write it as a product of a number between 1 and 10 and a power of 10. General form:* $M \times 10^{n}$, *where*

$M = $ a number between 1 and 10

$n = $ the exponent or power of the base number 10

---

## EXAMPLE 1

Write 325 in scientific notation.

$325 = 3.25 \times 10^{2}$. (Remember that $10^{2} = 10 \times 10 = 100$. Also, $3.25 \times 100 = 325$. This is the same as moving the decimal point two places.)

## EXAMPLE 2

Write 65,800 in scientific notation.

$$65,800 = 6.58 \times 10,000 = 6.58 \times (10 \times 10 \times 10 \times 10) = 6.58 \times 10^{4}$$

The following procedure should help you to write any decimal number in scientific notation.

1. Place a decimal point after the first nonzero digit reading from left to right.
2. Place a caret ($_\wedge$) at the position of the original decimal point.
3. If the decimal point is to the left of the caret, the exponent of 10 is the number of places from the caret to the decimal point.
   Example:     $83,662 = 8.3662_{\wedge} \times 10^{\underline{4}}$
4. If the decimal point is to the right of the caret, the exponent of 10 is the negative of the number of places from the caret to the decimal point.
   Example:     $0.00683 = {}_{\wedge}006.83 \times 10^{\underline{-3}}$
5. If the decimal point and the caret coincide, the exponent of 10 is zero.
   Example:     $5.12 = 5_{\wedge}12 \times 10^{0}$

## EXAMPLE 3

Write 0.0000002486 in scientific notation.

$$0.0000002486 = {}_{\wedge}0000002.486 \times 10^{\underline{-7}} = 2.486 \times 10^{-7}$$

## PROBLEMS

*Write each number in scientific notation.*

| | | |
|---|---|---|
| **1.** 326 | **2.** 798 | **3.** 2650 |
| **4.** 14,500 | **5.** 826.4 | **6.** 24.97 |
| **7.** 0.00413 | **8.** 0.00053 | **9.** 6.43 |
| **10.** 482,300 | **11.** 0.000065 | **12.** 0.00224 |
| **13.** 540,000 | **14.** 1,400,000 | **15.** 0.0000075 |
| **16.** 0.0000009 | **17.** 0.00000005 | **18.** 3,500,000,000 |
| **19.** 732,000,000,000,000,000 | | **20.** 0.00000000000000000618 |

Converting from scientific notation to decimal form requires the ability to multiply by powers of 10. To multiply by a positive power of 10, move the decimal point to the right the same number of places as is indicated by the exponent of 10. Supply zeros as needed. To multiply by a negative power of 10, move the decimal point to the left the same number of places as the exponent of 10 indicates, and supply zeros as needed.

EXAMPLE 4

Write $7.62 \times 10^2$ in decimal form.

$7.62 \times 10^2 = 762$     (Move the decimal point two places to the right.)

EXAMPLE 5

Write $3.15 \times 10^{-4}$ in decimal form.

Move the decimal point four places to the left and insert three zeros:

$$3.15 \times 10^{-4} = 0.000315$$

Since all calculators used in science and technology accept numbers entered in scientific notation and give some results in scientific notation, it is essential that you fully understand this topic before going to the next section. It may be necessary now to consult Appendix A: Section A-1 on signed numbers and Section A-2 on the powers of 10.

## PROBLEMS

*Write each number in decimal form.*

1. $8.62 \times 10^4$
2. $8.67 \times 10^2$
3. $6.31 \times 10^{-4}$
4. $5.41 \times 10^3$
5. $7.68 \times 10^{-1}$
6. $9.94 \times 10^1$
7. $7.77 \times 10^8$
8. $4.19 \times 10^{-6}$
9. $6.93 \times 10^1$
10. $3.78 \times 10^{-2}$
11. $9.61 \times 10^4$
12. $7.33 \times 10^3$
13. $1.4 \times 10^0$
14. $9.6 \times 10^{-5}$
15. $8.4 \times 10^{-6}$
16. $9 \times 10^8$
17. $7 \times 10^{11}$
18. $4.05 \times 10^0$
19. $7.2 \times 10^{-7}$
20. $8 \times 10^{-9}$
21. $4.5 \times 10^{12}$
22. $1.5 \times 10^{11}$
23. $5.5 \times 10^{-11}$
24. $8.72 \times 10^{-10}$

**1-4**
## LENGTH MEASUREMENT

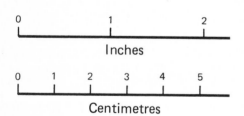

Length is one of the most basic measurements. Historically, human beings used their hands or feet or whatever was handy to make measurements. As civilization progressed, however, the need for everyone to use the same standards increased. Two systems are in use today. The English system has been historically used in the English-speaking countries and its basic units are the foot, pound, and second. The foot is the basic unit of length in the English system and it may be divided into 12 equal parts or *inches*.

The metric system is used in most all other countries and is used by scientists all over the world. Its basic unit of length is the *metre*.

1 foot (ft) = 12 inches (in.)

1 metre (m) = 100 centimetres (cm)

1 in. = 2.54 cm

1 m = 39.37 in. = 3.28 ft

1 yard (yd) = 3 ft

Turn to Table 3 in Appendix C for a listing of the English and metric unit conversions. Lines or distances may be measured in either system directly or indirectly.

To convert from one unit to another we use the fact that, when any quantity is multiplied by 1, its value is not changed. The correct conversion factor is a fraction whose numerator is equal to the denominator; that is, the fraction equals 1. The numerator should be expressed in the new units and the denominator should be expressed in the old units so that the old units divide (cancel out).

## EXAMPLE 1

Express 10 inches in centimetres.

$$1 \text{ in.} = 2.54 \text{ cm} \quad \text{so} \quad 10 \text{ in.} \times \frac{2.54 \text{ cm}}{1 \text{ in.}} = 25.4 \text{ cm}$$

It is very important that you use the correct units on all quantities and divide (cancel) units where possible. The answer will then appear in the desired units. This practice will aid you greatly in solving problems in later chapters.

The conversion factors you will need are to be found in Appendix C. The examples below will show you how to use these tables.

## EXAMPLE 2

Convert 15 miles to kilometres.

From Table 3 we find 1 mile listed in the left-hand column. Moving over to the fourth column, under the heading "km," we see that 1 mile (mi) = 1.61 km. Then we have

$$15 \text{ mi} \times \frac{1.61 \text{ km}}{1 \text{ mi}} = 24.15 \text{ km}$$

## EXAMPLE 3

Convert 220 centimetres to inches.

Find 1 centimetre in the left-hand column and move to the fifth column under the heading "in." We find that 1 centimetre = 0.394 in. Then

$$220 \text{ cm} \times \frac{0.394 \text{ in.}}{1 \text{ cm}} = 86.68 \text{ in.}$$

## EXAMPLE 4

Convert 3 yards to centimetres.

Since there is no direct conversion from yards to centimetres in the tables, we must first change yards to inches and then inches to centimetres:

$$3 \text{ yd} \times \frac{36 \text{ in.}}{1 \text{ yd}} \times \frac{2.54 \text{ cm}}{1 \text{ in.}} = 274.32 \text{ cm}$$

## PROBLEMS

*Use the technique of multiplying by 1 to do the following problems.*

1. The wheelbase of a certain automobile is 108 in. long. What is its length in yards? in feet? in centimetres?
2. The length of a connecting rod is 7 in. What is its length in centimetres?
3. The distance between two cities is 256 mi. What is the distance in kilometres?
4. Convert each length to metres.
   (a) 5 ft
   (b) 2.7 mi
   (c) 10.6 ft
   (d) 7 yd

5. Convert each length as indicated.
    (a) 5.94 m to feet
    (b) 7.1 cm to inches
    (c) 1.2 in. to centimetres

6. The turning radius of an auto is 20 ft. What is this in metres?

7. Would a wrench with an opening of 25 mm be larger or smaller than a 1-in. wrench?

8. What is the width in inches of 8-mm movie film?

9. How many reamers each 20 cm long can be cut from a bar 6 ft long, allowing 3 mm for each saw cut?

10. If 214 pieces each 47 cm long are ordered to be turned from $\frac{1}{4}$-in. round steel stock and $\frac{1}{8}$ in. waste is allowed on each piece, what length (in metres) of stock is required?

## 1-5
## TIME MEASUREMENT

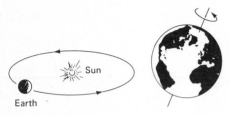

Revolution of Earth    Rotation about Axis

Oscillation of Balance Wheel

Airlines and other transportation systems run on time schedules that would be meaningless if we did not have a common unit for time measurement. All the common units for time measurement are the same in both the English and metric systems. These units are based on the motion of the earth and the moon. The year is approximately the time required for one complete revolution of the earth about the sun. The month is approximately the time for one complete revolution of the moon about the earth. The day is the time for one rotation of the earth about its axis.

The basic time unit is the *second* (s), which is defined to be 1/86,400 day. The second is not always convenient to use, so other units are necessary. The *minute* (min) is 60 seconds, the *hour* (h) is 60 minutes, and the *day* is 24 hours. The *year* is 365 days in length except for every fourth year, when it is 366 days long. This difference is necessary to keep the seasons at the same time each year, since one revolution of the earth about the sun takes $365\frac{1}{4}$ days. Other small variations are compensated for each year by adding or subtracting seconds from the last day of the year. This is done by international agreement.

Common devices for time measurement are the electric clock, the mechanical watch, and the quartz crystal watch. The accuracy of an electric clock depends on how accurately the 60-Hz (hertz = cycles per second) line voltage is controlled. In the United States this is controlled very accurately. Most mechanical watches have a balance wheel that oscillates near a given frequency, usually 18,000 to 36,000 vibrations per hour, and drives the hands of the watch. The quartz crystal in a watch is excited by a small power cell and vibrates 32,768 times per second. The accuracy of the watch depends on how well the frequency of oscillation is controlled.

## PROBLEMS

*In Problems 1 through 16, use conversion factors to convert each quantity.*

1. 12 min to seconds

2. 7800 s to minutes

3. 8 h to seconds

4. 2 days to minutes

5. 3 years to hours (use 1 yr = 365 days).

6. 1 day to seconds

7. $7.2 \times 10^{12}$ s to hours

8. 150 ms to seconds

9. $2.4 \times 10^{-8}$ s to microseconds

10. 250 $\mu$s to seconds

**11.** 75 $\mu$s to milliseconds

**12.** 4.5 ns to picoseconds

**13.** 18 ms to microseconds

**14.** $4 \times 10^5$ ps to microseconds

**15.** 15 $\mu$s to nanoseconds

**16.** 15 ms to nanoseconds

**17.** How many minutes does it take to turn out 1500 mainsprings if it takes 15 s to turn out one mainspring?

**18.** An arbor requires $1\frac{1}{2}$ h to machine. How many arbors can a department of 4 machinists turn out in a 5-day week if the workday is 8 h?

**1-6**

**MASS AND WEIGHT MEASUREMENT**

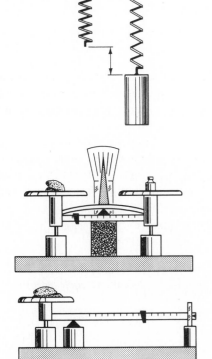

The *mass* of an object is a measure of the amount of matter in that object. The kilogram (kg), a unit of mass, is one of the seven basic SI metric units. It is defined as the amount of water contained in a litre (L). The litre of water fills a cube 10 cm on an edge.

The *weight* of an object is a measure of the gravitational force or pull acting on an object. The pound (lb), a unit of force, is one of the basic English system units. It is defined as the pull of the earth on a cylinder of a platinum–iridium alloy which is stored in a vault at the U.S. Bureau of Standards.

The *ounce* (oz) is another common unit of weight in the English system. The relationship between pounds and ounces is

$$1 \text{ lb} = 16 \text{ oz}$$

The corresponding weight unit in the metric system is the *newton* (N). The following relationships can be used for conversion between systems of units:

$$1 \text{ N} = 0.225 \text{ lb} \quad \text{or} \quad 1 \text{ lb} = 4.45 \text{ N}$$

Mass and weight and their units of measure are discussed later in more detail.

A common method of measuring weights is based on the *spring balance*. The basis of this method is the fact that the distance a spring stretches when a body is supported by it is proportional to the weight of the body. A pointer can be attached to the spring and a calibrated scale added so that the device will read directly in pounds or newtons. The common bathroom scale uses this principle to measure weights.

The other common device for measuring mass is the *equal arm balance*. Two platforms are connected by a horizontal rod which balances on a knife edge. This device compares the pull of gravity on objects which are on the two platforms. The platforms are at the same height only when the unknown mass of the object on the left is equal to the known mass placed on the right. It is also possible to use one platform and a mass that slides along a calibrated scale. Variations of this basic design are used in scales such as those found in some meat markets and truck scales.

EXAMPLE

The weight of the intake valve of an auto engine is 0.157 lb. What is the weight in ounces and in newtons?

To find the weight in ounces, we simply use the correct conversion factor as follows:

$$0.157 \text{ lb} \times \frac{16 \text{ oz}}{1 \text{ lb}} = 2.512 \text{ oz}$$

To find weight in newtons, we again use the correct conversion factor:

$$0.157 \, \cancel{lb} \times \frac{4.45 \, N}{1 \, \cancel{lb}} = 0.69865 \, N$$

## PROBLEMS

1. If the weight of a car is 3500 lb, what is its weight in newtons?
2. A certain bridge is designed to support 150,000 lb. What is the maximum supportable weight in newtons?
3. A man's weight is 200 lb. What is his weight in newtons?
4. Convert 80 lb to newtons.
5. Convert 2000 N to pounds.
6. Convert 2000 lb to newtons.
7. Convert 120 oz to pounds.
8. Convert 3.5 lb to ounces.
9. Convert 10 N to ounces.
10. Convert 8 oz to newtons.
11. Find the metric weight of a 94-lb bag of cement.
12. What is the weight in newtons of 500 blocks if each block weighs 3 lb?

**1-7**
**ELECTRICAL UNITS**

Later in this book we will study electricity and magnetism. *Electricity* is the flow of energy by charge transported through wires. The importance of electricity to our industrialized society cannot be underestimated. This is evidenced by the dependence of every household and every industry on it for the energy necessary to run appliances and large machinery.

For this study of electricity we will need to define several units. The *ampere* (A) is a basic unit and is a measure of the amount of electric current. The *coulomb* (C) is a measure of the amount of electrical charge. The *volt* (V) is a measure of electrical energy. The *watt* (W) is a measure of power. The *kilowatt-hour* (kWh) is a measure of work or electrical energy used.

**1-8**
**AREA**

To measure a surface area of an object, you must first decide upon a standard unit of area. Standard units of area are based on the square and are called square inches, square centimetres, square miles, or some square unit of measure. An area of 1 square centimetre (cm²) is the amount of area found within a square 1 cm on each side. An area of 1 square inch (in²) is the amount of area found within a square of 1 in. on each side.

The area of a figure is the number of square units that are contained in the figure. In general, when multiplying measurements of like units, multiply the numbers and then multiply the units as follows:

$$3 \text{ in.} \times 5 \text{ in.} = (3 \times 5)(\text{in.} \times \text{in.}) = 15 \text{ in}^2$$

$$2 \text{ cm} \times 4 \text{ cm} = (2 \times 4)(\text{cm} \times \text{cm}) = 8 \text{ cm}^2$$

$$1.4 \text{ m} \times 6.7 \text{ m} = (1.4 \times 6.7)(\text{m} \times \text{m}) = 9.38 \text{ m}^2$$

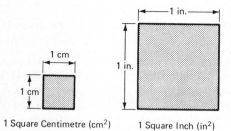

1 Square Centimetre (cm²)    1 Square Inch (in²)

## EXAMPLE 1

Find the area of a rectangle that is 6 in. long and 4 in. wide.

Each square is 1 in². To find the area of the rectangle, simply count the number of squares in the rectangle. Therefore, you find that the area = 24 in²; or, by using the formula $A = lw$,

$$A = lw$$
$$= (6 \text{ in.})(4 \text{ in.})$$
$$= 24 \text{ in}^2$$

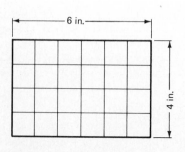

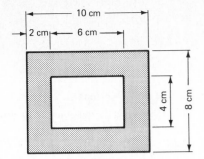

## EXAMPLE 2

What is the area of the metal plate shown?

To find the area of the metal plate, find the area of each of the two rectangles and then find the difference of their areas. The large rectangle is 10 cm long and 8 cm wide. The small rectangle is 6 cm long and 4 cm wide. The two areas are 80 cm² and 24 cm². Therefore, the area of the metal plate is 80 cm² − 24 cm² = 56 cm².

The surface that would be seen by cutting a geometric solid with a thin plate represents the cross-sectional area of a solid.

## EXAMPLE 3

Find a cross-sectional area of the following box.

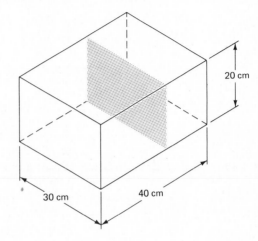

The indicated cross section of this box is a rectangle 30 cm long and 20 cm wide.

$$A = lw$$
$$= (30 \text{ cm})(20 \text{ cm})$$
$$= 600 \text{ cm}^2$$

The area of this rectangle is 600 cm², which represents the cross-sectional area of the box.

The formulas for finding the areas of other plane figures can be found in Table 18 of Appendix C.

To convert area or square units, use the same rule for finding a conversion factor as in Section 1-4. That is, the correct conversion factor will be in fractional form and equal to 1, with the numerator expressed in the units you wish to convert to and the denominator expressed in the units given. The conversion table for area is found in Table 4 of Appendix C.

## EXAMPLE 4

Convert 324 in² to yd².

$$324 \text{ in}^2 \times \frac{1 \text{ yd}^2}{1296 \text{ in}^2} = \frac{324}{1296} \text{yd}^2 = 0.25 \text{ yd}^2$$

## EXAMPLE 5

Convert 258 cm² to m².

$$258 \text{ cm}^2 \times \frac{1 \text{ m}^2}{10,000 \text{ cm}^2} = 0.0258 \text{ m}^2$$

Convert 28.5 m² to in².

$$28.5 \text{ m}^2 \times \frac{1550 \text{ in}^2}{1 \text{ m}^2} = 44,175 \text{ in}^2$$

## PROBLEMS

*Find the area of each figure.*

1.

5 cm

8 cm

2.

15 in.

28 in.

3.

8 in.

3 in.

5 in.

3 in.

4.

6 cm        3 cm

12 cm       7 cm

15 cm

5. Find the cross-sectional area of the I-beam shown.

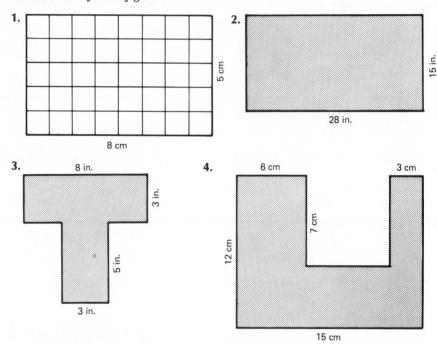

1 in.

1 in.

6 in.

1 in.

8 in.

6. Find the largest cross-sectional area of the following figure.

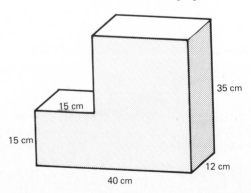

35 cm

15 cm

15 cm

12 cm

40 cm

7. Convert 5 yd² to ft².
8. Convert 15 cm² to mm².
9. How many m² are in 225 ft²?
10. Convert 15 ft² to cm².
11. How many ft² are there in a rectangle 15 m long and 12 m wide?
12. Convert 72 cm² to m².
13. Convert 99 in² to ft².
14. How many cm² are in 37,000 mm²?
15. How many in² are in 51 cm²?
16. How many in² are there in a square 11 yd on a side?
17. How many cm² are there in the cross section of a piece of metal stock that has an area of $1\frac{1}{2}$ in²?
18. How many m² are there in a door stoop whose area is 15 ft²?
19. How many cm² are there in a face plate whose area is 4 in²?
20. What is the area in cm² of a cross section of a rod whose area is 993 mm²?

**1-9**

**VOLUME**

Standard units of volume are based on the cube and are called cubic inches, cubic centimetres, cubic yards, or some other cubic unit of measure. A volume of 1 cubic centimetre (cm³) is the same as the amount of volume contained in a cube 1 cm on each side. One cubic inch (in³) is the volume contained in a cube 1 in. on each side.

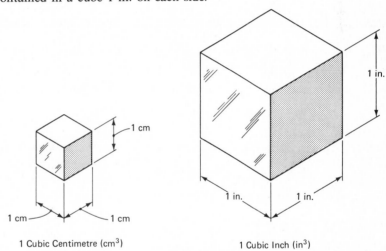

1 Cubic Centimetre (cm³)          1 Cubic Inch (in³)

The volume of a figure is the number of cubic units that are contained in the figure. *Note*: When multiplying measurements of like units, multiply the numbers and then multiply the units as follows:

$$3 \text{ in.} \times 5 \text{ in.} \times 4 \text{ in.} = (3 \times 5 \times 4)(\text{in.} \times \text{in.} \times \text{in.}) = 60 \text{ in}^3$$

$$2 \text{ cm} \times 4 \text{ cm} \times 1 \text{ cm} = (2 \times 4 \times 1)(\text{cm} \times \text{cm} \times \text{cm}) = 8 \text{ cm}^3$$

$$1.5 \text{ ft} \times 8.7 \text{ ft} \times 6 \text{ ft} = (1.5 \times 8.7 \times 6)(\text{ft} \times \text{ft} \times \text{ft}) = 78.3 \text{ ft}^3$$

### EXAMPLE 1

Find the volume of a rectangular prism that is 6 in. long, 4 in. wide, and 5 in. high.

Each cube is 1 in³. To find the volume of the rectangular solid, count the number of cubes in the bottom layer of the rectangular solid and then multiply that number by the number of layers that the solid can hold. Therefore, there are 5 layers of 24 cubes, which is 120 cubes or 120 cubic inches.

Or, by formula, $V = Bh$, where $B$ is the area of the base and $h$ is the height. However, the area of the base is found by $lw$, where $l$ is the length and $w$ is the

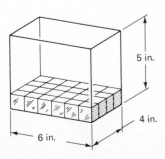

width of the rectangle. Therefore, the volume of a rectangular solid can be found by the formula

$$V = lwh$$
$$= (6 \text{ in.})(4 \text{ in.})(5 \text{ in.})$$
$$= 120 \text{ in}^3$$

## EXAMPLE 2

Find the volume of the figure.

$$V = lwh$$
$$= (8 \text{ cm})(4 \text{ cm})(5 \text{ cm})$$
$$= 160 \text{ cm}^3$$

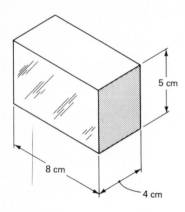

The relationship between the litre and the cubic centimetre deserves special mention. The litre is defined as the volume in 1 cubic decimetre ($dm^3$). That is, 1 litre of liquid fills a cube 1 dm (10 cm) on an edge.

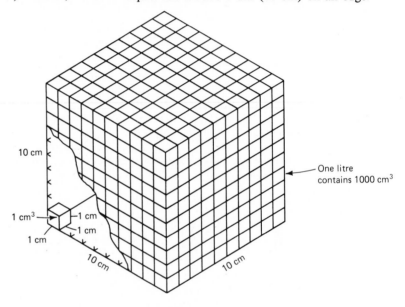

One litre (L)

The volume of this cube can be found by using the formula

$$V = lwh$$
$$= (10 \text{ cm})(10 \text{ cm})(10 \text{ cm})$$
$$= 1000 \text{ cm}^3$$

That is,

$$1 \text{ L} = 1000 \text{ cm}^3$$

Then

$$\frac{1}{1000} \text{ L} = 1 \text{ cm}^3$$

But

$$\frac{1}{1000} \text{ L} = 1 \text{ mL}$$

Therefore,

$$1 \text{ mL} = 1 \text{ cm}^3$$

The lateral (side) surface area of any geometric solid is the area of all the lateral faces. The total surface area of any geometric solid is the lateral surface area plus the area of the bases.

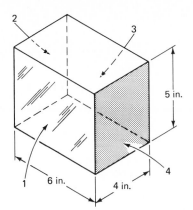

### EXAMPLE 3

Find the lateral surface area of the prism.

area of lateral face 1 = (6 in.)(5 in.) = 30 in²
area of lateral face 2 = (5 in.)(4 in.) = 20 in²
area of lateral face 3 = (6 in.)(5 in.) = 30 in²
area of lateral face 4 = (5 in.)(4 in.) = 20 in²
lateral surface area = 100 in²

### EXAMPLE 4

Find the total surface area of the prism in Example 3.

total surface area = lateral surface area + area of the bases
area of base = (6 in.)(4 in.) = 24 in²
area of both bases = 2(24 in²) = 48 in²
total surface area = 100 in² + 48 in² = 148 in²

Area formulas, volume formulas, and lateral surface area formulas can be found in Table 18 of Appendix C.

To convert volume units, use the same rule for finding a conversion factor. The conversion table for volume is Table 5 of Appendix C.

### EXAMPLE 5

Convert 24 ft³ to in³.

$$24 \text{ ft}^3 \times \frac{1728 \text{ in}^3}{1 \text{ ft}^3} = 41{,}472 \text{ in}^3$$

### EXAMPLE 6

Convert 56 in³ to cm³.

$$56 \text{ in}^3 \times \frac{16.4 \text{ cm}^3}{1 \text{ in}^3} = 918.4 \text{ cm}^3$$

### EXAMPLE 7

Convert 28 m³ to ft³.

$$28 \text{ m}^3 \times \frac{35.3 \text{ ft}^3}{1 \text{ m}^3} = 988.4 \text{ ft}^3$$

## PROBLEMS

*Find the volume in each figure.*

**1.**

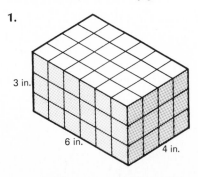

**2.**

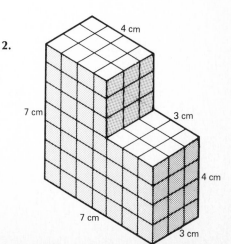

**3.**

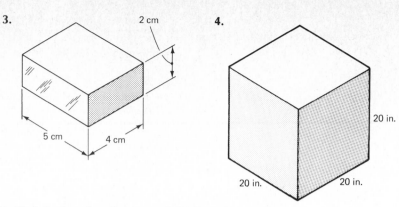

**4.**

5. Find the lateral surface area of the figure in Problem 3.
6. Find the lateral surface area of the figure in Problem 4.
7. Find the total surface area of the figure in Problem 3.
8. Find the total surface area of the figure in Problem 4.
9. How many mL of water would the figure in Problem 3 hold?
10. How many mL of water would the figure in Problem 2 hold?
11. Convert 19 yd³ to ft³.
12. Convert 5440 cm³ to m³.
13. How many in³ are there in 29 cm³?
14. How many yd³ are there in 23 m³?
15. How many cm³ are in 88 in³?
16. How many cm³ are there in 27 m³?
17. Convert 8 ft³ to in³.
18. Convert 79 ft³ to m³.
19. Convert 9 ft³ to cm³.
20. How many in³ are in 12 m³?
21. The volume of a casting is 38 in³. What is its volume in cm³?
22. How many castings of volume 14 cm³ could be made from a block of steel of volume 12 ft³?

**1-10
MEASUREMENT:
SIGNIFICANT DIGITS
AND ACCURACY**

Up to this time in your studies, probably all numbers and all measurements have been treated as exact numbers. An *exact number* is a number that has been determined as a result of counting, such as 24 students are enrolled in this class, or by some definition, such as 1 h = 60 min or 1 in. = 2.54 cm, a conversion definition agreed to by the world governments' bureaus of standards. Generally, the treatment of the addition, subtraction, multiplication, and division of exact numbers is the emphasis or main content of elementary mathematics.

However, nearly all data of a technical nature involve *approximate numbers;* that is, they have been determined as a result of some measurement process—some direct, as with a ruler, and some indirect, as with a surveying transit. Before studying how to perform the calculations with approximate numbers (measurements), we first must determine the "correctness" of an approximate number. First, we realize that no measurement can be found exactly. The length of the cover of this book can be found using many instruments. The better the measuring device used, the better the measurement.

A measurement may be expressed in terms of its accuracy or its precision. The *accuracy* of a measurement refers to the number of digits, called *significant digits,* which indicate the number of units that we are reasonably sure of having counted when making a measurement. The greater the number of significant digits given in a measurement, the better the accuracy, and vice versa.

## EXAMPLE 1

The average distance between the moon and the earth is 385,000 km. This measurement indicates measuring 385 thousands of kilometres; its accuracy is indicated by three significant digits.

## EXAMPLE 2

A measurement of 0.025 cm indicates measuring 25 thousandths of a centimetre; its accuracy is indicated by two significant digits.

## EXAMPLE 3

A measurement of 0.0500 mg indicates measuring $50\overline{0}$ ten-thousandths of a milligram; its accuracy is indicated by three significant digits.

Notice that sometimes a zero is significant and sometimes it is not. To clarify this, we use the following rules for significant digits:

---

1. All nonzero digits are significant: 156.4 m has four significant digits (this measurement indicates 1564 tenths of metres).
2. All zeros between significant digits are significant: 306.02 km has five significant digits (this measurement indicates 30,602 hundredths of kilometres).
3. In a number greater than 1, a zero that is specially tagged, such as by a bar above it, is significant: $23\overline{0},000$ km has three significant digits (this measurement indicates $23\overline{0}$ thousands of kilometres).
4. All zeros to the right of a significant digit *and* a decimal point are significant: 86.10 cm has four significant digits (this measurement indicates $861\overline{0}$ hundredths of centimetres).
5. In whole-number measurements, zeros at the right that are not tagged are *not* significant: 2500 m has two significant digits (25 hundreds of metres).
6. In measurements of less than 1, zeros at the left are *not* significant: 0.00752 m has three significant digits (752 hundred-thousandths of a metre).

---

When a number is written in scientific notation, the decimal part indicates the number of significant digits. For example, $20\overline{0},000$ m would be written in scientific notation as $2.00 \times 10^5$ m.

In summary, to find the number of significant digits:

1. All nonzero digits are significant.
2. Zeros are significant when they
   (a) Are between significant digits;
   (b) Follow the decimal point and a significant digit; or
   (c) Are in a whole number and a bar is placed over the zero.

## EXAMPLE 4

Determine the accuracy (the number of significant digits) of each measurement.

| Measurement | Accuracy (significant digits) |
|---|---|
| (a) 2642 ft | 4 |
| (b) 2005 m | 4 |
| (c) 2050 m | 3 |
| (d) 2500 m | 2 |
| (e) 25$\overline{0}$0 m | 3 |
| (f) 250$\overline{0}$ m | 4 |
| (g) 34,000 mi | 2 |
| (h) 15,670,000 lb | 4 |
| (i) 203.05 km | 5 |
| (j) 0.000345 kg | 3 |
| (k) 75 V | 2 |
| (l) 2.3 A | 2 |
| (m) 0.02700 g | 4 |
| (n) 2.40 cm | 3 |
| (o) 4.050 $\mu$A | 4 |
| (p) 100.050 km | 6 |
| (q) 0.004 s | 1 |
| (r) $2.03 \times 10^4$ m² | 3 |
| (s) $1.0 \times 10^{-3}$ A | 2 |
| (t) $5 \times 10^6$ kg | 1 |
| (u) $3.060 \times 10^8$ m³ | 4 |

## PROBLEMS

*Determine the accuracy (the number of significant digits) of each measurement.*

**1.** 536 V                 **2.** 307.3 mi

**3.** 5007 m               **4.** 5.00 cm

**5.** 0.0070 in.          **6.** 6.010 cm

**7.** 84$\overline{0}$0 km               **8.** 30$\overline{0}$0 ft

**9.** 187.40 m           **10.** 5$\overline{0}$0 g

**11.** 0.00700 in.       **12.** 10.30 cm

**13.** 376.52 m         **14.** 3.05 mi

**15.** 4087 kg          **16.** 35.00 mm

**17.** 0.0160 in.        **18.** 37$\overline{0}$ lb

**19.** 4$\overline{0}$00 $\Omega$                **20.** 5010 ft³

**21.** 7 A               **22.** 32,000 tons

**23.** 70.00 m²         **24.** 0.007 m

**25.** $2.4 \times 10^3$ kg     **26.** $1.20 \times 10^{-5}$ ms

**27.** $3.00 \times 10^{-4}$ A    **28.** $4.0 \times 10^6$ ft

**29.** $5.106 \times 10^7$ V     **30.** $1 \times 10^{-9}$ m

**1-11**
**MEASUREMENT:**
**PRECISION**

The *precision* of a measurement refers to the smallest unit with which a measurement is made, that is, the position of the last significant digit.

### EXAMPLE 1

The precision of the measurement 385,000 km is 1000 km. (The position of the last significant digit is in the thousands place.)

### EXAMPLE 2

The precision of the measurement 0.025 cm is 0.001 cm. (The position of the last significant digit is in the thousandths place.)

EXAMPLE 3

The precision of the measurement 0.0500 mg is 0.0001 mg. (The position of the last significant digit is in the ten-thousandths place.)

Unfortunately, the terms "accuracy" and "precision" have several different common meanings. Here we will use each term consistently as we have defined them. A measurement of 0.0004 cm has good precision and poor accuracy when compared with the measurement 378.0 cm, which has much better accuracy (one versus four significant digits) and poorer precision (0.0001 cm versus 0.1 cm).

EXAMPLE 4

Determine the precision of each measurement given in Example 4 of Section 1-10.

| Measurement | Precision | Accuracy (significant digits) |
|---|---|---|
| (a) 2642 ft | 1 ft | 4 |
| (b) 2005 m | 1 m | 4 |
| (c) 2050 m | 10 m | 3 |
| (d) 2500 m | 100 m | 2 |
| (e) 25$\overline{0}$0 m | 10 m | 3 |
| (f) 250$\overline{0}$ m | 1 m | 4 |
| (g) 34,000 mi | 1000 mi | 2 |
| (h) 15,670,000 lb | 10,000 lb | 4 |
| (i) 203.05 km | 0.01 km | 5 |
| (j) 0.000345 kg | 0.000001 kg | 3 |
| (k) 75 V | 1 V | 2 |
| (l) 2.3 A | 0.1 A | 2 |
| (m) 0.02700 g | 0.00001 g | 4 |
| (n) 2.40 cm | 0.01 cm | 3 |
| (o) 4.050 $\mu$A | 0.001 $\mu$A | 4 |
| (p) 100.050 km | 0.001 km | 6 |
| (q) 0.004 s | 0.001 s | 1 |
| (r) $2.03 \times 10^4$ m² | $0.01 \times 10^4$ m² or 100 m² | 3 |
| (s) $1.0 \times 10^{-3}$ A | $0.1 \times 10^{-3}$ A or 0.0001 A | 2 |
| (t) $5 \times 10^6$ kg | $1 \times 10^6$ kg or 1,000,000 kg | 1 |
| (u) $3.060 \times 10^8$ m³ | $0.001 \times 10^8$ m³ or $1 \times 10^5$ m³ or 100,000 m³ | 4 |

## PROBLEMS

*Determine the precision of each measurement.*

**1.** 536 V

**2.** 307.3 mi

**3.** 5007 m

**4.** 5.00 cm

**5.** 0.0070 in.

**6.** 6.010 cm

**7.** 84$\overline{0}$0 km

**8.** 30$\overline{0}$0 ft

**9.** 187.40 m

**10.** 5$\overline{0}$0 g

**11.** 0.00700 in.

**12.** 10.30 cm

**13.** 376.52 m

**14.** 3.05 mi

**15.** 4087 kg

**16.** 35.00 mm

**17.** 0.0160 in.

**18.** 37$\overline{0}$ lb

**19.** 4$\overline{0}$00 Ω

**20.** 5010 ft³

**21.** 7 A

**22.** 32,000 tons

**23.** 70.00 m²

**24.** 0.007 m

**25.** $2.4 \times 10^3$ kg

**26.** $1.20 \times 10^{-5}$ ms

**27.** $3.00 \times 10^{-4}$ A

**28.** $4.0 \times 10^6$ ft

**29.** $5.106 \times 10^7$ V

**30.** $1 \times 10^{-9}$ m

*In each set of measurements, find the measurement that is (a) the most accurate, and (b) the most precise.*

**31.** 15.7 in.;   0.018 in.;   0.07 in.

**32.** 368 ft;   600 ft;   180 ft

**33.** 0.734 cm;   0.65 cm;   16.01 cm

**34.** 3.85 m;   8.90 m;   7.00 m

**35.** 0.0350 A;   0.025 A;   0.00040 A;   0.051 A

**36.** 125.00 g;   8.50 g;   9.000 g;   0.05 g

**37.** 27,0$\overline{0}$0 L;   350 L;   27.6 L;   4.75 L

**38.** 8.4 m;   15 m;   180 m;   0.40 m

**39.** 500 Ω;   10,000 Ω;   500,000 Ω;   50 Ω

**40.** 7.5 mA;   14.2 mA;   10.5 mA;   120.0 mA

*In each set of measurements, find the measurement that is (a) the least accurate, and (b) the least precise.*

**41.** 16.4 in.;   0.075 in.;   0.05 in.

**42.** 475 ft;   300 ft;   360 ft

**43.** 27.5 m;   0.65 m;   12.02 m

**44.** 5.7 kg;   120 kg;   0.025 kg

**45.** 0.0250 g;   0.015 g;   0.00005 g;   0.75 g

**46.** 185.0 m;   6.75 m;   5.000 m;   0.09 m

**47.** 45,000 V;   250 V;   16.8 V;   0.25 V;   3 V

**48.** 2.50 kg;   42.0 kg;   15$\overline{0}$ kg;   0.500 kg

**49.** 20$\overline{0}$0 Ω;   10,$\overline{0}$00 Ω;   40$\overline{0}$,000 Ω;   20 Ω

**50.** 80 V;   250 V;   12,550 V;   26$\overline{0}$0 V

**1-12**
**ARITHMETIC OPERATIONS**
**WITH MEASUREMENTS**

If one person measured the length of one of two parts of a shaft with a micrometer calibrated in 0.01 mm as 42.28 mm and another person measured the second part with a ruler calibrated in mm as 54 mm, would the total length be 96.28 mm? Note that the sum 96.28 mm indicates a precision of 0.01 mm. The precision of the ruler is 1 mm, which means that the measurement 54 mm with the ruler could actually be anywhere between 53.50 mm and 54.50 mm using the micrometer (which has a precision of 0.01 mm). That is, using the ruler, any measurement between 53.50 mm and 54.50 mm can only be read as 54 mm. Of course, this means that the tenths and hundredths digits in the sum 96.28 mm are really meaningless. In other words, the sum or difference of measurements can be no more precise than the least precise measurement. That is,

---

*TO ADD OR SUBTRACT MEASUREMENTS:*

**1.** Make certain that all the measurements are expressed in the same units. If they are not, convert them all to the same units.

**2.** Round each measurement to the same precision as the least precise measurement.

**3.** Add or subtract.

---

## EXAMPLE 1

Add the measurements: 16.6 mi; 124 mi; 3.05 mi; 0.837 mi.

All measurements are in the same units, so round each measurement to the same precision as the least precise measurement, which is 124 mi. Then add.

$$
\begin{array}{rcr}
16.6 \;\; \text{mi} & \rightarrow & 17 \; \text{mi} \\
124 \;\;\;\;\; \text{mi} & \rightarrow & 124 \; \text{mi} \\
3.05 \;\; \text{mi} & \rightarrow & 3 \; \text{mi} \\
0.837 \; \text{mi} & \rightarrow & \underline{1 \; \text{mi}} \\
& & 145 \; \text{mi}
\end{array}
$$

## EXAMPLE 2

Add the measurements: 1370 cm; 1575 mm; 2.374 m; 8.63 m.

First, convert all measurements to the same units, say m.

$$1370 \text{ cm} = 13.7 \text{ m}$$

$$1575 \text{ mm} = 1.575 \text{ m}$$

Next, round each measurement to the same precision as the least precise measurement, which is 13.7 m. Then add.

$$
\begin{array}{rcr}
13.7 \;\;\;\; \text{m} & \rightarrow & 13.7 \; \text{m} \\
1.575 \; \text{m} & \rightarrow & 1.6 \; \text{m} \\
2.374 \; \text{m} & \rightarrow & 2.4 \; \text{m} \\
8.63 \;\;\; \text{m} & \rightarrow & \underline{8.6 \; \text{m}} \\
& & 26.3 \; \text{m}
\end{array}
$$

## EXAMPLE 3

Subtract the measurements: 3457.8 g − 2.80 kg.

First, convert both measurements to the same unit, say g.

$$2.80 \text{ kg} = 28\overline{0}0 \text{ g}$$

Next, round each measurement to the same precision as the least precise measurement, which is $28\overline{0}0$ g. Then subtract.

$$
\begin{array}{rcr}
3457.8 \; \text{g} & \rightarrow & 3460 \; \text{g} \\
28\overline{0}0 \;\;\;\; \text{g} & \rightarrow & \underline{28\overline{0}0 \; \text{g}} \\
& & 660 \; \text{g}
\end{array}
$$

Now suppose that you wish to find the area of the base of a rectangular building. You measure its length as 54.7 m and its width as 21.5 m. Its area is then

$$A = lw$$

$$A = (54.7 \text{ m})(21.5 \text{ m})$$

$$= 1176.05 \text{ m}^2$$

Note that the result contains six significant digits, whereas each of the original measurements contains only three significant digits. To rectify this inconsistency, we say that the product or quotient of measurements can be no more accurate than the least accurate measurement. That is,

---

*TO MULTIPLY OR DIVIDE MEASUREMENTS:*

1. Multiply or divide the measurements as given.
2. Round the result to the same number of significant digits as the measurement with the least number of significant digits.

---

Using the preceding rules the area of the base of the rectangular building is 1180 m².

*Note*: We assume throughout that you are using a calculator to do all calculations.

## EXAMPLE 4

Multiply the measurements: 124 ft × 187 ft.

$$124 \text{ ft} \times 187 \text{ ft} = 23{,}188 \text{ ft}^2$$

Round this product to three significant digits, which is the accuracy of the least accurate measurement (and also the accuracy of each measurement in the example). That is,

$$124 \text{ ft} \times 187 \text{ ft} = 23{,}200 \text{ ft}^2$$

## EXAMPLE 5

Multiply the measurements: (2.75 m)(1.25 m)(0.75 m).

$$(2.75 \text{ m})(1.25 \text{ m})(0.75 \text{ m}) = 2.578125 \text{ m}^3$$

Round this product to two significant digits, which is the accuracy of the least accurate measurement (0.75 m). That is,

$$(2.75 \text{ m})(1.25 \text{ m})(0.75 \text{ m}) = 2.6 \text{ m}^3$$

## EXAMPLE 6

Divide the measurements: 144,000 ft³ ÷ 108 ft.

$$144{,}000 \text{ ft}^3 \div 108 \text{ ft} = 1333.333 \ldots \text{ ft}^2$$

Round this quotient to three significant digits, which is the accuracy of the least accurate measurement (which is the accuracy of both measurements is this example). That is,

$$144{,}000 \text{ ft}^3 \div 108 \text{ ft} = 1330 \text{ ft}^2$$

## EXAMPLE 7

Find the value of $\dfrac{68 \text{ ft} \times 10{,}\overline{0}00 \text{ lb}}{95.6 \text{ s}}$.

$$\frac{68 \text{ ft} \times 10{,}\overline{0}00 \text{ lb}}{95.6 \text{ s}} = 7112.9707 \ldots \frac{\text{ft lb}}{\text{s}}$$

Round this result to two significant digits, which is the accuracy of the least accurate measurement (68 ft). That is,

$$\frac{68 \text{ ft} \times 10{,}\overline{0}00 \text{ lb}}{95.6 \text{ s}} = 7100 \frac{\text{ft lb}}{\text{s}}$$

## EXAMPLE 8

Find the value of $\dfrac{(58.0 \text{ kg})(2.40 \text{ m/s})^2}{5.40 \text{ m}}$.

$$\frac{(58.0 \text{ kg})(2.40 \text{ m/s})^2}{5.40 \text{ m}} = 61.8666 \ldots \frac{\text{kg m}}{\text{s}^2}$$

Carefully simplify the units:

$$\frac{(\text{kg})(\text{m/s})^2}{\text{m}} = \frac{(\text{kg})(\cancel{\text{m}}^{\;\text{m}}/\text{s}^2)}{\cancel{\text{m}}} = \frac{\text{kg m}}{\text{s}^2}$$

Round this result to three significant digits, which is the accuracy of the least accurate measurement (which is the accuracy of all measurements in this example). That is,

$$\frac{(58.0 \text{ kg})(2.40 \text{ m/s})^2}{5.40 \text{ m}} = 61.9 \ \frac{\text{kg m}}{\text{s}^2}$$

There are even more sophisticated methods for dealing with the calculations of measurements. The method one uses, and indeed whether one should even follow any given procedure, depends on the number of measurements and the sophistication needed for a particular situation.

In this text, we generally follow the customary practice of expressing measurements in terms of three significant digits, which is the accuracy used in most engineering and design work.

## PROBLEMS

*Use the rules for addition of measurements to add each set of measurements.*

| **1.** 3847 ft | **2.** 8,560 m | **3.** 42.8 cm | **4.** 0.456 g |
|---|---|---|---|
| 5800 ft | 84,000 m | 16.48 cm | 0.93 g |
| 4520 ft | 18,476 m | 1.497 cm | 0.402 g |
| | 12,500 m | 12.8 cm | 0.079 g |
| | | 9.69 cm | 0.964 g |

**5.** 39,000 V;   19,600 V;   8470 V;   2500 V

**6.** 6800 ft;   2760 ft;   4$\bar{0}$00 ft;   20$\bar{0}$0 ft

**7.** 467 m;   970 cm;   12$\bar{0}$0 cm;   1352 cm;   30$\bar{0}$ m

**8.** 36.8 m;   147.5 cm;   1.967 m;   125.0 m;   98.3 cm

**9.** 12 A;   1.004 A;   0.040 A;   3.9 A;   0.87 A

**10.** 160,000 V;   84,200 V;   4300 V;   239,000 V;   18,450 V

*Use the rules for subtraction of measurements to subtract each second measurement from the first.*

| **11.** 2876 kg | **12.** 14.73 m | **13.** 45.585 g | **14.** 34,500 kg |
|---|---|---|---|
| 2400 kg | 9.378 m | 4.6 g | 9,5$\bar{0}$0 kg |

**15.** 4200 km − 975 km

**16.** 64.73 g − 9.4936 g

**17.** 1,600,000 V − 685,000 V

**18.** 170 mm − 10.2 cm

**19.** 3.00 m − 26$\bar{0}$ cm

**20.** 1.40 mA − 0.708 mA

*Use the rules for multiplication of measurements to multiply each set of measurements.*

**21.** 125 m × 39 m

**22.** 470 ft × 1200 ft

**23.** (1637 km)(857 km)

**24.** (9100 m)(6$\bar{0}$0 m)

**25.** (18.70 m)(39.45 m)

**26.** (565 cm)(180 cm)

**27.** 14.5 cm × 18.7 cm × 20.5 cm

**28.** (0.046 m)(0.0317 m)(0.0437 m)

**29.** (45$\bar{0}$ in.)(315 in.)(205 in.)

**30.** (18.7 kg)(217 m)

*Use the rules for division of measurements to divide.*

**31.** 360 ft³ ÷ 12 ft²

**32.** 125 m² ÷ 3.0 m

**33.** 275 cm² ÷ 90.0 cm

**34.** 185 mi ÷ 4.5 h

**35.** $\dfrac{347 \text{ km}}{4.6 \text{ h}}$

**36.** $\dfrac{2700 \text{ m}^3}{9\bar{0}0 \text{ m}^2}$

**37.** $\dfrac{8800 \text{ V}}{8.5 \text{ A}}$

**38.** $\dfrac{4960 \text{ ft}}{2.95 \text{ s}}$

*Use the rules for multiplication and division of measurements to find the value of each of the following:*

**39.** $\dfrac{(18 \text{ ft})(290 \text{ lb})}{4.6 \text{ s}}$

**40.** $\dfrac{18.5 \text{ kg} \times 4.65 \text{ m}}{19.5 \text{ s}}$

**41.** $\dfrac{4500 \text{ V}}{12.3 \text{ A}}$

**42.** $\dfrac{48.9 \text{ kg}}{(1.5 \text{ m})(3.25 \text{ m})}$

**43.** $\dfrac{(48.7 \text{ m})(68.5 \text{ m})(18.4 \text{ m})}{(35.5 \text{ m})(40.0 \text{ m})}$

**44.** $\frac{1}{2}(270 \text{ kg})(16.4 \text{ m/s})^2$

**45.** $\dfrac{(115 \text{ V})^2}{25 \text{ }\Omega}$

**46.** $\dfrac{(45.2 \text{ kg})(13.7 \text{ m})}{(2.65 \text{ s})^2}$

**47.** $\dfrac{(85.7 \text{ kg})(25.7 \text{ m/s})^2}{12.5 \text{ m}}$

**48.** $\dfrac{(120 \text{ V})^2}{275 \text{ }\Omega}$

**49.** $\frac{4}{3}\,\pi(13.5 \text{ m})^3$

**50.** $\dfrac{140 \text{ g}}{(3.4 \text{ cm})(2.8 \text{ cm})(5.6 \text{ cm})}$

# CHAPTER
# *2*

# *Problem Solving*

**2-1**
FORMULAS A formula is an equation, usually expressed in letters (called variables in algebra) and numbers.

## EXAMPLE 1

The formula $s = vt$ states that the distance traveled, $s$, equals the product of the velocity, $v$, and the time, $t$.

## EXAMPLE 2

The formula $I = \dfrac{Q}{t}$ states that the current, $I$, equals the quotient of the charge, $Q$, and the time, $t$.

To solve a formula for a given letter means to express the given letter in terms of all the remaining letters. That is, by using the equation solving principles in Section A-3 of Appendix A, rewrite the formula so that the given letter appears on one side of the equation by itself and all the other letters appear on the other side.

## EXAMPLE 3

Solve $s = vt$ for $v$.

$$s = vt$$

$$\frac{s}{t} = \frac{vt}{t} \qquad \text{Divide both sides by } t.$$

$$\frac{s}{t} = v$$

## EXAMPLE 4

Solve $I = Q/t$
(a) for $Q$
(b) for $t$

(a)
$$I = \frac{Q}{t}$$

$$I(t) = \left(\frac{Q}{t}\right)t \qquad \text{Multiply both sides by } t.$$

$$It = Q$$

(b) Starting with $It = Q$, we obtain

$$\frac{It}{I} = \frac{Q}{I} \qquad \text{Divide both sides by } I.$$

$$t = \frac{Q}{I}$$

## EXAMPLE 5

Solve $V = E - Ir$ for $r$.

*Method 1:*
$$V = E - Ir$$
$$V - E = E - Ir - E \qquad \text{Subtract } E \text{ from both sides.}$$
$$V - E = -Ir$$
$$\frac{V - E}{-I} = \frac{-Ir}{-I} \qquad \text{Divide both sides by } -I.$$
$$\frac{V - E}{-I} = r$$

*Method 2:*
$$V = E - Ir$$
$$V + Ir = E - Ir + Ir \qquad \text{Add } Ir \text{ to both sides.}$$
$$V + Ir = E$$
$$V + Ir - V = E - V \qquad \text{Subtract } V \text{ from both sides.}$$
$$Ir = E - V$$
$$\frac{Ir}{I} = \frac{E - V}{I} \qquad \text{Divide both sides by } I.$$
$$r = \frac{E - V}{I}$$

Note that the two results are equivalent. Take the first result,

$$\frac{V - E}{-I}$$

and multiply numerator and denominator by $-1$. That is,

$$\left(\frac{V - E}{-I}\right)\left(\frac{-1}{-1}\right) = \frac{-V + E}{I} = \frac{E - V}{I}$$

which is the same as the second result.

It is often convenient to use the same quantity in more than one way in a formula. For example, we may wish to use a certain measurement of a quantity, such as velocity, at a given time, say at $t = 0$ s. Then use the velocity at a later time, say at $t = 6$ s. To write out these desired values of the velocity is rather awkward. We simplify this written statement by using *subscripts* (small letters or numbers printed a half space below the printed line and to the right of the quantity referred to) to shorten what we must write.

From the example given, $v$ at time $t = 0$ s will be written as $v_i$ (initial velocity); $v$ at time $t = 6$ s will be written as $v_f$ (final velocity). Mathematically, $v_i$ and $v_f$ are two different quantities, which in most cases are unequal. That is, $v_i$ and $v_f$ cannot be added as like terms or multiplied as numbers having the same base. The sum of $v_i$ and $v_f$ is written as $v_i + v_f$. The product of $v_i$ and $v_f$ is written as $v_i v_f$.

The subscript notation is used only to distinguish the general quantity, $v$, velocity, from the measure of that quantity at certain specified times.

## EXAMPLE 6

Solve the formula $x = x_i + v_i t + \frac{1}{2}at^2$ for $v_i$.

*Method 1:*

$$x = x_i + v_i t + \frac{1}{2}at^2$$

$$x - v_i t = x_i + v_i t + \frac{1}{2}at^2 - v_i t \qquad \text{Subtract } v_i t \text{ from both sides.}$$

$$x - v_i t = x_i + \frac{1}{2}at^2$$

$$x - v_i t - x = x_i + \frac{1}{2}at^2 - x \qquad \text{Subtract } x \text{ from both sides.}$$

$$-v_i t = x_i + \frac{1}{2}at^2 - x$$

$$\frac{-v_i t}{-t} = \frac{x_i + \frac{1}{2}at^2 - x}{-t} \qquad \text{Divide both sides by } -t.$$

$$v_i = \frac{x_i + \frac{1}{2}at^2 - x}{-t}$$

*Method 2:*

$$x = x_i + v_i t + \frac{1}{2}at^2$$

$$x - x_i - \frac{1}{2}at^2 = x_i + v_i t + \frac{1}{2}at^2 - x_i - \frac{1}{2}at^2 \qquad \textit{Subtract } x_i \textit{ and } \frac{1}{2}at^2 \textit{ from both sides.}$$

$$x - x_i - \frac{1}{2}at^2 = v_i t$$

$$\frac{x - x_i - \frac{1}{2}at^2}{t} = \frac{v_i t}{t} \qquad \text{Divide both sides by } t.$$

$$\frac{x - x_i - \frac{1}{2}at^2}{t} = v_i$$

## EXAMPLE 7

Solve the formula $v_{avg} = \frac{1}{2}(v_f + v_i)$ for $v_f$ (avg is a subscript meaning average).

$$v_{avg} = \frac{1}{2}(v_f + v_i)$$

$$2v_{avg} = v_f + v_i \qquad \text{Multiply both sides by 2.}$$

$$2v_{avg} - v_i = v_f \qquad \text{Subtract } v_i \text{ from both sides.}$$

## EXAMPLE 8

Solve $A = \pi d^2/4$ for $d$.

$$A = \frac{\pi d^2}{4}$$

$$4A = \left(\frac{\pi d^2}{4}\right)(4) \qquad \text{Multiply both sides by 4.}$$

$$4A = \pi d^2$$

$$\frac{4A}{\pi} = \frac{\pi d^2}{\pi} \qquad \text{Divide both sides by } \pi.$$

$$\frac{4A}{\pi} = d^2$$

$$\pm\sqrt{\frac{4A}{\pi}} = d \qquad \text{Take the square root of both sides.}$$

In this case, a negative diameter has no physical meaning, so the result is

$$d = \sqrt{\frac{4A}{\pi}}$$

*Solve each formula for the quantity given.*

**1.** $v = \dfrac{s}{t}$ for $s$

**2.** $a = \dfrac{v}{t}$ for $v$

**3.** $w = mg$ for $m$

**4.** $F = ma$ for $a$

**5.** $PE = mgh$ for $g$

**6.** $PE = mgh$ for $h$

**7.** $KE = \frac{1}{2} mv^2$ for $m$

**8.** $KE = \frac{1}{2} mv^2$ for $v^2$

**9.** $P = \dfrac{w}{t}$ for $t$

**10.** $p = \dfrac{F}{A}$ for $A$

**11.** $W = Fs$ for $s$

**12.** $v_f = v_i + at$ for $a$

**13.** $V = E - Ir$ for $I$

**14.** $v_2 = v_1 + at$ for $t$

**15.** $v_{avg} = \frac{1}{2}(v_f + v_i)$ for $v_i$

**16.** $2a(s - s_i) = v^2 - v_i^2$ for $a$

**17.** $2a(s - s_i) = v^2 - v_i^2$ for $s$

**18.** $Ft = m(V_2 - V_1)$ for $V_1$

**19.** $Q = \dfrac{I^2 Rt}{J}$ for $R$

**20.** $x = x_i + v_i t + \frac{1}{2} at^2$ for $x_i$

**21.** $A = \pi r^2$ for $r$

**22.** $V = \pi r^2 h$ for $r$

**2-2**

**SUBSTITUTING DATA INTO FORMULAS**

An important part of problem solving is being able to substitute the given data into the appropriate formula and to find the value of the unknown quantity.

Basically, there are two ways of substituting data into formulas to solve for the unknown quantity:

1. Solve the formula for the unknown quantity and then make the substitution of the data; or
2. Substitute the data into the formula first, and then solve for the unknown quantity.

When using a calculator, the first way is more useful. We will be using this way most of the time in this text.

EXAMPLE 1

Given the formula $A = bh$, $A = 120$ m², and $b = 15$ m. Find $h$.
First, solve for $h$.

$$A = bh$$

$$\frac{A}{b} = \frac{bh}{b} \qquad \text{Divide both sides by } b.$$

$$\frac{A}{b} = h$$

Then substitute the data:

$$h = \frac{A}{b} = \frac{120 \text{ m}^2}{15 \text{ m}} = 8.0 \text{ m}$$

(Remember to follow the rules of measurement discussed in Chapter 1. We use them consistently throughout.)

EXAMPLE 2

Given the formula $P = 2a + 2b$, $P = 824$ cm, and $a = 292$ cm. Find $b$.
First solve for $b$.

$$P = 2a + 2b$$

$$P - 2a = 2a + 2b - 2a \qquad \text{Subtract } 2a \text{ from both sides}$$

$$P - 2a = 2b$$

$$\frac{P - 2a}{2} = \frac{2b}{2} \qquad \text{Divide both sides by 2.}$$

$$\frac{P - 2a}{2} = b \qquad \left(\text{or } b = \frac{P}{2} - a\right)$$

Then substitute the data:

$$b = \frac{P - 2a}{2} = \frac{824 \text{ cm} - 2(292 \text{ cm})}{2}$$

$$= \frac{824 \text{ cm} - 584 \text{ cm}}{2}$$

$$= \frac{24\overline{0} \text{ cm}}{2} = 12\overline{0} \text{ cm}$$

## EXAMPLE 3

Given the formula $A = \left(\dfrac{a+b}{2}\right)h$, $A = 15\overline{0}$ m², $b = 18.0$ m, and $h = 10.0$ m. Find $a$.

First, solve for $a$.

$$A = \left(\frac{a+b}{2}\right)h$$

$$2A = \left[\left(\frac{a+b}{2}\right)h\right](2) \quad \text{Multiply both sides by 2.}$$

$$2A = (a+b)h$$

$$2A = ah + bh \qquad \text{Remove the parentheses}$$

$$2A - bh = ah + bh - bh \qquad \text{Subtract } bh \text{ from both sides.}$$

$$2A - bh = ah$$

$$\frac{2A - bh}{h} = \frac{ah}{h} \qquad \text{Divide both sides by } h.$$

$$\frac{2A - bh}{h} = a \qquad \left(\text{or } a = \frac{2A}{h} - b\right)$$

Then substitute the data:

$$a = \frac{2A - bh}{h}$$

$$= \frac{2(15\overline{0} \text{ m}^2) - (18.0 \text{ m})(10.0 \text{ m})}{10.0 \text{ m}}$$

$$= \frac{30\overline{0} \text{ m}^2 - 18\overline{0} \text{ m}^2}{10.0 \text{ m}}$$

$$= \frac{12\overline{0} \text{ m}^2}{10.0 \text{ m}} = 12.0 \text{ m}$$

## EXAMPLE 4

Given the formula $V = \frac{1}{3}\pi r^2 h$, $V = 64,400$ mm³, and $h = 48.0$ mm. Find $r$.

First, solve for $r$.

$$V = \frac{1}{3}\pi r^2 h$$

$$3V = (\frac{1}{3}\pi r^2 h)(3) \qquad \text{Multiply both sides by 3.}$$

$$3V = \pi r^2 h$$

$$\frac{3V}{\pi h} = \frac{\pi r^2 h}{\pi h} \qquad \text{Divide both sides by } \pi h.$$

$$\frac{3V}{\pi h} = r^2$$

$$\pm\sqrt{\frac{3V}{\pi h}} = r \qquad \text{Take the square root of both sides.}$$

In this case, a negative radius has no physical meaning, so the result is

$$r = \sqrt{\frac{3V}{\pi h}}$$

Then substitute the data:

$$r = \sqrt{\frac{3(64,400 \text{ mm}^3)}{\pi(48.0 \text{ mm})}}$$

$$= 35.8 \text{ mm}$$

## PROBLEMS

*In each of the following, (a) solve for the indicated letter and then (b) substitute the given values to find the value of the indicated letter. Follow the rules of calculations with measurements.*

| Formula | Data | Find |
|---|---|---|
| 1. $A = bh$ | $b = 14.5$ ft, $h = 11.2$ ft | $A$ |
| 2. $V = lwh$ | $l = 16.7$ m, $w = 10.5$ m, $h = 25.2$ m | $V$ |
| 3. $A = bh$ | $A = 34.5$ cm², $h = 4.60$ cm | $b$ |
| 4. $P = 4b$ | $P = 42\bar{0}$ in. | $b$ |
| 5. $P = a + b + c$ | $P = 48.5$ cm, $a = 18.2$ cm, $b = 24.3$ cm | $c$ |
| 6. $C = \pi d$ | $C = 495$ ft | $d$ |
| 7. $C = 2\pi r$ | $C = 68.5$ yd | $r$ |
| 8. $A = \frac{1}{2} bh$ | $A = 468$ m², $b = 36.0$ m | $h$ |
| 9. $P = 2(a + b)$ | $P = 88.7$ km, $a = 11.2$ km | $b$ |
| 10. $V = \pi r^2 h$ | $r = 61.0$ m, $h = 125.3$ m | $V$ |
| 11. $V = \pi r^2 h$ | $V = 368$ m³, $r = 4.38$ m | $h$ |
| 12. $A = 2\pi rh$ | $A = 51\bar{0}$ cm², $r = 14.0$ cm | $h$ |
| 13. $V = Bh$ | $V = 2185$ m³, $h = 14.2$ m | $B$ |
| 14. $A = \pi r^2$ | $A = 463.5$ m² | $r$ |
| 15. $A = b^2$ | $A = 465$ in² | $b$ |
| 16. $V = \frac{1}{3}\pi r^2 h$ | $V = 2680$ m³, $h = 14.7$ m | $r$ |
| 17. $C = 2\pi r$ | $r = 19.36$ m | $C$ |
| 18. $V = \frac{4}{3}\pi r^3$ | $r = 25.65$ m | $V$ |
| 19. $V = \frac{1}{3} Bh$ | $V = 19,850$ ft³, $h = 486.5$ ft | $B$ |
| 20. $A = \left(\dfrac{a + b}{2}\right) h$ | $A = 205.2$ m², $a = 16.50$ m, $b = 19.50$ m | $h$ |

**2-3**
**'ROBLEM-SOLVING**
**METHOD**

Problem solving in technical fields is more than plugging into formulas. It is necessary that you develop skill in taking data, analyzing the problems present, and finding the solution in an orderly manner.

Understanding the principle involved in a problem is more important than blindly substituting into a formula. By following an orderly procedure for problem solving, we hope to develop an approach to problem solving that you can use in your studies and on the job.

In all problems in the remainder of the book, the following method described will be applied to all problems where appropriate.

*Problem-solving method*

1. *Read the problem carefully.* This might appear obvious, but it is the most important step in solving a problem. As a matter of habit, you should read the problem at least twice.

(a) The first time you should read the problem straight through from beginning to end. Do not stop to think about setting up an equation or formula. You are only trying to get a general overview of the problem during this first reading.

(b) Read through a second time slowly and *completely,* beginning to think ahead to the following steps.

2. *Make a sketch.* All problems may not lend themselves to a sketch. However, make a sketch whenever it is possible. Many times, seeing a sketch of the problem will show if you have forgotten important parts of the problem and may suggest the solution. This is a *very important* part of problem solving and is often overlooked.

3. *Write down all given information.* This is necessary to get all essential facts in mind before looking for the solution. There are some common phrases that have understood physical meanings. For example, the term "from rest" means the initial velocity equals zero or $v_0 = 0$; the term "smooth surface" means assume that no friction is present.

4. *Write down the unknown or quantity asked for in the problem.* Many students have difficulty solving problems because they don't know what they are looking for and solve for the wrong quantity.

5. *Write down the basic equation or formula that relates the known and unknown quantities.* We find the basic formula or equation to use by studying what we are given and what we are asked to find. Then look for a formula or equation that relates these quantities. Sometimes, we may need to use more than one equation or formula in working a problem.

6. *Find a working equation by solving the basic equation or formula for the unknown quantity.*

7. *Substitute the data in the working equation, including the appropriate units.* It is important that you *carry the units all the way through the problem* as a check that you have solved the problem correctly. For example, if you are asked to find the weight of an object in newtons and the units of your answer work out to be metres, you need to review your solution for the error. (When the unit analysis is not obvious, we will go through it step by step in a box nearby.)

8. *Perform the indicated operations and work out the solution.* Although this will be your final written step in the solution, in every case you should ask yourself, "Is my answer reasonable?" Here and on the job you will be dealing with practical problems. A quick estimate will many times reveal an error in your calculations.

To help you recall the procedure detailed earlier, with most every problem set that follows you will find the figure shown here. This figure is not meant to be complete, but is only an outline to assist you in remembering and following the procedure for solving problems. *You should follow this outline in solving all problems in this course.*

This problem-solving method will be demonstrated in terms of relationships and formulas with which you are probably familiar.

The formulas for finding area and volume can be found in Table 18 of Appendix C.

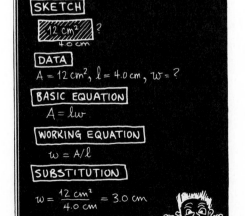

## EXAMPLE 1

Find the volume of concrete required to fill a rectangular bridge abutment whose dimensions are $l = 6.00$ ft, $w = 3.00$ ft, and $h = 15.0$ ft.

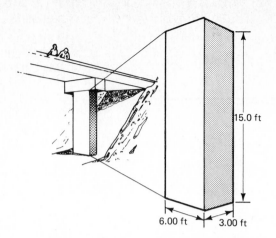

DATA:

$l = 6.00$ ft ⎫
$w = 3.00$ ft ⎬ This is a listing of the infor-
$h = 15.0$ ft ⎭ mation that is known.

$V = ?$    This identifies the unknown.

BASIC EQUATION:

$$V = lwh$$

WORKING EQUATION:   Same

SUBSTITUTION:

$$V = (6.00 \text{ ft})(3.00 \text{ ft})(15.0 \text{ ft})$$
$$= 27\bar{0} \text{ ft}^3$$

*Note*: ft × ft × ft = ft³.

## EXAMPLE 2

A rectangular holding tank of length 24.0 m and of width 15.0 m is used to store water for short periods of time in an industrial plant. If 2880 m³ of water is pumped into the tank, what is the depth of the water?

SKETCH:

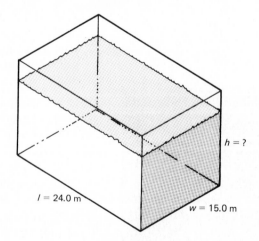

DATA:

$$V = 2880 \text{ m}^3$$
$$l = 24.0 \text{ m}$$
$$w = 15.0 \text{ m}$$
$$h = ?$$

BASIC EQUATION:

$$V = lwh$$

WORKING EQUATION:

$$h = \frac{V}{lw}$$

SUBSTITUTION:

$$h = \frac{2880 \text{ m}^3}{(24.0 \text{ m})(15.0 \text{ m})}$$

$$= 8.00 \text{ m} \qquad \boxed{\frac{\text{m}^3}{\text{m} \times \text{m}} = \text{m}}$$

## EXAMPLE 3

A storage bin in the shape of a cylinder contains 814 m³ of storage space. If its radius is 6.00 m, find its height.

SKETCH:

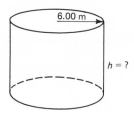

DATA:

$$V = 814 \text{ m}^3$$

$$r = 6.00 \text{ m}$$

$$h = ?$$

If your calculator does not have a button for $\pi$, use 3.14.

BASIC EQUATION:

$$V = \pi r^2 h$$

WORKING EQUATION:

$$h = \frac{V}{\pi r^2}$$

SUBSTITUTION:

$$h = \frac{814 \text{ m}^3}{\pi (6.00 \text{ m})^2}$$

$$= 7.20 \text{ m} \qquad \boxed{\frac{\text{m}^3}{\text{m}^2} = \text{m}}$$

## EXAMPLE 4

A rectangular piece of sheet metal measures 45.0 cm by 75.0 cm. A 10.0-cm square is then cut from each corner. The metal is then folded to form a boxlike container without a top. Find the volume of the container.

Sketch:

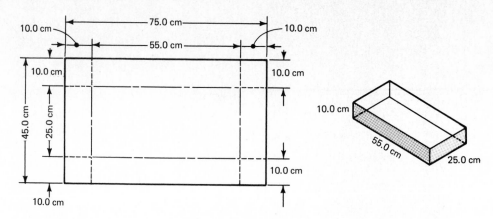

Data:

$$l = 55.0 \text{ cm}$$
$$w = 25.0 \text{ cm}$$
$$h = 10.0 \text{ cm}$$
$$V = ?$$

Basic Equation:

$$V = lwh$$

Working Equation:   Same

Substitution:

$$V = (55.0 \text{ cm})(25.0 \text{ cm})(10.0 \text{ cm})$$

$$= 13{,}800 \text{ cm}^3 \qquad \boxed{\text{cm} \times \text{cm} \times \text{cm} = \text{cm}^3}$$

## EXAMPLE 5

The cross-sectional area of a hole is to be $72\bar{0}$ cm². Find the radius of the hole.

Sketch:

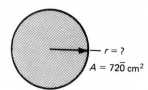

$r = ?$
$A = 72\bar{0} \text{ cm}^2$

Data:

$$A = 72\bar{0} \text{ cm}^2$$
$$r = ?$$

Basic Equation:

$$A = \pi r^2$$

Working Equation:

$$r = \sqrt{\frac{A}{\pi}}$$

Substitution:

$$r = \sqrt{\frac{72\bar{0} \text{ cm}^2}{\pi}}$$

$$= 15.1 \text{ cm} \qquad \boxed{\sqrt{\text{cm}^2} = \text{cm}}$$

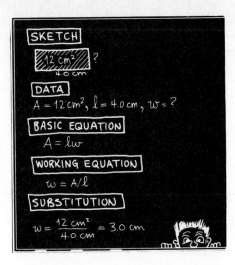

SKETCH

$\boxed{12 cm^2}$ ?
4.0 cm

DATA
$A = 12\ cm^2,\ \ell = 4.0\ cm,\ w = ?$

BASIC EQUATION
$A = \ell w$

WORKING EQUATION
$w = A/\ell$

SUBSTITUTION

$w = \dfrac{12\ cm^2}{4.0\ cm} = 3.0\ cm$

## PROBLEMS

*Use the problem-solving method to work each problem. (Here, as throughout the text, follow the rules of calculations with measurements.)*

**1.** Find the volume of the box.

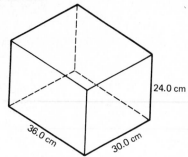

36.0 cm  30.0 cm  24.0 cm

**2.** Find the volume of a cylinder whose height is 7.50 in. and diameter is 4.20 in.

**3.** Find the volume of a cone whose height is 9.30 cm if the radius of the base is 5.40 cm.

The cylinder in an engine of a road grader is 11.40 cm in diameter and 24.00 cm high.

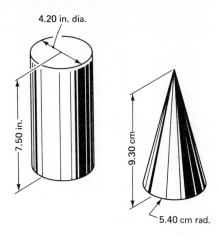

4.20 in. dia.

7.50 in.

9.30 cm

5.40 cm rad.

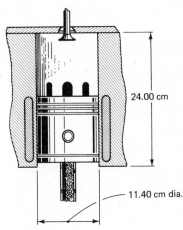

24.00 cm

11.40 cm dia.

**4.** Find the volume of the cylinder.

**5.** Find the cross-sectional area of the cylinder.

**6.** Find the lateral surface area of the cylinder.

**7.** Find the volume of the figure shown at left.

**8.** Find the cross-sectional area of the concrete dam shown.

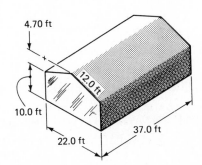

4.70 ft

12.0 ft

10.0 ft

22.0 ft

37.0 ft

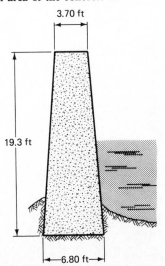

3.70 ft

19.3 ft

6.80 ft

9. Find the volume of a rectangular room 9.00 ft by 12.0 ft by 8.00 ft.

10. Find the cross-sectional area of a piston head with a diameter of 3.25 in.

11. Find the area of a right triangle that has legs of 4.00 cm and 6.00 cm.

12. Find the length of the hypotenuse of the right triangle in Problem 11.

13. Find the cross-sectional area of a pipe whose outer diameter is 3.50 cm and inner diameter is 3.20 cm.

14. Find the volume of a spherical water tank whose radius is 8.00 m.

15. The area of a rectangular parking lot is to be $90\bar{0}$ m². If the length is to be 25.0 m, what should be its width?

16. The volume of a rectangular crate is 192 ft³. If its length is 8.00 ft and its width is 4.00 ft, what is its height?

17. A cylindrical silo has a circumference of 29.5 m. What is its diameter?

18. If the silo in Problem 17 has a capacity of $100\bar{0}$ m³, what is its height?

19. A wheel 30.0 cm in diameter moving along level ground makes 145 complete rotations. How many metres did the wheel travel?

20. The side of the silo in Problems 17 and 18 needs to be painted. If each litre of paint covers 5.0 m², how many litres of paint will be needed? (Round up to the nearest litre.)

21. You are asked to design a cylindrical water tank that holds $50\bar{0},000$ gal. Its radius is to be 18.0 ft. What will be its height? (1 ft³ = 7.50 gal.)

22. If the height of the water tank in Problem 21 would be 42.0 ft, what would be its radius?

23. A ceiling is 12.0 ft by 15.0 ft. How many suspension panels 1.00 ft by 3.00 ft are needed to cover the ceiling?

24. Find the cross-sectional area of the dovetail slide shown.

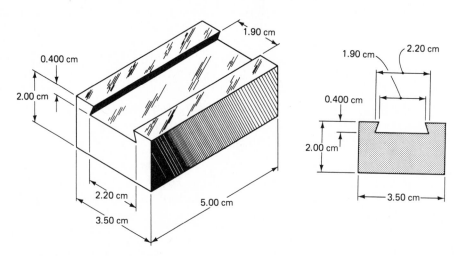

25. Find the volume of the storage bin shown at left.

26. The volume of a spherical propane storage tank is 4.00 m³. Will it fit into a 2.00-m-wide trailer?

27. How many cubic yards of concrete would be required to pour a patio 12.0 ft × 20.0 ft and 6.00 in. thick?

28. What length of sidewalk 4.00 in. thick and 4.00 ft wide could be poured with 2.00 yd³ of concrete?

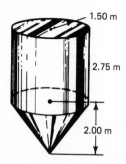

# CHAPTER
# 3

# *Motion*

**3-1**
INTRODUCTION
TO MECHANICS

The part of physics that is concerned with motion is called *mechanics*. The study of motion is very important in almost every area of science and technology.

Automotive technicians are concerned not only with the motion of the entire auto, but also with the motion of pistons, valves, driveshafts, and so on. Obviously, the motion of each of the internal parts has a direct and very important effect on the motion of the entire automobile.

The highway engineer must determine the correct banking angle of a curve if he or she is to design a safe road. This angle is determined from several laws of motion that we will soon study.

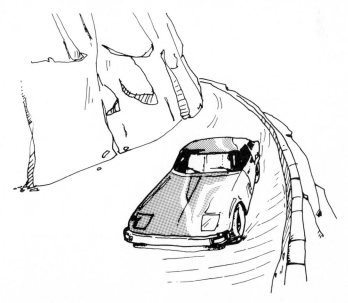

In the next few chapters we will develop the skills necessary for you to understand the basic aspects of motion. You will find this knowledge very helpful if you should decide to pursue a career in automotive mechanics, construction technology, electronics technology, mechanical technology, microprecision technology, or other similar fields.

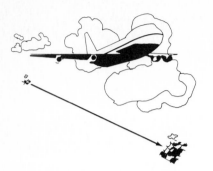

**3-2**
**DISPLACEMENT**

*Motion* can be defined as a change of position. An airplane is in motion when it flies through the air because its position is changing as it flies from one city to another. To describe the change of position of an object, such as an airplane, it is necessary to introduce the term *displacement*. Displacement, as a result of motion, is a change of position.

Suppose that a friend asked you how to reach your home from school. If you replied that he should walk four blocks, you would not have given him enough information. Obviously, you need to tell him which direction to go. If you had replied, "four blocks north," your friend could then find your home.

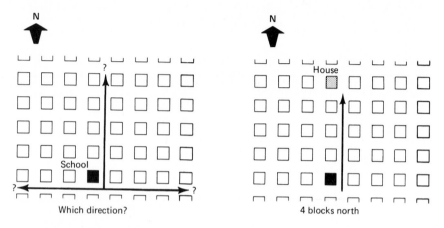

Which direction?  |  4 blocks north

Displacement involves all the necessary information about a change in position; that is, it includes both *distance and direction*. It does not contain any information about the path that has been followed. *The units of displacement are length units,* such as feet, metres, or miles. If your friend decides to walk one block west, four blocks north, and then one block east, he will still arrive at your house. This resultant displacement is the same as if he had walked four blocks north.

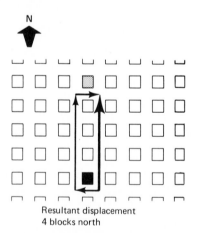

Resultant displacement
4 blocks north

**3-3**
**VECTORS AND SCALARS**

Displacement problems are examples of a certain type of problem that is easily solved by graphical methods. To solve this type of problem we need to examine the difference between what we call scalar and vector quantities. In Chapter 1 we discussed quantities of length, time, and volume. All these quantities can be expressed by a number with the appropriate units. For example, the area of a floor may be expressed as 150 ft². *Quantities such as this which can be completely described by a number and a unit are called scalar quantities or scalars.* They show magnitude only, not direction.

Many quantities, such as displacement, force, and velocity, must have their direction specified in addition to a number with a unit. To describe such quantities we use vectors. A vector is a quantity that has both size (magnitude) and direction. The magnitude of the displacement vector "$2\bar{0}$ miles NE" is $2\bar{0}$ miles. *Thus, a vector has both magnitude and direction.*

To represent a vector in our diagrams we draw an arrow that points in the correct direction. The magnitude of the vector is indicated by the length of the arrow. We usually choose a scale, such as 1.0 cm = 25 mi, for this purpose. Thus, a displacement of $10\bar{0}$ mi west would be drawn as an arrow (pointing west) 4.0 cm long since

$$10\bar{0} \text{ mi} \times \frac{1.0 \text{ cm}}{25 \text{ mi}} = 4.0 \text{ cm}$$

One end of the vector is called the initial end and the other is called the terminal end, as shown.

Displacements of $5\bar{0}$ mi east and $5\bar{0}$ mi north using the same scale are also shown in the figure above.

## EXAMPLE 1

Using the scale 1.0 cm = $5\bar{0}$ km, draw the displacement vector 275 km at 45° north of west.

First, find the length of the vector.

$$275 \text{ km} \times \frac{1.0 \text{ cm}}{5\bar{0} \text{ km}} = 5.5 \text{ cm}$$

Then draw the vector at an angle 45° north of west.

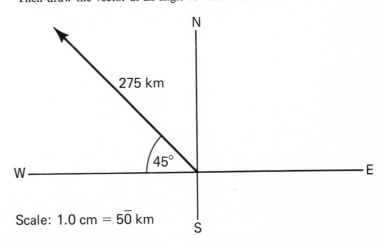

Scale: 1.0 cm = $5\bar{0}$ km

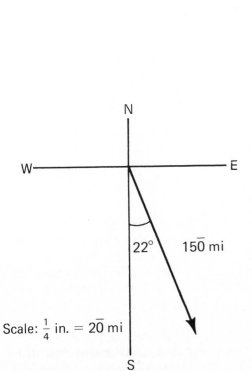

Scale: $\frac{1}{4}$ in. = $2\bar{0}$ mi

## EXAMPLE 2

Using the scale $\frac{1}{4}$ in. = $2\bar{0}$ mi, draw the displacement vector $15\bar{0}$ mi at 22° east of south.

First, find the length of the vector.

$$15\bar{0} \text{ mi} \times \frac{\frac{1}{4} \text{ in.}}{2\bar{0} \text{ mi}} = 1\frac{7}{8} \text{ in.}$$

Then draw the vector at 22° east of south as shown at the left.

*Using the scale of 1.0 cm = 25 mi, find the length of the vectors that represent the following displacements.*

**1.** Displacement 75 mi north                          length = _____ cm

**2.** Displacement $1\overline{0}0$ mi west                          length = _____ cm

**3.** Displacement 35 mi at 45° south of east          length = _____ cm

**4.** Displacement 160 mi at 35° west of north        length = _____ cm

**5.** Displacement $9\overline{0}$ mi at $1\overline{0}$° east of north          length = _____ cm

**6.** Displacement $6\overline{0}$ mi at 55° east of south        length = _____ cm

**7–12.** Draw the vectors in Problems 1 through 6 using the scale indicated there. Be sure to include the direction.

*Using the scale 1.0 cm = $5\overline{0}$ km, find the length of the vectors that represent the following displacements.*

**13.** Displacement $10\overline{0}$ km east                        length = _____ cm

**14.** Displacement 125 km south                          length = _____ cm

**15.** Displacement $14\overline{0}$ km at 45° east of south      length = _____ cm

**16.** Displacement $26\overline{0}$ km at $3\overline{0}$° south of west      length = _____ cm

**17.** Displacement 315 km at 65° north of east        length = _____ cm

**18.** Displacement 187 km at 17° north of west        length = _____ cm

**19–24.** Draw the vectors in Problems 13 through 18 using the scale indicated there.

*Using the scale $\frac{1}{4}$ in. = $2\overline{0}$ mi, find the length of the vectors that represent the following displacements.*

**25.** Displacement $10\overline{0}$ mi west                        length = _____ in.

**26.** Displacement $17\overline{0}$ mi north                        length = _____ in.

**27.** Displacement $21\overline{0}$ mi at 45° south of west      length = _____ in.

**28.** Displacement 145 mi at $6\overline{0}$° north of east        length = _____ in.

**29.** Displacement 75 mi at 25° west of north        length = _____ in.

**30.** Displacement $16\overline{0}$ mi at 72° west of south      length = _____ in.

**31–36.** Draw the vectors in Problems 25 through 30 using the scale indicated there.

**3-4**
GRAPHICAL ADDITION
OF VECTORS

Vectors may be denoted by a single letter with a small arrow above, such as $\vec{A}$, $\vec{v}$, or $\vec{R}$. This notation is especially useful when writing vectors on paper or on a chalkboard.

In this book we use the traditional boldface type to denote vectors, such as **A**, **v**, or **R**.

Any given displacement can be the result of many different combinations of displacements.

In the following diagram, the displacement represented by the arrow, labeled **R**, resultant, is the result of either of the two paths shown. This vector is called the *resultant* of the vectors that make up either path 1 or path 2. *The resultant vector is the result (sometimes called the sum) of a set of vectors.* The resultant vector, **R**, in the diagram is the sum of the vectors **A, B, C** and **D.** It is also the sum of the vectors **E** and **F.**

To solve a vector addition problem such as displacement, use the following procedure:

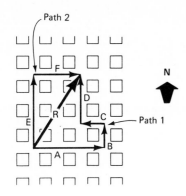

1. Choose a suitable scale and calculate the length of each vector.
2. Draw the north-south reference line. Graph paper should be used.
3. Using a ruler and protractor, draw the first vector and then draw the other vectors so that the initial end of each vector is placed at the terminal end of the previous vector.
4. Draw the sum or resultant vector from the initial end of the first vector to the terminal end of the last vector.
5. Measure the length of the resultant and use the scale to find the magnitude of the vector. Use a protractor to measure the angle of the resultant.

## EXAMPLE 1

Find the resultant displacement of an airplane that flies $2\overline{0}$ mi due east, then $3\overline{0}$ mi due north, and then $1\overline{0}$ mi at $6\overline{0}°$ west of south.

We choose a scale of 1.0 cm = 5.0 mi so that the vectors are large enough to be accurate and small enough to fit on the paper. The length of the first vector is

$$2\overline{0} \text{ mi} \times \frac{1.0 \text{ cm}}{5.0 \text{ mi}} = 4.0 \text{ cm}$$

The length of the second vector is

$$3\overline{0} \text{ mi} \times \frac{1.0 \text{ cm}}{5.0 \text{ mi}} = 6.0 \text{ cm}$$

The length of the third vector is

$$1\overline{0} \text{ mi} \times \frac{1.0 \text{ cm}}{5.0 \text{ mi}} = 2.0 \text{ cm}$$

We draw the north-south reference line and draw the first vector as shown in diagram 1. The second and third vectors are then drawn as shown in diagrams 2 and 3.

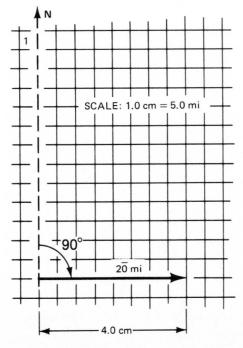

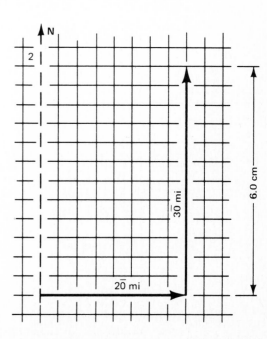

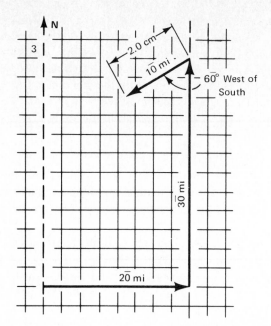

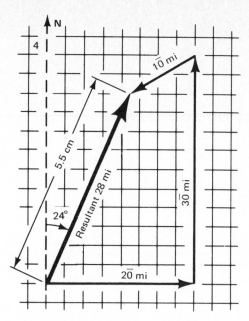

The length of the resultant is measured to be 5.5 cm, as in diagram 4. Since 1.0 cm = 5.0 mi, this represents a displacement with magnitude:

$$5.5 \text{ cm} \times \frac{5.0 \text{ mi}}{1.0 \text{ cm}} = 28 \text{ mi}$$

The angle is measured to be 24°, so the resultant is 28 mi at 24° east of north.

## EXAMPLE 2

Find the resultant of the following displacements: 15̄0 km due west, then 20̄0 km due east, and then 125 km due south.

Choose a scale of 1.0 cm = 5̄0 km.

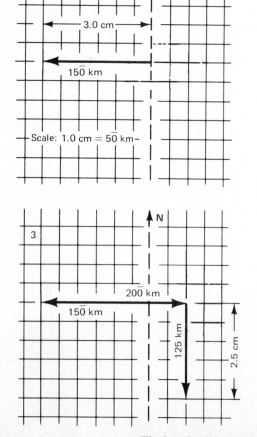

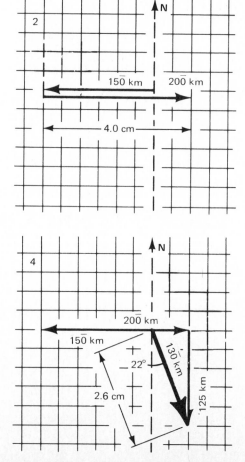

The length of the resultant is 2.6 cm, which represents 130 km at 22° east of south.

*Using graph paper, find the resultant of the following displacement pairs.*

**1.** 35 km due east, then $5\bar{0}$ km due north.

**2.** $6\bar{0}$ km due west, then $9\bar{0}$ km due south.

**3.** $5\bar{0}0$ mi at 75° east of north, then $15\bar{0}0$ mi at $2\bar{0}°$ west of south.

**4.** $2\bar{0}$ mi at 3° north of east, then 17 mi at 9° west of south.

**5.** 67 km at 55° north of west, then 46 km at 25° south of east.

**6.** 4.0 km at 25° west of south, then 2.0 km at 15° north of east.

*Using graph paper, find the resultant of the following combinations of displacements.*

**7.** $6\bar{0}$ km due south, then $9\bar{0}$ km at 15° north of west, and then 75 km at 45° north of east.

**8.** 110 km at $5\bar{0}°$ north of east, then 170 km at $3\bar{0}°$ east of south, and then 145 km at $2\bar{0}°$ north of east.

**9.** 1700 mi due north, then 2400 mi at 10° north of east, and then $2\bar{0}00$ mi at $2\bar{0}°$ south of west.

**10.** $9\bar{0}$ mi at $1\bar{0}°$ west of north, then 75 mi at $3\bar{0}°$ west of south, and then 55 mi at $2\bar{0}°$ east of south.

**11.** 75 km at 25° north of east, then 75 km at 65° south of west, and then 75 km due south.

**12.** 17 km due north, then $1\bar{0}$ km at 7° south of east, and then 15 km at $1\bar{0}°$ west of south.

**13.** 12 mi at 58° north of east, then 16 mi at 78° north of east, then $1\bar{0}$ mi at 45° north of east, and then 14 mi at $1\bar{0}°$ north of east.

**14.** $1\bar{0}$ km at 15° west of south, then 27 km at 35° north of east, then 31 km at 5° north of east, and then 22 km at $2\bar{0}°$ west of north.

**3-5**
**VELOCITY**

When an automobile travels a certain distance, we are interested in how fast the displacement took place. *The distance traveled per unit of time is called the speed.* Speed is a scalar (showing only magnitude, not direction) since it is described by a number and a unit. The unit of time is usually an hour or a second, so that the units of speed are miles per hour (mi/h), feet per second (ft/s), kilometres per hour (km/h), or metres per second (m/s).

The speed equals the total distance traveled divided by the total time. That is,

$$\text{speed} = \frac{\text{total distance traveled or displacement}}{\text{total time}}$$

This relationship may be expressed by the equation

$$v = \frac{s}{t}$$

or

$$\boxed{s = vt}$$

where

$s =$ displacement

$v =$ velocity

$t =$ time

We are often interested in the direction of travel of an automobile or an airplane. *Velocity is a vector that gives the direction of travel and the distance traveled per unit of time.* The units of velocity are the same as those of speed. (A useful conversion is $6\bar{0}$ mi/h $= 88$ ft/s.)

## EXAMPLE 1

Find the speed of an automobile that travels 160 km in 2.0 h.

SKETCH: None needed

DATA:

$$s = 160 \text{ km}$$
$$t = 2.0 \text{ h}$$
$$v = ?$$

BASIC EQUATION:

$$s = vt$$

WORKING EQUATION:

$$v = \frac{s}{t}$$

SUBSTITUTION:

$$v = \frac{160 \text{ km}}{2.0 \text{ h}}$$
$$= 8\bar{0} \text{ km/h}$$

## EXAMPLE 2

An airplane flies $35\bar{0}0$ mi in 5.00 h. What is its speed?

SKETCH: None needed

DATA:

$$s = 35\bar{0}0 \text{ mi}$$
$$t = 5.00 \text{ h}$$
$$v = ?$$

BASIC EQUATION:

$$s = vt$$

WORKING EQUATION:

$$v = \frac{s}{t}$$

SUBSTITUTION:

$$v = \frac{35\bar{0}0 \text{ mi}}{5.00 \text{ h}}$$
$$= 70\bar{0} \text{ mi/h}$$

## EXAMPLE 3

Find the velocity of a plane that travels $60\bar{0}$ km due north in 3 h 15 min.

SKETCH: None needed

DATA:

$$s = 60\bar{0} \text{ km}$$
$$t = 3 \text{ h } 15 \text{ min} = 3.25 \text{ h}$$
$$v = ?$$

BASIC EQUATION:

$$s = vt$$

WORKING EQUATION:

$$v = \frac{s}{t}$$

SUBSTITUTION:

$$v = \frac{60\bar{0} \text{ km}}{3.25 \text{ h}}$$

$$= 185 \text{ km/h}$$

The direction is north.
Thus, the velocity is 185 km/h due north.

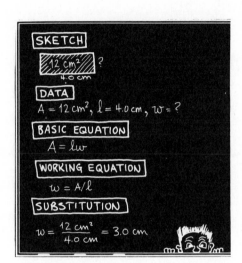

## PROBLEMS

*Find the speed of an auto that travels the following distances in the given times.*

1. Distance of $8\bar{0}$ mi in 1.0 h      speed = ____ mi/h
2. Distance of $12\bar{0}$ ft in 2.0 s      speed = ____ ft/s
3. Distance of $27\bar{0}$ km in 3.0 h      speed = ____ km/h
4. Distance of 190 m in 8.5 s      speed = ____ m/s
5. Distance of 150 mi in 3.0 h      speed = ____ mi/h
6. Distance of 45 km in 0.50 h      speed = ____ km/h
7. Distance of 8550 m in 6 min 35 s    speed = ____ m/s
8. Distance of 785 ft in 11.5 s      speed = ____ ft/s
9. Find the speed (in mi/h) of a racing car that turns a lap on a 1.00-mi oval track in 30.0 s.
10. While driving at $9\bar{0}$ km/h, how far can you travel in 3.5 h?
11. While driving at $9\bar{0}$ km/h, how far (in metres) do you travel in 1.0 s?

*An automobile is traveling at 55 mi/h. Find its speed*

12. in ft/s
13. in m/s
14. in km/h

*An automobile is traveling at 22.0 m/s. Find its speed*

15. in km/h
16. in mi/h
17. in ft/s

*Find the velocity for the following displacements and times.*

18. $1\bar{0}0$ km north in 3.0 h.
19. 160 km east in 2.0 h.
20. 31.0 mi west in 0.500 h.
21. $10\bar{0}0$ mi south in 8.00 h.
22. 426 km at 45° north of west in 2.75 h.
23. 275 km at $3\bar{0}$° south of east in 4.50 h.
24. 870 mi at 68° west of north in 3.5 h.

**3-6
ACCELERATION**
When a dragster travels down a quarter-mile track, its velocity changes.

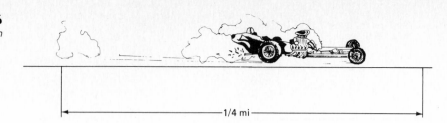

—— 1/4 mi ——

Its velocity in the last few feet of the race is much greater than its velocity at the start. The faster the velocity of the dragster changes, the faster its time will be. And the faster the velocity changes, the larger the acceleration will be. *Acceleration is the change in velocity per unit time.* That is,

$$\text{acceleration} = \frac{\text{change in velocity}}{\text{elapsed time}}$$

$$= \frac{\text{final velocity} - \text{initial velocity}}{\text{time}}$$

This relationship may be expressed by the equation

$$a = \frac{\Delta v}{t}$$

or

$$\boxed{\Delta v = at}$$

where: $\Delta v =$ change in velocity
$a =$ acceleration
$t =$ time

## EXAMPLE 1

A dragster starts from rest (velocity = 0 ft/s) and attains a velocity of $15\overline{0}$ ft/s in 10.0 seconds. Find its acceleration.

SKETCH:   None needed

DATA:

$$\Delta v = 15\overline{0} \text{ ft/s} - 0 \text{ ft/s} = 15\overline{0} \text{ ft/s}$$
$$t = 10.0 \text{ s}$$
$$a = ?$$

BASIC EQUATION:

$$\Delta v = at$$

WORKING EQUATION:

$$a = \frac{\Delta v}{t}$$

SUBSTITUTION:

$$a = \frac{15\overline{0} \text{ ft/s}}{10.0 \text{ s}}$$

$$= 15.0 \frac{\text{ft/s}}{\text{s}} \quad \text{or} \quad 15.0 \text{ feet per second per second}$$

Recall from arithmetic that to simplify fractions in the form

$$\frac{\dfrac{a}{b}}{\dfrac{c}{d}}$$

we divide by the denominator, that is, invert and multiply:

$$\frac{\dfrac{a}{b}}{\dfrac{c}{d}} = \frac{a}{b} \div \frac{c}{d} = \frac{a}{b} \cdot \frac{d}{c} = \frac{ad}{bc}$$

Use this idea to simplify the units from above:

$$\frac{\dfrac{15.0 \text{ ft}}{s}}{\dfrac{s}{1}} = \frac{15 \text{ ft}}{s} \div \frac{s}{1} = \frac{15.0 \text{ ft}}{s} \cdot \frac{1}{s} = \frac{15.0 \text{ ft}}{s^2} \quad \text{or} \quad 15.0 \text{ ft/s}^2$$

The units of acceleration are usually ft/s² or m/s².

## EXAMPLE 2

A car accelerates from 45 km/h to $8\bar{0}$ km/h in 3.00 s. Find its acceleration (in m/s²).

SKETCH:   None needed

DATA:

$$\Delta v = 8\bar{0} \text{ km/h} - 45 \text{ km/h} = 35 \text{ km/h}$$

$$t = 3.00 \text{ s}$$

$$a = ?$$

BASIC EQUATION:

$$\Delta v = at$$

WORKING EQUATION:

$$a = \frac{\Delta v}{t}$$

SUBSTITUTION:

$$a = \frac{35 \text{ km/h}}{3.00 \text{ s}} \times \frac{1000 \text{ m}}{1 \text{ km}} \times \frac{1 \text{ h}}{3600 \text{ s}}$$

$$= 3.2 \text{ m/s}^2$$

Note the use of the conversion factors to change the units km/h/s to m/s².

## EXAMPLE 3

A plane accelerates at 8.5 m/s² for 4.5 s. What is its increase in velocity (in m/s)?

SKETCH:   None needed

DATA:

$$a = 8.5 \text{ m/s}^2$$

$$t = 4.5 \text{ s}$$

$$\Delta v = ?$$

BASIC EQUATION:

$$\Delta v = at$$

WORKING EQUATION: Same

SUBSTITUTION:

$$\Delta v = (8.5 \text{ m/s}^2)(4.5 \text{ s})$$
$$= 38 \text{ m/s}$$

$$\boxed{\frac{\text{m}}{\text{s}^2} \times \text{s} = \frac{\text{m}}{\text{s}}}$$

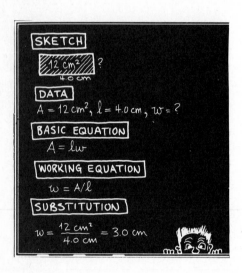

## PROBLEMS

*An automobile changes velocity as shown below. Find its acceleration (a) as indicated.*

| Velocity change | Time interval | Find a |
|---|---|---|
| **1.** From 0 to $1\overline{0}$ ft/s | 1.0 s | in ft/s² |
| **2.** From 0 to 18 m/s | 3.0 s | in m/s² |
| **3.** From $6\overline{0}$ ft/s to $7\overline{0}$ ft/s | 1.0 s | in ft/s² |
| **4.** From 45 m/s to 65 m/s | 2.0 s | in m/s² |
| **5.** From 25 km/h to $9\overline{0}$ km/h | 5.6 s | in m/s² |
| **6.** From $1\overline{0}$ mi/h to $5\overline{0}$ mi/h | 3.5 s | in ft/s² |

**7.** A dragster starts from rest and attains a velocity of 62.5 m/s in 10.0 s. Find its acceleration (in m/s²).

**8.** A car accelerates from 25 mi/h to 55 mi/h in 4.5 s. Find its acceleration (in ft/s²).

**9.** A train accelerates from $1\overline{0}$ km/h to $11\overline{0}$ km/h in 2 min 15 s. Find its acceleration (in m/s²).

*A plane accelerates at 30.0 ft/s² for 3.30 s. What is its increase in velocity*

**10.** in ft/s?

**11.** in mi/h?

*A rocket accelerates at 10.0 m/s² from rest for 20.0 s. What is its increase in velocity*

**12.** in m/s?

**13.** in km/h?

**14.** How long (in seconds) does it take for a rocket sled accelerating at 15.0 m/s² to change its velocity from 20.0 m/s to 65.0 m/s?

**15.** How long (in seconds) does it take for a truck accelerating at 1.50 m/s² to go from rest to 90.0 km/h?

**16.** How long (in seconds) does it take for a car accelerating at 3.50 m/s² to go from rest to $12\overline{0}$ km/h?

**3-7**
**MORE ON VELOCITY
AND ACCELERATION**

Every time a truck speeds up or slows down, its velocity changes. This change of velocity is called *acceleration.* Acceleration may be an increase or decrease in velocity. A negative (−) acceleration is commonly called *deceleration,* meaning the object is slowing down.

Note, however, that a driver does not always speed up or slow down at the same rate. If a child runs out in front of the truck, the driver may have to stop very quickly. The truck's acceleration is not uniform acceleration.

Because we lack the mathematical tools to study all kinds of motion, we must limit our study to one kind—uniformly accelerated motion. The most common example of this kind of motion is that of a freely falling body. Because of the complexity of this kind of problem, we must assume that falling bodies are unaffected by the resistance of the air, although, in fact, air resistance is an important factor in the design of machines that must move through the atmosphere. In learning to solve motion problems, we will assume air resistance to be negligible. Note also that for freely falling bodies the acceleration *(a)* due to gravity is $a = 32.2$ ft/s² (English system) or $a = 9.80$ m/s² (metric system).

A number of formulas and equations have been discovered that apply to freely falling bodies and uniformly accelerated motion in general.

$$s = v_{avg}t \qquad\qquad s = v_i t + \tfrac{1}{2} a_{avg} t^2$$

$$v_{avg} = \frac{v_f + v_i}{2} \qquad\qquad v_f = v_i + a_{avg} t$$

$$a_{avg} = \frac{v_f - v_i}{t} \qquad\qquad s = \tfrac{1}{2}(v_f + v_i)t$$

$$2a_{avg}s = v_f{}^2 - v_i{}^2$$

where:   $s =$ displacement      $v_{avg} =$ average velocity
$v_f =$ final velocity      $a_{avg} =$ average acceleration
$v_i =$ initial velocity      $t =$ time

We will now consider some problems using these equations, applying our problem-solving method.

## EXAMPLE 1

The average velocity of a rolling freight car is 7.00 ft/s. How long does it take for the car to roll 54.0 ft?

SKETCH:   None needed

DATA:

$$s = 54.0 \text{ ft}$$
$$v_{avg} = 7.00 \text{ ft/s}$$
$$t = ?$$

BASIC EQUATION:

$$s = v_{avg}t$$

WORKING EQUATION:

$$t = \frac{s}{v_{avg}}$$

SUBSTITUTION:

$$t = \frac{54.0 \text{ ft}}{7.00 \text{ ft/s}}$$
$$= 7.71 \text{ s}$$

$$\frac{\text{ft}}{\text{ft/s}} = \text{ft} \div \frac{\text{ft}}{\text{s}} = \text{ft} \cdot \frac{\text{s}}{\text{ft}} = \text{s}$$

## EXAMPLE 2

A dragster starting from a dead stop reaches a final velocity of 318 km/h. What is its average velocity?

SKETCH:   None needed

DATA:

$$v_i = 0$$
$$v_f = 318 \text{ km/h}$$
$$v_{avg} = ?$$

BASIC EQUATION:

$$v_{avg} = \frac{v_f + v_i}{2}$$

WORKING EQUATION:  Same

SUBSTITUTION:

$$v_{avg} = \frac{318 \text{ km/h} + 0 \text{ km/h}}{2}$$
$$= 159 \text{ km/h}$$

## EXAMPLE 3

A rock is thrown straight down from a cliff with an initial velocity of 10.0 ft/s. Its final velocity when it strikes the water below is $31\overline{0}$ ft/s. The acceleration due to gravity is 32.2 ft/s². How long is the rock in flight?

SKETCH:  None needed

DATA:

$$v_i = 10.0 \text{ ft/s}$$
$$a = 32.2 \text{ ft/s}^2$$
$$v_f = 31\overline{0} \text{ ft/s}$$
$$t = ?$$

Note the importance of listing all the data as an aid to finding the basic equation.

BASIC EQUATION:

$$v_f = v_i + a_{avg}t \quad \text{or} \quad a_{avg} = \frac{v_f - v_i}{t}$$

(two forms of same equation)

WORKING EQUATION:

$$t = \frac{v_f - v_i}{a_{avg}}$$

SUBSTITUTION:

$$t = \frac{31\overline{0} \text{ ft/s} - 10.0 \text{ ft/s}}{32.2 \text{ ft/s}^2}$$

$$= \frac{30\overline{0} \text{ ft/s}}{32.2 \text{ ft/s}^2}$$

$$= 9.32 \text{ s}$$

$$\boxed{\frac{\text{ft/s}}{\text{ft/s}^2} = \frac{\text{ft}}{\text{s}} \div \frac{\text{ft}}{\text{s}^2} = \frac{\text{ft}}{\text{s}} \cdot \frac{\text{s}^2}{\text{ft}} = \text{s}}$$

## EXAMPLE 4

A train slowing to a stop has an average acceleration of −3.00 m/s². [Note that a minus (−) acceleration is commonly called deceleration, meaning that the object is slowing down.] If its initial velocity is 30.0 m/s, how far does it travel in 4.00 s?

SKETCH:  None needed

DATA:

$$a_{avg} = -3.00 \text{ m/s}^2$$
$$v_i = 30.0 \text{ m/s}$$
$$t = 4.00 \text{ s}$$
$$s = ?$$

BASIC EQUATION:

$$s = v_i t + \tfrac{1}{2} a_{avg} t^2$$

WORKING EQUATION: Same

SUBSTITUTION:

$$s = (30.0 \text{ m/s})(4.00 \text{ s}) + \tfrac{1}{2}(-3.00 \text{ m/s}^2)(4.00 \text{ s})^2$$
$$= 12\bar{0} \text{ m} - 24.0 \text{ m}$$
$$= 96 \text{ m}$$

### EXAMPLE 5

An automobile accelerates from 67.0 km/h to 96.0 km/h in 7.80 s. What is its acceleration (in m/s²)?

SKETCH: None needed

DATA:

$$v_f = 96.0 \text{ km/h}$$
$$t = 7.80 \text{ s}$$
$$v_i = 67.0 \text{ km/h}$$
$$a = ?$$

BASIC EQUATION:

$$a_{avg} = \frac{v_f - v_i}{t}$$

WORKING EQUATION: Same

SUBSTITUTION:

$$a_{avg} = \frac{96.0 \text{ km/h} - 67.0 \text{ km/h}}{7.80 \text{ s}}$$
$$= \frac{29.0 \text{ km/h}}{7.80 \text{ s}}$$
$$= \frac{29.0 \, \dfrac{\cancel{km}}{\cancel{h}} \times \dfrac{10^3 \text{ m}}{1 \, \cancel{km}} \times \dfrac{1 \, \cancel{h}}{3600 \text{ s}}}{7.80 \text{ s}}$$
$$= 1.03 \text{ m/s}^2$$

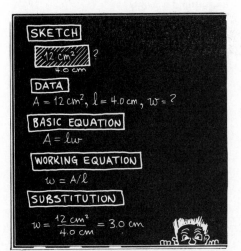

## PROBLEMS

*Substitute in the given equation and find the unknown quantity.*

**1.** Given: $v_{avg} = \dfrac{v_f + v_i}{2}$
$v_f = 6.20 \text{ ft/s}$
$v_i = 3.90 \text{ ft/s}$
$v_{avg} = ?$

**2.** Given: $a_{avg} = \dfrac{v_f - v_i}{t}$
$a_{avg} = 3.07 \text{ m/s}^2$
$v_f = 16.8 \text{ m/s}$
$t = 4.10 \text{ s}$
$v_i = ?$

3. Given:  $s = v_i t + \frac{1}{2} a_{avg} t^2$
$t = 3.00$ s
$a_{avg} = 6.40$ ft/s$^2$
$v_i = 33.0$ ft/s
$s = ?$

4. Given:  $2 a_{avg} s = v_f^2 - v_i^2$
$a_{avg} = 8.41$ m/s$^2$
$s = 4.81$ m
$v_i = 1.24$ m/s
$v_f = ?$

5. Given:  $v_f = v_i + a_{avg} t$
$v_f = 10.4$ ft/s
$v_i = 4.01$ ft/s
$t = 3.00$ s
$a_{avg} = ?$

6. The average velocity of a mini-bike is 15.0 km/h. How long does it take for the bike to go 35.0 m?

7. A sprinter starting from rest reaches a final velocity of 18.0 mi/h. What is her average velocity?

8. A coin is dropped with no initial velocity. Its final velocity when it strikes the earth below is 50.0 ft/s. The acceleration of gravity is 32.2 ft/s$^2$. How long does it take to strike the earth?

9. A rocket lifting off from earth has an average acceleration of 44.0 ft/s$^2$. Its initial velocity is zero. How far into the atmosphere does it travel during the first 5.00 s, assuming that it goes straight up?

10. The final velocity of a truck is 74.0 ft/s. If it accelerates at a rate of 2.00 ft/s$^2$ from an initial velocity of 5.00 ft/s, how long is required for it to attain its final velocity?

11. A truck can be accelerated from 85 km/h to $12\bar{0}$ km/h in 9.2 s. What is its acceleration in m/s$^2$?

12. How long does it take a rock to drop 95.0 m from rest?

13. What final velocity does the rock in Problem 12 attain?

14. A ball is thrown downward from the top of a 43.0-ft building with an initial velocity of 62.0 ft/s. What is its final velocity as it strikes the ground?

15. A car is traveling at $7\bar{0}$ km/h. It then uniformly decelerates to a complete stop in 12 s. Find its acceleration (in m/s$^2$).

16. A car is traveling at $6\bar{0}$ km/h. It then accelerates at 3.6 m/s$^2$ to $9\bar{0}$ km/h.
    (a) How long does it take to reach the new speed?
    (b) How far does it travel while accelerating?

17. A rock is dropped from a bridge to the water below. It takes 2.40 s for the rock to hit the water.
    (a) What is the speed (in m/s) of the rock as it hits the water?
    (b) How high (in metres) is the bridge above the water?

18. A bullet is fired vertically from a gun and reaches a height of $70\bar{0}0$ ft.
    (a) What was its initial velocity?
    (b) How long does it take to reach its maximum height?
    (c) How long is it in flight?

19. A bullet is fired vertically from a gun with an initial velocity of $25\bar{0}$ m/s.
    (a) How high does it go?
    (b) How long does it take to reach its maximum height?
    (c) How long is it in flight?

20. A worm accelerates from rest to a final velocity of 0.070 cm/s. If it takes 0.70 s to attain final velocity, what is its acceleration?

# *4*

# *Forces in One Dimension*

**4-1
FORCE
AND THE LAW
OF INERTIA**

To understand the causes of the various types of motion studied in technical programs, we need to study forces. Many types of forces are responsible for the motion of an automobile. The force produced by a hot expanding gas on each piston causes it to move.

When a structural engineer designs the supports for a bridge, he or she must allow for the weight of the vehicles on it and also the weight of the bridge itself. These forces do not cause motion but are still very important.

*A force is a push or a pull that tends to cause motion. Force is a vector quantity and thus has both magnitude and direction.* Some forces, such as the weight of the bridge shown, do not cause motion because they are balanced by other forces. The downward force of the bridge's weight is balanced by the upward force supplied by the supports. If the supports were weakened and could not supply this force, the downward force would no longer be balanced and the bridge would move; that is, it would collapse.

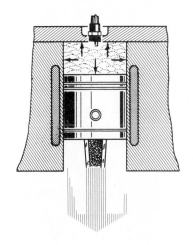

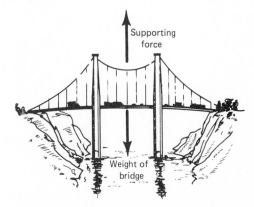

The units for measuring force are the pound (lb) in the English system and the newton (N) in the metric system. The conversion factor is

$$1 \text{ lb} = 4.45 \text{ N}.$$

We now want to examine the relationship between forces and motion. There are three relationships or laws that were discovered by Isaac Newton during the late seventeenth century. The three laws are often called Newton's laws. The first of these is the following:

53

*LAW OF INERTIA*
*A body that is in motion continues in motion with the same velocity (at constant speed and in a straight line), and a body at rest continues at rest unless an unbalanced force acts upon it.*

Inertia is the property of a body that causes it to remain at rest if it is at rest or to continue moving with a constant velocity unless an unbalanced force acts upon it.

When the accelerating force of an automobile engine is no longer applied to a moving car, it will slow down. This is not a violation of the law of inertia because there are forces being applied to the car through air resistance, friction in the bearings, and the rolling resistance of the tires. If these forces could be removed, the auto would continue moving with a constant velocity.

Anyone who has tried to stop quickly on ice knows the effect of the law of inertia when frictional forces are small.

Some objects tend to resist changes in their motion more than others. It is much easier to push a small automobile than to push a large truck into motion. *Mass is a measure of the resistance a body has to change in its motion.*

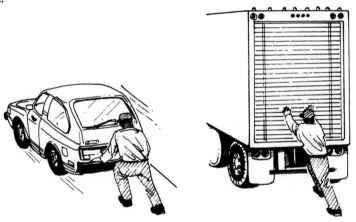

The common unit of mass is the slug in the English system and the kilogram (kg) in the metric system.

**4-2**
**FORCE**
**AND THE LAW**
**OF ACCELERATION**

The second law of motion, which we will call the law of acceleration, relates the applied force and the acceleration of an object:

*LAW OF ACCELERATION*
*The total force acting on a body is equal to the mass of the body times its acceleration.*

In equation form this is

$$F = ma$$

where   $F$ = the total force
$m$ = the mass
$a$ = the acceleration

Let us determine what force is necessary to give an object of mass 1 slug an acceleration of 1 ft/s². The law of acceleration gives us

$$F = ma$$
$$= 1 \text{ slug} \times 1 \text{ ft/s}^2$$
$$= 1 \text{ slug ft/s}^2$$

We would expect that the answer should be in the force unit, lb. The units will be correct if we use the fact that the slug is not a basic unit, and it can be expressed in terms of the basic units as

$$1 \text{ slug} = 1 \frac{\text{lb s}^2}{\text{ft}}$$

Used as a conversion factor, this equation becomes

$$F = 1 \text{ slug} \frac{\text{ft}}{\text{s}^2} \times \frac{1 \text{ lb s}^2}{1 \text{ slug ft}}$$
$$= 1 \text{ lb}$$

So a force of 1 lb gives a mass of 1 slug an acceleration of 1 ft/s².
This conversion factor is often useful when written in the form

$$1 \text{ lb} = 1 \frac{\text{slug ft}}{\text{s}^2}$$

In the metric system the conversion is

$$1 \text{ kg} = 1 \frac{\text{N s}^2}{\text{m}}$$

This conversion factor is often useful when written in the form

$$1 \text{ N} = 1 \frac{\text{kg m}}{\text{s}^2}$$

When the same force is applied to two different masses, these masses will have different accelerations. For example, a much smaller force is required to accelerate a baseball from rest to 90 km/h than to accelerate an automobile from rest to 90 km/h in the same time period. The reason for this is that the automobile has a much larger mass.

When the same amount of force is applied to two different masses, the smaller in mass will be accelerated more than the larger in mass. Compare Samples 1 and 2 and Samples 3 and 4 in the following computer printout, which illustrates these principles.

**SAMPLE 1:**

**FORCE = 80.0 N**
**MASS = 4.00 kg**
**ACCELERATION = 20.0 m/s²**

| TIME (s) | POSITION (m) | VELOCITY (m/s) |
|----------|--------------|----------------|
| 0.000 | 0.00 | 0.00 |
| 0.100 | 0.10 | 2.00 |
| 0.200 | 0.40 | 4.00 |
| 0.300 | 0.90 | 6.00 |
| 0.400 | 1.60 | 8.00 |
| 0.500 | 2.50 | 10.0 |
| 0.600 | 3.60 | 12.0 |
| 0.700 | 4.90 | 14.0 |
| 0.800 | 6.40 | 16.0 |
| 0.900 | 8.10 | 18.0 |
| 1.00 | 10.0 | 20.0 |
| 1.10 | 12.1 | 22.0 |
| 1.20 | 14.4 | 24.0 |
| 1.30 | 16.9 | 26.0 |
| 1.40 | 19.6 | 28.0 |
| 1.50 | 22.5 | 30.0 |

**SAMPLE 2:**

**FORCE = 80.0 N**
**MASS = 20$\bar{0}$ kg**
**ACCELERATION = 0.400 m/s²**

| TIME (s) | POSITION (m) | VELOCITY (m/s) |
|----------|--------------|----------------|
| 0.000 | 0.000 | 0.000 |
| 0.100 | 0.002 | 0.040 |
| 0.200 | 0.008 | 0.080 |
| 0.300 | 0.018 | 0.120 |
| 0.400 | 0.032 | 0.160 |
| 0.500 | 0.050 | 0.200 |
| 0.600 | 0.072 | 0.240 |
| 0.700 | 0.098 | 0.280 |
| 0.800 | 0.128 | 0.320 |
| 0.900 | 0.162 | 0.360 |
| 1.00 | 0.200 | 0.400 |
| 1.10 | 0.242 | 0.440 |
| 1.20 | 0.288 | 0.480 |
| 1.30 | 0.338 | 0.520 |
| 1.40 | 0.392 | 0.560 |
| 1.50 | 0.450 | 0.600 |

**SAMPLE 3:**

**FORCE = 12$\bar{0}$0 N**
**MASS = 4.00 kg**
**ACCELERATION = 30$\bar{0}$ m/s²**

| TIME (s) | POSITION (m) | VELOCITY (m/s) |
|----------|--------------|----------------|
| 0.000 | 0.00 | 00.0 |
| 0.100 | 1.50 | 30.0 |
| 0.200 | 6.00 | 60.0 |
| 0.300 | 13.5 | 90.0 |
| 0.400 | 24.0 | 120. |
| 0.500 | 37.5 | 150. |
| 0.600 | 54.0 | 180. |
| 0.700 | 73.5 | 210. |
| 0.800 | 96.0 | 240. |
| 0.900 | 122. | 270. |
| 1.00 | 150. | 300. |
| 1.10 | 182. | 330. |
| 1.20 | 216. | 360. |
| 1.30 | 254. | 390. |
| 1.40 | 294. | 420. |
| 1.50 | 338. | 450. |

**SAMPLE 4:**

**FORCE = 12$\overline{0}$0 N**
**MASS = 30$\overline{0}$0 kg**
**ACCELERATION = 0.400 m/s²**

| TIME (s) | POSITION (m) | VELOCITY (m/s) |
|---|---|---|
| 0.000 | 0.00000 | 0.0000 |
| 0.100 | 0.00200 | 0.0400 |
| 0.200 | 0.00800 | 0.0800 |
| 0.300 | 0.0180 | 0.120 |
| 0.400 | 0.0320 | 0.160 |
| 0.500 | 0.0500 | 0.200 |
| 0.600 | 0.0720 | 0.240 |
| 0.700 | 0.0980 | 0.280 |
| 0.800 | 0.128 | 0.320 |
| 0.900 | 0.162 | 0.360 |
| 1.00 | 0.200 | 0.400 |
| 1.10 | 0.242 | 0.440 |
| 1.20 | 0.288 | 0.480 |
| 1.30 | 0.338 | 0.520 |
| 1.40 | 0.392 | 0.560 |
| 1.50 | 0.450 | 0.600 |

## EXAMPLE 1

What total force is necessary to produce an acceleration of 2.00 ft/s² on a mass of 3.00 slugs?

SKETCH:

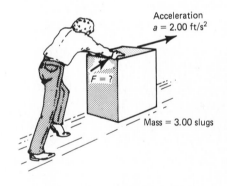

DATA:

$$m = 3.00 \text{ slugs}$$

$$a = 2.00 \frac{\text{ft}}{\text{s}^2}$$

$$F = ?$$

BASIC EQUATION:

$$F = ma$$

WORKING EQUATION:     Same

SUBSTITUTION:

$$F = 3.00 \text{ slugs} \times 2.00 \frac{\text{ft}}{\text{s}^2}$$

$$= 6.00 \frac{\text{slug ft}}{\text{s}^2}$$

$$= 6.00 \frac{\cancel{\text{slug ft}}}{\cancel{\text{s}^2}} \times \frac{1 \text{ lb } \cancel{\text{s}^2}}{1 \cancel{\text{slug ft}}}$$

$$= 6.00 \text{ lb}$$

(*Note*: We must use a conversion factor to obtain force units.)

What is the acceleration produced by a total force of $50\overline{0}$ N applied to a mass of 20.0 kg?

SKETCH:     None needed

DATA:

$$F = 50\overline{0} \text{ N}$$
$$m = 20.0 \text{ kg}$$
$$a = ?$$

BASIC EQUATION:

$$F = ma$$

WORKING EQUATION:

$$a = \frac{F}{m}$$

SUBSTITUTION:

$$a = \frac{50\overline{0} \text{ N}}{20.0 \text{ kg}}$$

$$= 25.0 \frac{\text{N}}{\text{kg}}$$

$$= 25.0 \frac{\cancel{\text{N}}}{\cancel{\text{kg}}} \times \frac{1 \cancel{\text{kg}} \text{ m}}{1 \cancel{\text{N}} \text{ s}^2}$$

$$= 25.0 \text{ m/s}^2$$

(*Note*: We must use a conversion factor to obtain acceleration units.)

## PROBLEMS

*Find the total force necessary to give the following masses the given acceleration.*

**1.** $m = 15.0$ kg, $a = 2.00$ m/s²    $F = \rule{1cm}{0.4pt}$ N
**2.** $m = 4.00$ slugs, $a = 0.500$ ft/s²    $F = \rule{1cm}{0.4pt}$ lb
**3.** $m = 111$ slugs, $a = 6.70$ ft/s²    $F = \rule{1cm}{0.4pt}$ lb
**4.** $m = 91.0$ kg, $a = 6.00$ m/s²    $F = \rule{1cm}{0.4pt}$ N
**5.** $m = 28.0$ slugs, $a = 9.00$ ft/s²    $F = \rule{1cm}{0.4pt}$ lb
**6.** $m = 42.0$ kg, $a = 3.00$ m/s²    $F = \rule{1cm}{0.4pt}$ N

*Find the acceleration of the following masses with the given total force.*

**7.** $m = 7.00$ slugs, $F = 12.0$ lb    $a = \rule{1cm}{0.4pt}$ ft/s².
**8.** $m = 19\overline{0}$ kg, $F = 76.0$ N    $a = \rule{1cm}{0.4pt}$ m/s².
**9.** $m = 3.60$ slugs, $F = 42.0$ lb    $a = \rule{1cm}{0.4pt}$ ft/s².
**10.** $m = 0.790$ slugs, $F = 13.0$ lb    $a = \rule{1cm}{0.4pt}$ ft/s².
**11.** $m = 11\overline{0}$ kg, $F = 57.0$ N    $a = \rule{1cm}{0.4pt}$ m/s².
**12.** $m = 84.0$ kg, $F = 33.0$ N    $a = \rule{1cm}{0.4pt}$ m/s².

**13.** Find the total force (in newtons) necessary to give an automobile of mass 1750 kg an acceleration of 3.00 m/s².

**14.** Find the acceleration produced by a total force of 93.0 N on a mass of 6.00 kg.

**15.** Find the total force (in pounds) necessary to give an automobile of mass $12\overline{0}$ slugs an acceleration of 11.0 ft/s².

**16.** Find the total force (in pounds) necessary to give a rocket of mass $25,\overline{0}00$ slugs an acceleration of 28.0 ft/s².

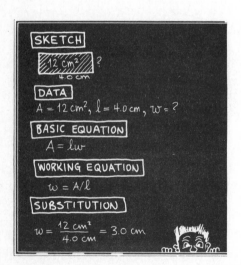

**17.** Find the acceleration produced by a total force of $30\overline{0}$ lb on a mass of 0.750 slug.

**18.** Find the mass of an object that has an acceleration of 15.0 m/s² when an unbalanced force of 90.0 N acts on it.

**19.** An automobile has a mass of $10\overline{0}$ slugs. The passengers it carries have a mass of 7.00 slugs each.

    (a) Find the acceleration of the auto and one passenger if the total force acting is $15\overline{0}0$ lb.

    (b) Find the acceleration of the auto and six passengers if the total force is again $15\overline{0}0$ lb.

**20.** Find the acceleration produced by a force of $6.75 \times 10^6$ N on a rocket of mass $5.27 \times 10^5$ kg.

**4-3**
**FRICTION**

When two objects slide across each other, a force that resists the motion is produced. This force is called *friction*. Friction is caused by the irregularities of the two surfaces, which tend to catch on each other. Severe engine damage can be caused by friction if proper lubricants are not used. Friction makes it hard to push objects along the floor.

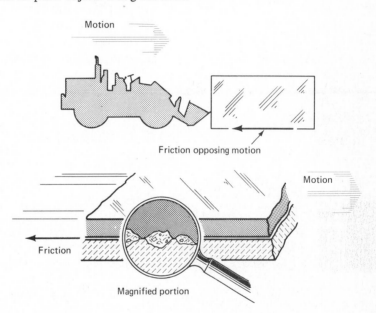

The frictional force depends on the materials and the smoothness of the surfaces involved. In general, the more polished the surfaces, the smaller the frictional force will be.

Friction is both a necessity and a hindrance to our everyday living. Consider:

Walking without friction between your shoes and the ground:

Driving on a road without friction between the road and your tires:

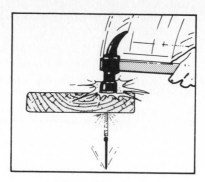

Driving a nail into wood with
no friction between the nail
and wood:

Cutting your steak with no
friction between the steak and
plate and table:

It is difficult to describe friction in terms of simple formulas or laws.
So let us study the following general properties of friction:

1. *Friction is a force that always acts parallel to the surfaces which are in
contact and opposite to the motion.* If there is no motion, friction acts
opposite any force that tends to produce motion. (See the first two figures
in this section.)

2. *Friction between rough surfaces is greater than between smooth surfaces.*
It is much easier to slide a crate on ice than on a wooden floor.

3. *Starting friction is greater than moving friction.* If you have ever pushed
a car by hand, you probably noticed that it took more force to start
the car moving than it did to keep it moving.

4. *Friction increases as the force between the surfaces increases.* It is much
easier to slide a light crate across the floor than a heavy one.

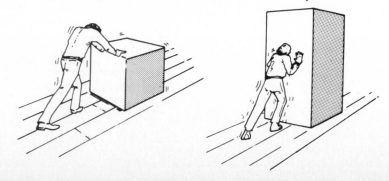

This property can be written as an equation:

$$F_f = \mu F_N$$

where:   $F_f$ = frictional force

$F_N$ = normal force (force acting perpendicular to the contact surfaces)

$\mu$ = coefficient of friction

Values for the coefficients of friction for some surfaces are given in the table.

| | Coefficient of friction | |
|---|---|---|
| Material | Starting friction | Sliding friction |
| Steel on steel | 0.58 | 0.20 |
| Glass on glass | 0.95 | 0.40 |
| Hardwood on hardwood | 0.40 | 0.25 |
| Steel on concrete | | 0.30 |
| Aluminum on aluminum | 1.9 | |
| Rubber on dry concrete | 2.0 | 1.0 |
| Rubber on wet concrete | 1.5 | 0.97 |
| Aluminum on wet snow | 0.4 | 0.02 |
| Teflon on Teflon | 0.04 | 0.04 |

In general, the following will help to reduce friction:

1. Use smoother surfaces.
2. Lubrication provides a thin film between surfaces and thus reduces friction.
3. Teflon greatly reduces friction between surfaces when an oil lubricant is not desirable, such as in electric motors.
4. Substitute rolling friction for sliding friction. Using ball bearings and roller bearings greatly reduces friction.

EXAMPLE

A force of 170 N is needed to keep a 530-N wooden box sliding on a wooden floor. What is the coefficient of sliding friction?

SKETCH:

DATA:

$$F_f = 170 \text{ N}$$
$$F_N = 530 \text{ N}$$
$$\mu = ?$$

BASIC EQUATION:

$$F_f = \mu F_N$$

WORKING EQUATION:

$$\mu = \frac{F_f}{F_N}$$

$$\mu = \frac{170\ N}{530\ N}$$

$$= 0.32$$

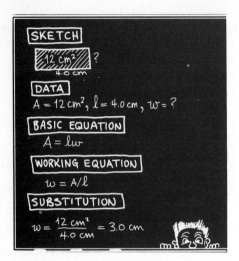

## PROBLEMS

1. A cart on wheels weighs 2400 N. The coefficient of rolling friction between the wheels and floor is 0.16. What force is needed to keep the cart rolling uniformly?

2. A wooden crate weighs 780 lb. What force is needed to start the crate sliding on a wooden floor when the coefficient of starting friction is 0.40?

3. A piano weighs 4700 N. What force is needed to start the piano rolling across the floor when the coefficient of starting friction is 0.23?

4. A force of 850 N is needed to keep the piano in Problem 3 rolling uniformly. What is the coefficient of rolling friction?

5. A dog sled weighing 750 lb is pulled over level snow at a uniform speed by a dog team exerting a force of $6\bar{0}$ lb. Find the coefficient of friction.

6. A horizontal conveyor belt system has a coefficient of moving friction of 0.65. The motor driving the system can deliver a maximum force of $2.5 \times 10^6$ N. What maximum total weight can be placed on the conveyor system?

**4-4**

**TOTAL FORCES IN ONE DIMENSION**

In the examples used to illustrate the law of acceleration, we discussed total forces only. We need to remember that forces are vectors and have magnitude and direction. The total force acting on an object is the resultant of the separate forces. *When forces act in the same or opposite directions (in one dimension), the total force can be found by adding the forces which act in one direction and subtract from that the forces which act in the opposite direction.* It is useful to draw the forces as vectors (arrows) in the sketch before working the problem.

### EXAMPLE 1

Two workers push in the same direction (to the right) on a crate. The force exerted by one worker is $15\bar{0}$ lb. The force by the other is 175 lb. Find the net force.

SKETCH:

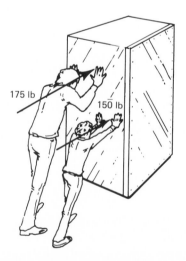

175 lb

$15\bar{0}$ lb

Both forces act in the same direction, so the total force is the sum of the two.

$$\mathbf{F} = 15\bar{0}\ lb + 175\ lb$$

$$= 325\ lb \quad \text{to the right}$$

EXAMPLE 2

The same two workers push the crate to the right and the motion is opposed by a frictional force of $30\overline{0}$ lb. Find the net force.

SKETCH:

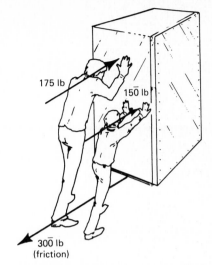

175 lb

$15\overline{0}$ lb

$30\overline{0}$ lb
(friction)

The workers push in one direction and friction pushes in the opposite direction, so we add the forces exerted by the worker and subtract the frictional force.

$$\mathbf{F} = 175\text{ lb} + 15\overline{0}\text{ lb} - 30\overline{0}\text{ lb}$$

$$= 25\text{ lb} \quad \text{to the right}$$

EXAMPLE 3

The crate in Example 2 has a mass of 5.00 slugs. What is its acceleration when the workers are pushing against the frictional force?

SKETCH: None needed

DATA:

$$F = 25\text{ lb} \qquad \text{(from Example 2)}$$
$$m = 5.00\text{ slugs}$$
$$a = ?$$

BASIC EQUATION:

$$F = ma$$

WORKING EQUATION:

$$a = \frac{25\text{ lb}}{5.00\text{ slugs}}$$

$$= 5.0\frac{\text{lb}}{\text{slug}}$$

$$= 5.0\frac{\cancel{\text{lb}}}{\cancel{\text{slug}}} \times \frac{1\ \cancel{\text{slug}}\ \text{ft}}{1\ \cancel{\text{lb}}\ \text{s}^2}$$

$$= 5.0\text{ ft/s}^2$$

(*Note*: We must use a conversion factor to obtain acceleration units.)

EXAMPLE 4

Two workers push in the same direction on a large pallet. The force exerted by one worker is $60\overline{0}$ N. The force by the other worker is $68\overline{0}$ N. The motion is opposed by a frictional force of $118\overline{0}$ N. Find the net force.

$$\mathbf{F} = 60\overline{0}\text{ N} + 68\overline{0}\text{ N} - 118\overline{0}\text{ N}$$

$$= 10\overline{0}\text{ N}$$

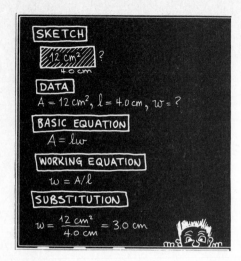

SKETCH

DATA
$A = 12\,cm^2,\ l = 4.0\,cm,\ w = ?$

BASIC EQUATION
$A = lw$

WORKING EQUATION
$w = A/l$

SUBSTITUTION
$w = \dfrac{12\,cm^2}{4.0\,cm} = 3.0\,cm$

## PROBLEMS

*Find the net force acting when the following forces act in the direction indicated.*

1. 17.0 lb to the left, 20.0 lb to the right.

   Net force = _____ lb

   The direction is _____

2. 265 N to the left, 85 N to the right.

   Net force = _____ N

   The direction is _____

3. 100.0 lb to the left, 75.0 lb to the right, and 10.0 lb to the right.

   Net force = _____ lb

   The direction is _____

4. $19\overline{0}$ lb to the left, 87 lb to the right, and 49 lb to the right.

   Net force = _____ lb

   The direction is _____

5. 346 N to the right, 247 N to the left, and 103 N to the left.

   Net force = _____ N

   The direction is _____

6. 37 N to the right, 24 N to the left, 65 N to the right, and 85 N to the right.

   Net force = _____ N

   The direction is _____

7. Find the acceleration of an automobile of mass $10\overline{0}$ slugs acted upon by a driving force of $50\overline{0}$ lb which is opposed by a frictional force of $10\overline{0}$ lb.

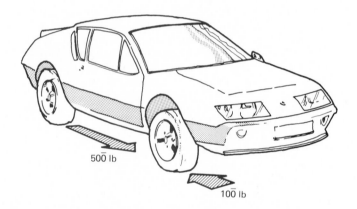

$50\overline{0}$ lb

$10\overline{0}$ lb

8. Find the acceleration of an automobile of mass $15\overline{0}0$ kg acted upon by a driving force of $22\overline{0}0$ N which is opposed by a frictional force of 450 N.

9. A truck of mass 13,100 kg is acted upon by a driving force of $89\overline{0}0$ N. The motion is opposed by a frictional force of 2230 N. Find the acceleration.

10. A speed boat of mass 30.0 slugs has a $30\overline{0}$-lb force applied by the propellers. The friction of the water on the hull is a force of $10\overline{0}$ lb. Find the acceleration.

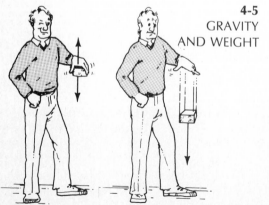

Upward force equals downward force

Downward force greater than upward force

**4-5**
**GRAVITY AND WEIGHT**

We have said that the weight of an object is the amount of gravitational pull exerted on an object by the earth. If this force is not balanced by other forces, an acceleration is produced. When you hold a brick in your hand, you exert an upward force on the brick that balances the downward force (weight). If you remove your hand, the brick moves downward due to the unbalanced force. The velocity of the falling brick changes but the acceleration (rate of change of the velocity) is constant.

The acceleration of all objects that are near the surface of the earth is the same if air resistance is ignored. We call this acceleration due to the gravitational pull of the earth—*g*. Its value is 32.2 ft/s² in the English system and 9.80 m/s² in the metric system.

The weight of an object is the force that gives the body the acceleration $g$. This force can be found using $F = ma$, where $a = 9.80$ m/s² $= 32.2$ ft/s² $= g$. If we abbreviate weight by $F_w$, the equation for weight is

$$F_w = mg$$

where:  $F_w$ = weight
$\phantom{where:}$  $m$ = mass
$\phantom{where:}$  $g$ = acceleration due to gravity

## EXAMPLE 1

Find the weight of 1.00 slug.

DATA:

$$m = 1.00 \text{ slug}$$
$$a = 32.2 \text{ ft/s}^2$$
$$F_w = ?$$

BASIC EQUATION:

$$F_w = mg$$

WORKING EQUATION:  Same

SUBSTITUTION:

$$F_w = 1.00 \text{ slug} \times 32.2 \text{ ft/s}^2$$
$$= 32.2 \frac{\text{slug ft}}{\text{s}^2}$$
$$= 32.2 \frac{\cancel{\text{slug ft}}}{\cancel{\text{s}^2}} \times \frac{1 \text{ lb } \cancel{\text{s}^2}}{\cancel{\text{slug ft}}}$$
$$= 32.2 \text{ lb}$$

(*Note*: We must use a conversion factor to obtain force units.)

## EXAMPLE 2

Find the weight of 1.00 kg.

DATA:

$$m = 1.00 \text{ kg}$$
$$a = 9.80 \text{ m/s}^2$$
$$F_w = ?$$

BASIC EQUATION:

$$F_w = mg$$

WORKING EQUATION:  Same

SUBSTITUTION:

$$F_w = 1.00 \text{ kg} \times 9.80 \text{ m/s}^2$$
$$= 9.80 \frac{\text{kg m}}{\text{s}^2}$$
$$= 9.80 \frac{\cancel{\text{kg m}}}{\cancel{\text{s}^2}} \times \frac{1 \text{ N } \cancel{\text{s}^2}}{\cancel{\text{kg m}}}$$
$$= 9.80 \text{ N}$$

(*Note*: We must use a conversion factor to obtain force units.)

Note that the mass of an object remains the same, but its weight varies according to the gravitational pull. For example, an astronaut of mass $7\overline{0}$ kg has a weight of

$$F_w = mg = (7\overline{0} \text{ kg})(9.80 \text{ m/s}^2) = 690 \text{ N}$$

on earth. As the spacecraft goes to the moon, we say the astronaut is "weightless." Actually, his or her weight is almost zero; the astronaut's mass remains $7\overline{0}$ kg. The following example will show the astronaut's weight on the moon.

## EXAMPLE 3

Find the $7\overline{0}$-kg astronaut's weight on the moon, where $g = 1.63$ m/s².

SKETCH: None needed

DATA:

$$m = 7\overline{0} \text{ kg}$$
$$g = 1.63 \text{ m/s}^2$$
$$F_w = ?$$

BASIC EQUATION:

$$F_w = mg$$

WORKING EQUATION: Same

SUBSTITUTION:

$$F_w = (7\overline{0} \text{ kg})(1.63 \text{ m/s}^2)$$
$$= 110 \text{ N}$$

Note that the astronaut's mass on the moon remains $7\overline{0}$ kg.

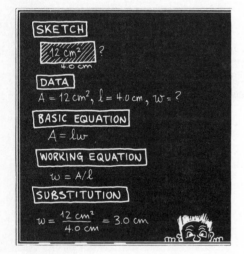

## PROBLEMS

*Find the unknown quantity in the following problems.*

1. $m = 10.0$ slugs    $F_w = \underline{\hspace{1cm}}$ lb
2. $m = 9.00$ kg    $F_w = \underline{\hspace{1cm}}$ N
3. $F_w = 17.0$ N    $m = \underline{\hspace{1cm}}$ kg
4. $F_w = 17.0$ N    $m = \underline{\hspace{1cm}}$ slug
5. $F_w = 21.0$ lb    $m = \underline{\hspace{1cm}}$ slug
6. $F_w = 170$ lb    $m = \underline{\hspace{1cm}}$ slug
7. $F_w = 170$ lb    $m = \underline{\hspace{1cm}}$ kg
8. $F_w = 835$ N    $m = \underline{\hspace{1cm}}$ kg
9. Find the weight (in N) of a $15\overline{0}0$-kg automobile.
10. Find the weight (in lb) of a $15\overline{0}0$-kg automobile.
11. Find the weight (in N) of a $15\overline{0}0$-kg automobile on the moon.
12. Find your weight (in N).
13. Find your mass (in kg).
14. Find your mass (in slugs).
15. Find your weight (in N) on the moon.
16. Find your mass (in kg) on the moon.

**4-6**

**LAW OF ACTION
AND REACTION**

When an automobile accelerates, we know that a force is being applied to it. What applies this force? You may think that the tires exert this force on the auto. This is not correct, since the tires move along with the auto and there must be a force applied to them also. The ground below the tires

Force of tires on ground (action)

Force of ground on tires (reaction)

$$F_1 = -F_2$$

supplies the force that accelerates the car. This force is called a *reaction* to the force exerted by the tires on the ground, which is called the *action force.*

The third law of motion, which is called the law of action and reaction, can be stated as follows:

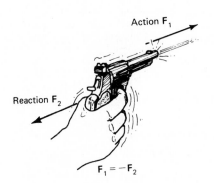

Action $F_1$

Reaction $F_2$

$F_1 = -F_2$

---

*LAW OF ACTION AND REACTION*
*For every force applied by object* A *to object* B *(action), there is a force exerted by object* B *on object* A *(reaction) which has the same magnitude but is opposite in direction.*

---

When a bullet is fired from a handgun (action), the recoil felt is the reaction. These forces are shown in the diagram. Note that the action and reaction forces *never* act on the same object.

**4-7**
**MOMENTUM**

We all know that if two automobiles are moving with the same velocity and one is heavier than the other, the heavier auto would cause more damage in a head-on collision. The lighter auto can cause as much or more damage if its velocity is greater than that of the heavier auto. *Momentum is a measure of the effect an object would have in a collision brought to rest in a certain amount of time. Momentum is equal to the mass times the velocity of an object.*

$$p = mv$$

where   $p$ = momentum
   $m$ = mass
   $v$ = velocity

The units of momentum are slug ft/s in the English system and kg m/s in the metric system.

Momentum is a vector quantity whose direction is the same as the velocity.

EXAMPLE 1

---

Find the momentum of an auto that has a mass of $11\overline{0}$ slugs and a velocity of 60.0 mi/h.

SKETCH: None needed

DATA:

$$m = 11\bar{0} \text{ slugs}$$
$$v = 60.0 \text{ mi/h} = 88.0 \text{ ft/s}$$
$$p = ?$$

BASIC EQUATION:

$$p = mv$$

WORKING EQUATION: Same

SUBSTITUTION:

$$p = 11\bar{0} \text{ slugs} \times 88.0 \text{ ft/s}$$
$$= 9680 \text{ slug ft/s}$$

## EXAMPLE 2

Find the momentum of an auto that has a mass of 1350 kg at speed of 75.0 km/h.

SKETCH: None needed

DATA:

$$m = 1350 \text{ kg}$$
$$v = 75.0 \frac{\text{km}}{\text{h}} \times \frac{1000 \text{ m}}{1 \text{ km}} \times \frac{1 \text{ h}}{3600 \text{ s}} = 20.8 \text{ m/s}$$
$$p = ?$$

BASIC EQUATION:

$$p = mv$$

WORKING EQUATION: Same

SUBSTITUTION:

$$p = (1350 \text{ kg})(20.8 \text{ m/s})$$
$$= 28,100 \text{ kg m/s}$$

## EXAMPLE 3

Find the velocity of a bullet of mass $1.00 \times 10^{-2}$ kg if it is to have the same momentum as a bullet of mass $1.80 \times 10^{-3}$ kg and a velocity of $30\bar{0}$ m/s.

SKETCH:

$m_1 = 1.00 \times 10^{-2}$ kg      $m_2 = 1.80 \times 10^{-3}$ kg

$v_1 = ?$          $v_2 = 30\bar{0}$ m/s

DATA:

Heavier bullet:            Lighter bullet:
$m_1 = 1.00 \times 10^{-2}$ kg      $m_2 = 1.80 \times 10^{-3}$ kg
$v_1 = ?$                  $v_2 = 30\bar{0}$ m/s
$p_1 = ?$                  $p_2 = ?$

BASIC EQUATION:

$$p_1 = m_1 v_1$$
$$p_2 = m_2 v_2$$

we want:

$$p_1 = p_2$$

or

$$m_1 v_1 = m_2 v_2$$

WORKING EQUATION:

$$v_1 = \frac{m_2}{m_1} v_2$$

SUBSTITUTION:

$$v_1 = \frac{1.80 \times 10^{-3} \text{ kg}}{1.00 \times 10^{-2} \text{ kg}} \times 30\overline{0} \text{ m/s}$$

$$= 54.0 \text{ m/s}$$

The *impulse* on an object is the product of the force and the time interval during which the force acts on the object. That is,

$$\boxed{\text{impulse} = Ft}$$

where   $F =$ force
        $t =$ time interval during which the force acts

How are impulse and momentum related? Recall that

$$a = \frac{v_f - v_i}{t}$$

If we substitute this equation into Newton's second law of motion,

$$F = ma$$

we have

$$F = m\left(\frac{v_f - v_i}{t}\right)$$

or

$$F = \frac{mv_f - mv_i}{t}$$

Multiply both sides by $t$:

$$Ft = mv_f - mv_i$$

Note that $mv_f$ is the final momentum and $mv_i$ is the initial momentum. That is,

$$\boxed{\text{impulse} = \text{change in momentum}}$$

A common example that illustrates this relationship is a golf club hitting a golf ball. During the time that the club and ball are in contact, the force of the swinging club is transmitting most of its momentum to the ball. The impulse given to the ball is directly related to the force with which the ball was hit and the length of time that the club and ball are in contact. After the ball leaves the club, the ball has a momentum equal to its mass times its velocity. This principle is the basis for the necessity of "follow-through" for long drives.

A 175-g bullet is fired at a muzzle velocity of $57\overline{0}$ m/s from a gun whose mass is 8.00 kg and whose barrel length is 75.0 cm.
(a) How long is the bullet in the barrel?
(b) What is the force of the bullet as it leaves the barrel?
(c) What is the impulse given to the bullet while it is in the barrel?
(d) What is the bullet's momentum as it leaves the barrel?

(a) SKETCH:

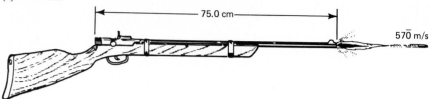

DATA:

$$s = 75.0 \text{ cm} = 0.750 \text{ m}$$
$$v_f = 57\overline{0} \text{ m/s}$$
$$v_i = 0 \text{ m/s}$$
$$v_{avg} = \frac{v_f + v_i}{2} = \frac{57\overline{0} \text{ m/s} + 0 \text{ m/s}}{2} = 285 \text{ m/s}$$
$$t = ?$$

BASIC EQUATION:

$$s = v_{avg} \, t$$

WORKING EQUATION:

$$t = \frac{s}{v_{avg}}$$

SUBSTITUTION:

$$t = \frac{0.750 \text{ m}}{285 \text{ m/s}}$$
$$= 0.00263 \text{ s}$$

*Note*: This is the length of time that the force is applied to the bullet.

(b) DATA:

$$t = 0.00263 \text{ s}$$
$$m = 175 \text{ g} = 0.175 \text{ kg}$$
$$v_f = 575 \text{ m/s}$$
$$v_i = 0 \text{ m/s}$$
$$F = ?$$

BASIC EQUATION:

$$Ft = mv_f - mv_i$$

WORKING EQUATION:

$$F = \frac{mv_f - mv_i}{t}$$

SUBSTITUTION:

$$F = \frac{(0.175 \text{ kg})(575 \text{ m/s}) - (0.175 \text{ kg})(0 \text{ m/s})}{0.00263 \text{ s}}$$
$$= 38{,}300 \text{ kg m/s}^2$$
$$= 38{,}300 \text{ N}$$

(c) DATA:

$$t = 0.00263 \text{ s}$$
$$F = 38,300 \text{ N}$$
$$\text{impulse} = ?$$

BASIC EQUATION:

$$\text{impulse} = Ft$$

WORKING EQUATION:   Same

SUBSTITUTION:

$$\text{impulse} = (38,300 \text{ N})(0.00263 \text{ s})$$
$$= 101 \text{ N s}$$
$$= 101 \text{ (kg m/s}^2)(s) \quad [1 \text{ N} = 1 \text{ kg m/s}^2]$$
$$= 101 \text{ kg m/s}$$

(d) DATA:

$$m = 175 \text{ g} = 0.175 \text{ kg}$$
$$v = 575 \text{ m/s}$$
$$p = ?$$

BASIC EQUATION:

$$p = mv$$

WORKING EQUATION:   Same

SUBSTITUTION:

$$p = (0.175 \text{ kg})(575 \text{ m/s})$$
$$\doteq 101 \text{ kg m/s}$$

*Note*: The impulse equals the change in momentum.

## PROBLEMS

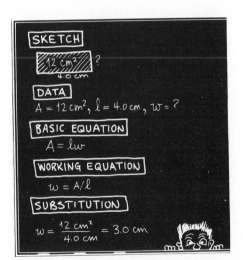

*Find the momentum for the following objects.*

1. $m = 2.00$ kg, $v = 40.0$ m/s, $p =$ _____ kg m/s
2. $m = 5.00$ slugs, $v = 90.0$ ft/s, $p =$ _____ slug ft/s
3. $m = 17.0$ slugs, $v = 45.0$ ft/s, $p =$ _____ slug ft/s
4. $m = 38.0$ kg, $v = 97.0$ m/s, $p =$ _____ kg m/s
5. $m = 3.8 \times 10^5$ kg, $v = 2.5 \times 10^3$ m/s, $p =$ _____ kg m/s
6. $m = 3.84$ kg, $v = 1.6 \times 10^5$ m/s, $p =$ _____ kg m/s
7. $F_w = 1.50 \times 10^5$ N, $v = 4.50 \times 10^4$ m/s, $p =$ _____ kg m/s
8. $F_w = 3200$ lb, $v = 6\overline{0}$ mi/h (change to ft/s), $p =$ slug ft/s
9. (a) Find the momentum of a heavy automobile of mass $18\overline{0}$ slugs traveling with velocity 70.0 ft/s.
   (b) Find the velocity (in km/h) of a light auto of mass 80.0 slugs if it is to have the same momentum as the auto in part (a).
   (c) Find the weight (in lb) of each of the autos in parts (a) and (b).
10. (a) Find the momentum of a bullet of mass $1.00 \times 10^{-3}$ slug and velocity $70\overline{0}$ ft/s.
    (b) Find the velocity of a bullet of mass $5.00 \times 10^{-4}$ slug if it is to have the same momentum as the bullet in part (a).
11. (a) Find the momentum of a heavy automobile of mass 2630 kg traveling at a velocity of 21.0 m/s.
    (b) Find the velocity (in km/h) of a light auto of mass 1170 kg if it is to have the same momentum as the auto in part (a).

12. A cannon is mounted on a railroad car. The cannon shoots a 1.75-kg ball with a muzzle velocity of $30\overline{0}$ m/s. The cannon and the railroad car together have a mass of $45\overline{0}0$ kg. If the ball, cannon, and railroad car are initially at rest, what is the recoil velocity of the car and cannon?

A 75.0-g bullet is fired at a muzzle velocity of $46\overline{0}$ m/s from a gun whose mass is 3.75 kg and whose barrel length is 66.5 cm.

13. How long is the bullet in the barrel?

14. What is the force of the bullet as it leaves the barrel?

15. What is the impulse given to the bullet while it is in the barrel?

16. What is the bullet's momentum as it leaves the barrel?

**4-8**
**CONSERVATION OF MOMENTUM**

One of the most important laws of physics is the following:

> *LAW OF CONSERVATION OF MOMENTUM*
> **When no outside forces are acting on a system of moving objects, the total momentum of the system remains constant.**

Let us look at some examples. Consider a 35-kg boy and a 75-kg man standing next to each other on ice skates on "frictionless ice." The man pushes on the boy, which gives the boy a velocity of 0.40 m/s. What happens to the man? Initially, the total momentum was zero because the initial velocity of each was zero. Because of the law of conservation of momentum, the

$v_b = 0.40$ m/s    $v_m = 0.19$ m/s

total momentum must still be zero. That is,

$$m_{boy}\, v_{boy} + m_{man}\, v_{man} = 0$$
$$(35 \text{ kg})(0.40 \text{ m/s}) + (75 \text{ kg})\, v_{man} = 0$$
$$v_{man} = -0.19 \text{ m/s}$$

*Note*: The minus sign indicates that the man's velocity and the boy's velocity are in opposite directions.

Rocket propulsion is another illustration of conservation of momentum. Like the skaters, the total momentum of a rocket on the launch pad is zero. When the rocket engines are fired, hot exhaust gases (actually gas molecules) are expelled downward through the rocket nozzle at tremendous speeds. As the rocket takes off, the sum of the total momentums of the rocket and the gas particles must remain zero. The total momentum of the gas particles is the sum of the products of each mass and its corresponding velocity and is directed down. The momentum of the rocket is the product of its mass and its velocity and is directed up.

When the rocket is in space, its propulsion works in the same manner. The conservation of momentum is still valid except that when the rocket engines are fired, the total momentum is a nonzero constant. This is because the rocket has velocity.

Actually, repair work is more difficult in space than it is on earth because of the conservation of momentum and the "weightlessness" of objects in orbit. On earth, when a hammer is swung, the person is coupled to the earth by frictional forces, so that the person's mass includes that of the earth. In space orbit, because the person is weightless, there is no friction to couple him or her to the spaceship and the person's mass alone must equal the magnitude of the momentum of the hammer to keep the momentum of the system conserved. A person in space has roughly the same problem driving a nail as a person on earth would have wearing a pair of "frictionless" roller skates.

# CHAPTER
# 5

# Vectors and Trigonometry

To this point we have discussed vectors in terms of a north-south reference. To study vectors more thoroughly, trigonometry of the right triangle is needed.

A right triangle is a triangle with one right angle (90°), two acute angles (less than 90°), two legs, and a hypotenuse (the side opposite the right angle).

When it is necessary to label a triangle, the vertices are labeled using capital letters and the sides opposite the vertices are labeled using the corresponding lowercase letter.

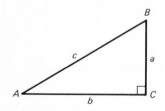

*Note:* The side opposite angle $A$ is $a$.
The side opposite angle $B$ is $b$.
The side opposite angle $C$ is c.

If we consider a certain acute angle of a right triangle, the two legs can be identified as the side opposite or the side adjacent to an acute angle.

The side opposite angle $A$ is $a$.
The side adjacent to angle $A$ is $b$.
The side opposite angle $B$ is $b$.
The side adjacent to angle $B$ is $a$.
The hypotenuse is $c$.

*Note:* The side opposite angle $A$ is the same as the side adjacent to angle $B$.
The side adjacent to angle $A$ is the same as the side opposite angle $B$.
The side opposite angle $B$ is the same as the side adjacent to angle $A$.
The side adjacent to angle $B$ is the same as the side opposite angle $A$.

74

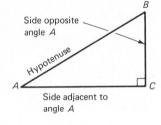

*Use right triangle ABC at the left to fill in each blank.*

1. The side opposite angle $A$ is _____.
2. The side opposite angle $B$ is _____.
3. The hypotenuse is _____.
4. The side adjacent to angle $A$ is _____.
5. The side adjacent to angle $B$ is _____.
6. The angle opposite side $a$ is _____.
7. The angle opposite side $b$ is _____.
8. The angle opposite side $c$ is _____.
9. The angle adjacent to side $a$ is _____.
10. The angle adjacent to side $b$ is _____.

A ratio is a comparison by division of two quantities. In a right triangle there are three ratios that are very important. Consider the right triangle shown.

These ratios are:

$$\frac{\text{side opposite angle } A}{\text{hypotenuse}}$$ is called the sine $A$ (abbreviated sin $A$)

$$\frac{\text{side adjacent to angle } A}{\text{hypotenuse}}$$ is called the cosine $A$ (abbreviated cos $A$)

$$\frac{\text{side opposite angle } A}{\text{side adjacent to angle } A}$$ is called the tangent $A$ (abbreviated tan $A$)

$$\sin A = \frac{\text{side opposite angle } A}{\text{hypotenuse}}$$

$$\cos A = \frac{\text{side adjacent to angle } A}{\text{hypotenuse}}$$

$$\tan A = \frac{\text{side opposite angle } A}{\text{side adjacent to angle } A}$$

The ratios can be defined similarly for angle $B$.

$$\sin B = \frac{\text{side opposite angle } B}{\text{hypotenuse}}$$

$$\cos B = \frac{\text{side adjacent to angle } B}{\text{hypotenuse}}$$

$$\tan B = \frac{\text{side opposite angle } B}{\text{side adjacent to angle } B}$$

EXAMPLE 1

Find the three trigonometric ratios of angle $A$.

$$\sin A = \frac{\text{side opposite angle } A}{\text{hypotenuse}} = \frac{3}{5} = 0.6$$

$$\cos A = \frac{\text{side adjacent to angle } A}{\text{hypotenuse}} = \frac{4}{5} = 0.8$$

$$\tan A = \frac{\text{side opposite angle } A}{\text{side adjacent to angle } A} = \frac{3}{4} = 0.75$$

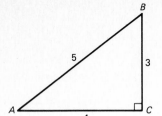

## EXAMPLE 2

Find the three trigonometric ratios of angle $B$.

$$\sin B = \frac{\text{side opposite angle } B}{\text{hypotenuse}} = \frac{4}{5} = 0.8$$

$$\cos B = \frac{\text{side adjacent to angle } B}{\text{hypotenuse}} = \frac{3}{5} = 0.6$$

$$\tan B = \frac{\text{side opposite angle } B}{\text{side adjacent to angle } B} = \frac{4}{3} = 1.33$$

*Note:* Every acute angle has three trigonometric ratios associated with it. In Table 20 of Appendix C, these angles and their trigonometric ratios are given.

In this text we assume that you will be using a calculator. (Table 20 is to be used as a reference and when you do not have access to a calculator.) When calculations involve a trigonometric ratio, we will use the following generally accepted rule of thumb for significant digits:

| Angle expressed to nearest: | Length of side contains: |
| --- | --- |
| 1° | Two significant digits |
| 0.1° | Three significant digits |
| 0.01° | Four significant digits |

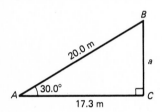

## EXAMPLE 3

Find angle $B$ and side $a$.

Angle $B$ can be found directly by using the fact that the sum of the angles of a triangle is 180°.

$$30.0° + 90° + B = 180°$$
$$B = 60.0°$$

To find side $a$ we must use a trigonometric ratio. Note that we are looking for the side opposite angle $A$ and that the hypotenuse is given. The trigonometric ratio having these two quantities is the sine.

$$\sin A = \frac{\text{side opposite angle } A}{\text{hypotenuse}}$$

$$\sin 30.0° = \frac{a}{20.0 \text{ m}}$$

Since $\sin 30.0° = 0.5000$,

$$0.5000 = \frac{a}{20.0 \text{ m}}$$

$$(0.5000)(20.0 \text{ m}) = \left(\frac{a}{20.0 \text{ m}}\right)(20.0 \text{ m}) \quad \text{Multiply both sides by 20.0 m.}$$

$$10.0 \text{ m} = a$$

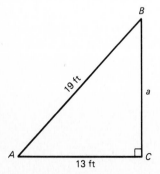

## EXAMPLE 4

Find angle $A$, angle $B$, and side $a$.

First, find angle $A$. The side adjacent to angle $A$ and the hypotenuse are given. Therefore, we use $\cos A$ to find angle $A$ because the $\cos A$ uses these two quantities:

$$\cos A = \frac{\text{side adjacent to } A}{\text{hypotenuse}}$$

$$\cos A = \frac{13 \text{ ft}}{19 \text{ ft}} = 0.684$$

Using a calculator, we find that $A = 47°$.

To find angle *B,* we use the fact that the sum of the angles of a triangle equals 180°.

$$90° + 47° + B = 180°$$
$$137° + B = 180°$$
$$B = 43°$$

To find side *a* we use sin *A* because the hypotenuse is given and side *a* is opposite angle *A.*

$$\sin A = \frac{\text{side opposite angle } A}{\text{hypotenuse}}$$

$$\sin 47° = \frac{a}{19 \text{ ft}}$$

Since sin 47° = 0.731,

$$0.731 = \frac{a}{19 \text{ ft}}$$

$$(0.731)(19 \text{ ft}) = \left(\frac{a}{19 \text{ ft}}\right)(19 \text{ ft}) \qquad \text{Multiply both sides by 19 ft.}$$

$$14 \text{ ft} = a$$

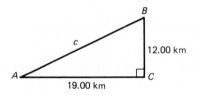

## EXAMPLE 5

Find angle *A,* angle *B,* and the hypotenuse in the right triangle shown.
To find angle *A,* use tan *A:*

$$\tan A = \frac{\text{side opposite angle } A}{\text{side adjacent to angle } A}$$

$$\tan A = \frac{12.00 \text{ km}}{19.00 \text{ km}} = 0.6316$$

$$A = 32.28°$$

To find angle *B,*

$$90° + 32.28° + B = 180°$$
$$122.28° + B = 180°$$
$$B = 57.72°$$

To find the hypotenuse, use sin *A.*

$$\sin A = \frac{\text{side opposite angle } A}{\text{hypotenuse}}$$

$$\sin 32.28° = \frac{12.00 \text{ km}}{c}$$

$$(\sin 32.28°)(c) = \left(\frac{12.00 \text{ km}}{c}\right)(c) \qquad \text{Multiply both sides by } c.$$

$$(\sin 32.28°)(c) = 12.00 \text{ km}$$

$$\frac{c(\sin 32.28°)}{\sin 32.28°} = \frac{12.00 \text{ km}}{\sin 32.28°} \qquad \begin{array}{l}\text{Divide both sides by} \\ \sin 32.28°.\end{array}$$

$$c = \frac{12.00 \text{ km}}{\sin 32.28°}$$

$$= \frac{12.00 \text{ km}}{0.5341}$$

$$= 22.47 \text{ km}$$

*Find the following trigonometric ratios using a calculator. Round each ratio to four significant digits.*

**1.** sin 71°
**2.** cos 4$\overline{0}$°
**3.** tan 61°

**4.** tan 41.2°
**5.** cos 11.5°
**6.** sin 79.0°

**7.** cos 49.63°
**8.** tan 53.45°
**9.** tan 17.04°

**10.** cos 34°
**11.** sin 27.5°
**12.** cos 58.72°

*Find each angle rounded to the nearest whole degree.*

**13.** sin $A = 0.2678$
**14.** cos $B = 0.1046$
**15.** tan $A = 0.9237$

**16.** sin $B = 0.9253$
**17.** cos $B = 0.6742$
**18.** tan $A = 1.351$

*Find each angle rounded to the nearest tenth degree.*

**19.** sin $B = 0.5963$
**20.** cos $A = 0.9406$
**21.** tan $B = 1.053$

**22.** sin $A = 0.9083$
**23.** cos $A = 0.8660$
**24.** tan $B = 0.9433$

*Find each angle rounded to the nearest hundredth degree.*

**25.** sin $A = 0.3792$
**26.** cos $B = 0.06341$
**27.** tan $B = 0.3010$

**28.** sin $A = 0.4540$
**29.** cos $B = 0.8141$
**30.** tan $A = 2.369$

*Solve each triangle (find the missing angles and sides) using trigonometric ratios.*

**31.**

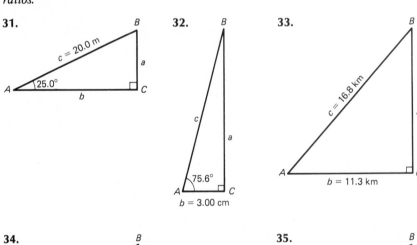

**32.**

**33.**

**34.**

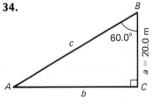

**35.**

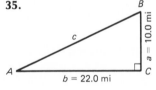

**36.**

**37.**

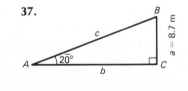

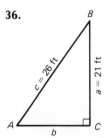

**38.**

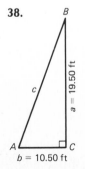

**39.**

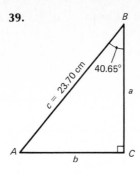

**40.**

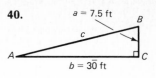

**41.** Answer the following questions about the round taper shown in the figure.
  (a) What is the value of $\angle BAC$?
  (b) What length is $BC$?
  (c) What is the diameter of end $x$?

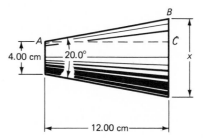

**42.** Across the flats *(a)* a hexagonal nut is $\frac{3}{4}$ in. Calculate the distance across the corners *(b)*.

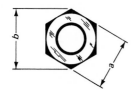

**43.** What are the distances $C$ and $D$ between the holes of the plate shown in the figure?

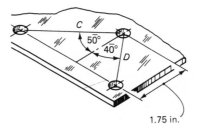

**5-2**
**PYTHAGOREAN THEOREM**

When given the two legs of a right triangle, the hypotenuse can be found without using trigonometric ratios.

From geometry, the sum of the squares of the legs of a right triangle is equal to the square of the hypotenuse *(Pythagorean theorem):*

$$a^2 + b^2 = c^2$$

or, taking the square root of each side of the equation,

$$\boxed{c = \sqrt{a^2 + b^2}}$$

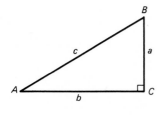

**EXAMPLE 1**

Find the hypotenuse of the triangle shown.

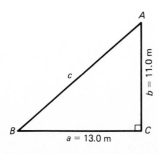

$$c = \sqrt{a^2 + b^2}$$
$$c = \sqrt{(13.0 \text{ m})^2 + (11.0 \text{ m})^2}$$
$$= \sqrt{169 \text{ m}^2 + 121 \text{ m}^2}$$
$$= \sqrt{29\overline{0} \text{ m}^2}$$
$$= 17.0 \text{ m}$$

Also, if one leg and the hypotenuse are given, the other leg can be found by

$$a = \sqrt{c^2 - b^2}$$
$$b = \sqrt{c^2 - a^2}$$

## EXAMPLE 2

Find side $b$ in the triangle shown.

$$b = \sqrt{c^2 - a^2}$$
$$b = \sqrt{(12.2 \text{ km})^2 - (7.30 \text{ km})^2}$$
$$= 9.77 \text{ km}$$

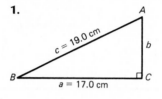

## PROBLEMS

*Find the missing side in each right triangle using the Pythagorean theorem.*

**1.**

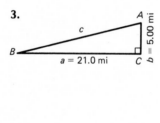

**2.**

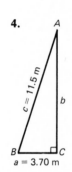

**3.**

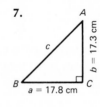

**4.**

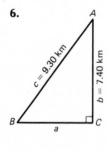

**5.**

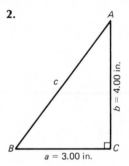

**6.**

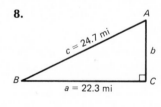

**7.**

**8.**

**9.**

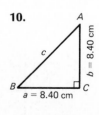

**10.**

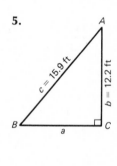

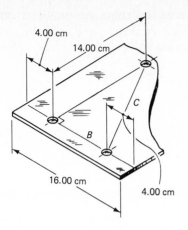

4.00 cm

14.00 cm

C

B

16.00 cm

4.00 cm

**11.** What are the distances between holes on the plate shown in the figure?

**12.** A piece of electrical conduit cuts across a corner of a room $24\bar{0}$ cm from the corner. It meets the adjoining wall $35\bar{0}$ cm from the corner. Find length *AB* of conduit needed.

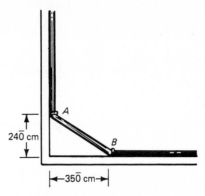

$24\bar{0}$ cm

A

B

$\leftarrow 35\bar{0}$ cm $\rightarrow$

**5-3**

**COMPONENTS OF A VECTOR**

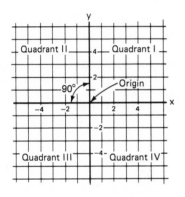

Before further study of vectors, we need to discuss components of vectors. This requires an understanding of the number plane. The number plane is determined by a horizontal line called the *x*-axis and a vertical line called the *y*-axis intersecting at right angles as shown. These two lines divide the number plane into four quadrants which we will label as quadrants I, II, III, and IV as illustrated.

Each axis has a scale, and the intersection of the two axes is called the *origin*. The *x*-axis contains positive numbers to the right of the origin and negative numbers to the left of the origin. The *y*-axis contains positive numbers above the origin and negative numbers below the origin.

When a vector is expressed graphically as a sum of vectors, the vectors are called *components* of the resultant vector.

The components of vector **R** are vectors **A**, **B**, and **C**.

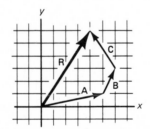
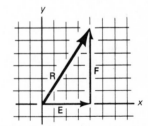

*Note:* A vector may have more than one set of component vectors. The components of vector **R** are vectors **E** and **F**.

We are interested in the components of a vector which are perpendicular to each other and which are on or parallel to the *x*- and *y*-axes. In particular, we are interested in the type of component vectors we found in the preceding figure (component vectors **E** and **F**). The component vector that lies on or is parallel to the *x*-axis is called the *x component*. The component vector that lies on or is parallel to the *y*-axis is called the *y component*. Here are three examples.

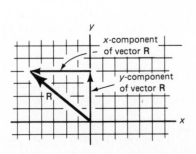
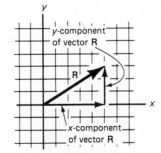
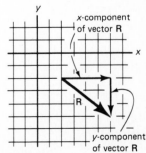

Thus far we have considered the $x$ and $y$ components as vectors. However, they can also be thought of as signed numbers. The sign of the number corresponds to the direction of the components as follows:

| *x component* | *y component* |
|---|---|
| +, if right | +, if up |
| −, if left | −, if down |

The absolute value of the signed number corresponds to the magnitude of the vector.

## EXAMPLE 1

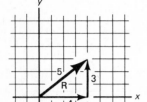

Find the $x$ and $y$ components of vector **R**.

$$x \text{ component of } \mathbf{R} = +4$$
$$y \text{ component of } \mathbf{R} = +3$$

## EXAMPLE 2

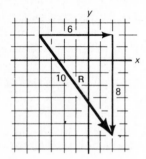

Find the $x$ and $y$ components of vector **R**.

$$x \text{ component of } \mathbf{R} = +6$$
$$y \text{ component of } \mathbf{R} = -8$$

($y$ component points in a negative direction)

## EXAMPLE 3

Find the $x$ and $y$ components of vector **R**.

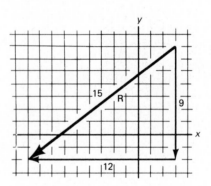

$x$ component of **R** = −12

$y$ component of **R** = −9

(both $x$ and $y$ components point in a negative direction)

Now that we have considered the $x$ and $y$ components as signed numbers, we can find the resultant vector of several vectors using arithmetic and graphing. To find the resultant vector of several vectors, find the $x$ component of each vector and find the sum of the $x$ components. Then find the $y$ component of each vector and find the sum of the $y$ components. The two sums are the $x$ and $y$ components of the resultant vector. This is shown in the following example.

## EXAMPLE 4

Find the $x$ and $y$ components of vector **R** using the $x$ and $y$ components of vectors **A** and **B**.

$$\mathbf{A} + \mathbf{B} = \mathbf{R}$$

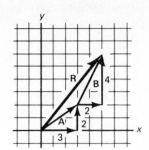

*Note:* The *x* component of **A** is +3
The *x* component of **B** is +2
The *x* component of **R** is (+3) + (+2) = +5

The *y* component of **A** is +2
The *y* component of **B** is +4
The *y* component of **R** is (+2) + (+4) = +6

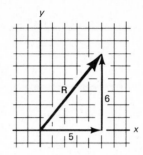

## EXAMPLE 5

Find the *x* and *y* components of vector **R** using the *x* and *y* components of vectors **A**, **B**, and **C**.

$$A + B + C = R$$

| Vector | x component | y component |
|--------|-------------|-------------|
| **A** | +7 | +2 |
| **B** | −2 | +1 |
| **C** | −1 | +4 |
| **R** | +4 | +7 |

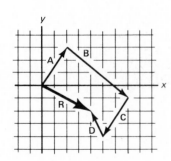

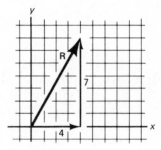

## EXAMPLE 6

Find the *x* and *y* components of vector **R** using the *x* and *y* components of vectors **A**, **B**, **C**, and **D**.

$$A + B + C + D = R$$

| Vector | x component | y component |
|--------|-------------|-------------|
| **A** | +2 | +3 |
| **B** | +5 | −4 |
| **C** | −2 | −3 |
| **D** | −1 | +2 |
| **R** | +4 | −2 |

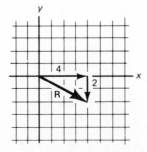

*Find the* x *and* y *components of the following vectors. (Express them as signed numbers and graph them as vectors.)*

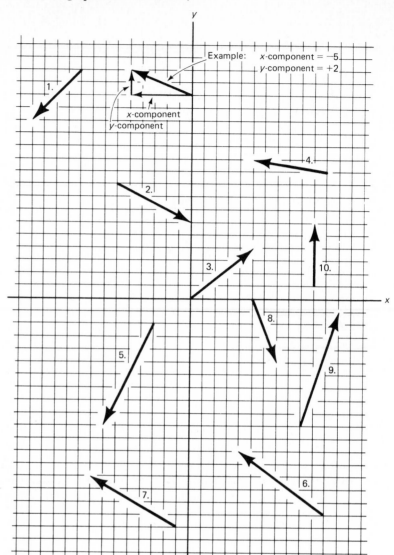

*Find the* x *and* y *components of the resultant vector* ***R*** *and graph the resultant vector* ***R***.

| | Vector | x component | y component |
|---|---|---|---|
| **11.** | **A** | +2 | +3 |
| | **B** | +7 | +2 |
| | **R** | | |
| **12.** | **A** | +9 | −5 |
| | **B** | −4 | −6 |
| | **R** | | |
| **13.** | **A** | −2 | +13 |
| | **B** | −11 | +1 |
| | **C** | +3 | −4 |
| | **R** | | |
| **14.** | **A** | +10 | −5 |
| | **B** | −13 | −9 |
| | **C** | +4 | +3 |
| | **R** | | |

| | | | |
|---|---|---|---|
| **15.** | **A** | +17 | +7 |
| | **B** | −14 | +11 |
| | **C** | +7 | +9 |
| | **D** | −6 | −15 |
| **16.** | **A** | +1 | +7 |
| | **B** | +9 | −4 |
| | **C** | −4 | +13 |
| | **D** | −11 | −4 |
| | **R** | | |
| **17.** | **A** | +1.5 | −1.5 |
| | **B** | −3 | −2 |
| | **C** | +7.5 | −3 |
| | **D** | +2 | +2.5 |
| | **R** | | |
| **18.** | **A** | +1 | −1 |
| | **B** | −4 | −2 |
| | **C** | +2 | +4 |
| | **D** | +5 | −3 |
| | **E** | +3 | +5 |
| | **R** | | |
| **19.** | **A** | +1.5 | +2.5 |
| | **B** | −2 | −3 |
| | **C** | +3.5 | −7.5 |
| | **D** | −4 | +6 |
| | **E** | −5.5 | +2 |
| | **R** | | |
| **20.** | **A** | −7 | +15 |
| | **B** | +13.5 | −17.5 |
| | **C** | −7.5 | −20 |
| | **D** | +6 | +13.5 |
| | **E** | +2.5 | +2.5 |
| | **F** | −11 | +11.5 |
| | **R** | | |

**5-4**
**VECTORS**
**IN STANDARD POSITION**

Any vector may be placed in any position in the number plane as long as its magnitude and direction are not changed. A vector is in *standard position* when its initial point is at the origin of the number plane. A vector in standard position is expressed in terms of its length and its angle $\theta$, where $\theta$ *is measured counterclockwise from the positive x-axis to the vector.* The vectors shown are in standard position.

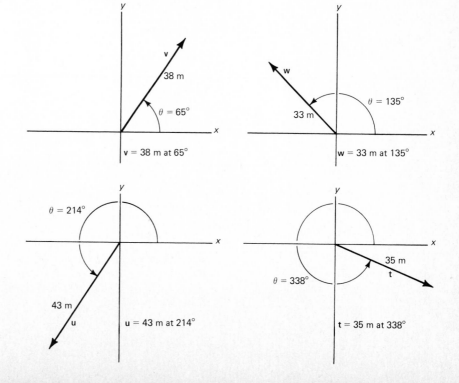

$\mathbf{v} = 38$ m at $65°$

$\mathbf{w} = 33$ m at $135°$

$\mathbf{u} = 43$ m at $214°$

$\mathbf{t} = 35$ m at $338°$

Find the $x$ and $y$ components of the vector 10.0 m at 60.0°.

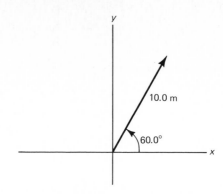

To find the $x$ and $y$ components, draw a right triangle where the legs represent the $x$ and $y$ components of the vector.

The absolute value of the $x$ component of the vector is the length of the side adjacent to the 60.0° angle. Therefore, to find the $x$ component

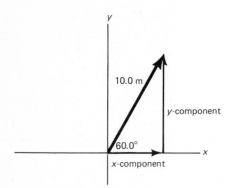

$$\cos 60.0° = \frac{\text{side adjacent to } 60.0°}{\text{hypotenuse}}$$

Since $\cos 60.0° = 0.5000$,

$$\text{hypotenuse} = 10.0 \text{ m}$$
$$\text{side adjacent to } 60.0° = x \text{ component}$$

we have

$$0.5000 = \frac{x \text{ component}}{10.0 \text{ m}}$$

$$(0.5000)(10.0 \text{ m}) = \left(\frac{x \text{ component}}{10.0 \text{ m}}\right)(10.0 \text{ m}) \qquad \text{Multiply both sides by 10.0 m.}$$

$$5.00 \text{ m} = x \text{ component}$$

Since the $x$ component is pointing in the positive $x$ direction, the $x$ component is +5.00 m.

The absolute value of the $y$ component of the vector is the length of the side opposite the 60.0° angle. Therefore, to find the $y$ component,

$$\sin 60.0° = \frac{\text{side opposite } 60.0°}{\text{hypotenuse}}$$

Since, $\sin 60.0° = 0.8660$,

$$\text{hypotenuse} = 10.0 \text{ m}$$
$$\text{side opposite } 60.0° = y \text{ component}$$

we have

$$0.8660 = \frac{y \text{ component}}{10.0 \text{ m}}$$

$$(0.8660)(10.0 \text{ m}) = \left(\frac{y \text{ component}}{10.0 \text{ m}}\right)(10.0 \text{ m}) \qquad \text{Multiply both sides by 10.0 m.}$$

$$8.66 \text{ m} = y \text{ component}$$

Since the $y$ component is pointing in the positive $y$ direction, the $y$ component is +8.66 m.

Find the $x$ and $y$ components of the vector 13.0 km at 220.0°.

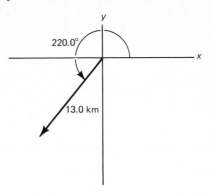

Complete a right triangle with the $x$ and $y$ components being the two legs. First, find angle $A$ as follows:

$$180° + A = 220.0°$$

$$A = 40.0°$$

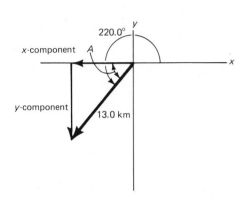

The absolute value of the $x$ component is the length of the side adjacent to angle $A$. Therefore, to find the $x$ component,

$$\cos A = \frac{\text{side adjacent to } A}{\text{hypotenuse}}$$

Since $\cos 40.0° = 0.7660$,

$$\text{hypotenuse} = 13.0 \text{ km}$$

$$\text{side adjacent to } A = x \text{ component}$$

we have

$$0.7660 = \frac{x \text{ component}}{13.0 \text{ km}}$$

$$(0.7660)(13.0 \text{ km}) = \left(\frac{x \text{ component}}{13.0 \text{ km}}\right)(13.0 \text{ km}) \qquad \text{Multiply both sides by 13.0 km.}$$

$$9.96 \text{ km} = x \text{ component}$$

Since the $x$ component is pointing in the negative $x$ direction, the $x$ component is $-9.96$ km.

The absolute value of the $y$ component of the vector is the length of the side opposite angle $A$. Therefore, to find the $y$ component,

$$\sin A = \frac{\text{side opposite } A}{\text{hypotenuse}}$$

Since $\sin 40.0° = 0.6428$,

$$\text{hypotenuse} = 13.0 \text{ km}$$

$$\text{side opposite } A = y \text{ component}$$

we have

$$0.6428 = \frac{y \text{ component}}{13.0 \text{ km}}$$

$$(0.6428)(13.0 \text{ km}) = \left(\frac{y \text{ component}}{13.0 \text{ km}}\right)(13.0 \text{ km}) \qquad \text{Multiply both sides by 13.0 km.}$$

$$8.36 \text{ km} = y \text{ component}$$

Since the $y$ component is pointing in the negative $y$ direction, the $y$ component is $-8.36$ km.

Find the $x$ and $y$ components of the vector 27.0 ft/s at 125.0°.

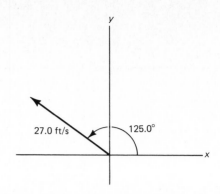

To find the $x$ and $y$ components, draw a right triangle, with the $x$ and $y$ components being the two legs.

First, find angle $A$ as follows:

$$A + 125.0° = 180°$$
$$A = 55.0°$$

To find the $x$ component, find the length of the side adjacent to angle $A$. We then use

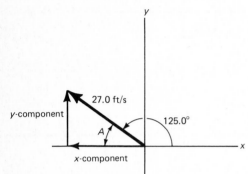

$$\cos A = \frac{\text{side adjacent to } A}{\text{hypotenuse}}$$

Since $\cos 55.0° = 0.5736$,

$$\text{hypotenuse} = 27.0 \text{ ft/s}$$
$$\text{side adjacent to } A = x \text{ component}$$

we have

$$0.5736 = \frac{x \text{ component}}{27.0 \text{ ft/s}}$$

$$(0.5736)(27.0 \text{ ft/s}) = \left(\frac{x \text{ component}}{27.0 \text{ ft/s}}\right)(27.0 \text{ ft/s})$$

$$15.5 \text{ ft/s} = x \text{ component}$$

Since the $x$ component is pointing in the negative $x$ direction, the $x$ component is $-15.5$ ft/s.

To find the $y$ component, find the length of the side opposite angle $A$. We then use

$$\sin A = \frac{\text{side opposite } A}{\text{hypotenuse}}$$

Since $\sin 55.0° = 0.8192$,

$$\text{hypotenuse} = 27.0 \text{ ft/s}$$
$$\text{side opposite } A = y \text{ component}$$

we have

$$0.8192 = \frac{y \text{ component}}{27.0 \text{ ft/s}}$$

$$(0.8192)(27.0 \text{ ft/s}) = \left(\frac{y \text{ component}}{27.0 \text{ ft/s}}\right)(27.0 \text{ ft/s})$$

$$22.1 \text{ ft/s} = y \text{ component}$$

Since the $y$ component is pointing in the positive $y$ direction, the $y$ component is $+22.1$ ft/s.

In general:

> To find the $x$ and $y$ components of a vector when it is given in standard position:
>
> **1.** Complete the right triangle with the legs being the $x$ and $y$ components of the vector.
> **2.** Find the lengths of the legs of the right triangle.
> **3.** Determine the sign of the $x$ and $y$ components.

## EXAMPLE 4

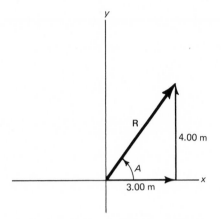

Find vector **R** in standard position whose $x$ component is +3.00 m and $y$ component is +4.00 m.

First, find angle $A$ as follows:

$$\tan A = \frac{\text{side opposite } A}{\text{side adjacent to } A}$$

Since

$$\text{side opposite } A = 4.00 \text{ m}$$
$$\text{side adjacent to } A = 3.00 \text{ m}$$

we have

$$\tan A = \frac{4.00 \text{ m}}{3.00 \text{ m}} = 1.333$$

and then

$$A = 53.1° = \theta$$

The magnitude can be found by using the Pythagorean theorem:

$$c = \sqrt{a^2 + b^2}$$

Thus,

$$\mathbf{R} = \sqrt{(3.00 \text{ m})^2 + (4.00 \text{ m})^2}$$
$$= 5.00 \text{ m}$$

That is, **R** is 5.00 m at 53.1°.

## EXAMPLE 5

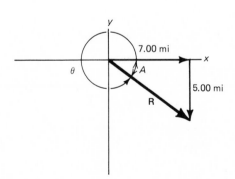

Find vector **R** in standard position whose $x$ component is +7.00 mi and $y$ component is −5.00 mi.

First, find angle $A$ as follows:

$$\tan A = \frac{\text{side opposite } A}{\text{side adjacent to } A}$$

Since

$$\text{side opposite } A = 5.00 \text{ mi}$$
$$\text{side adjacent to } A = 7.00 \text{ mi}$$

we have

$$\tan A = \frac{5.00 \text{ mi}}{7.00 \text{ mi}} = 0.7143$$

and then

$$A = 35.5°$$

But

$$\theta = 360° - A$$
$$= 360° - 35.5°$$
$$= 324.5°$$

The magnitude can be found by using the Pythagorean theorem:

$$c = \sqrt{a^2 + b^2}$$

Thus,

$$\mathbf{R} = \sqrt{(7.00 \text{ mi})^2 + (5.00 \text{ mi})^2}$$
$$= 8.60 \text{ mi}$$

That is, **R** is 8.60 mi at 324.5°.

## EXAMPLE 6

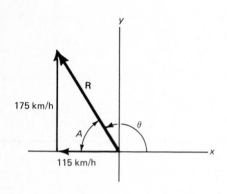

Find vector **R** in standard position whose *x* component is −115 km/h and *y* component is +175 km/h.

First, find angle *A* as follows:

$$\tan A = \frac{\text{side opposite } A}{\text{side adjacent to } A}$$

Since

$$\text{side opposite } A = 175 \text{ km/h}$$
$$\text{side adjacent to } A = 115 \text{ km/h}$$

we have

$$\tan A = \frac{175 \text{ km/h}}{115 \text{ km/h}} = 1.522$$

and then

$$A = 56.7°$$

But

$$\theta = 180° - A$$
$$= 180° - 56.7°$$
$$= 123.3°$$

The magnitude can be found by using the Pythagorean theorem:

$$c = \sqrt{a^2 + b^2}$$
$$\mathbf{R} = \sqrt{(115 \text{ km/h})^2 + (175 \text{ km/h})^2}$$
$$= 209 \text{ km/h}$$

That is, **R** is 209 km/h at 123.3°.

In general:

> To find a vector in standard position when the *x* and *y* components are given:
>
> 1. Complete the right triangle with the legs being the *x* and *y* components of the vector.
> 2. Find the acute angle *A* of the right triangle whose vertex is at the origin by using tan *A*.

**3.** Find angle $\theta$ in standard position as follows:

$$\theta = A \qquad (\theta \text{ in first quadrant})$$
$$\theta = 180° - A \qquad (\theta \text{ in second quadrant})$$
$$\theta = 180° + A \qquad (\theta \text{ in third quadrant})$$
$$\theta = 360° - A \qquad (\theta \text{ in fourth quadrant})$$

**4.** Find the magnitude of the vector by using the Pythagorean theorem:

$$c = \sqrt{a^2 + b^2}$$

## EXAMPLE 7

A plane is flying due north (at 90°) at 475 km/h. Suddenly there is a wind from the east (at 180°) at 55.0 km/h. What is the plane's velocity with respect to the ground in standard position?

First, find angle $A$ as follows:

$$\tan A = \frac{\text{side opposite } A}{\text{side adjacent to } A}$$

Since

$$\text{side opposite } A = 475 \text{ km/h}$$
$$\text{side adjacent to } A = 55.0 \text{ km/h}$$

we have

$$\tan A = \frac{475 \text{ km/h}}{55.0 \text{ km/h}} = 8.636$$

Then

$$A = 83.4°$$

But

$$\theta = 180° - 83.4 = 96.6°$$

The magnitude of the velocity (ground speed) can be found by using the Pythagorean theorem:

$$c = \sqrt{a^2 + b^2}$$
$$\mathbf{R} = \sqrt{(55.0 \text{ km/h})^2 + (475 \text{ km/h})^2}$$
$$= 478 \text{ km/h}$$

That is, the velocity of the plane is 478 km/h at 96.6°.

## EXAMPLE 8

A plane is flying northwest (at 135.0°) at 325 km/h. Suddenly there is a wind from 30.0° south of west (at 30.0°) at 65.0 km/h. What is the plane's velocity with respect to the ground in standard position?

First, find the $x$ and $y$ components of each vector:

*Plane:* See figure at left.

$$A = 180° - 135.0° = 45.0°$$

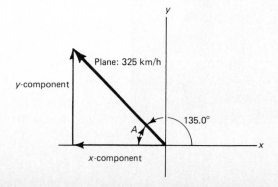

To find:

| x component | y component |
|---|---|
| $\cos A = \dfrac{\text{side adjacent to } A}{\text{hypotenuse}}$ | $\sin A = \dfrac{\text{side opposite } A}{\text{hypotenuse}}$ |
| $\cos 45.0° = \dfrac{x \text{ component}}{325 \text{ km/h}}$ | $\sin 45.0° = \dfrac{y \text{ component}}{325 \text{ km/h}}$ |
| $(0.7071)(325 \text{ km/h}) = x \text{ component}$ | $(0.7071)(325 \text{ km/h}) = y \text{ component}$ |
| $23\bar{0} \text{ km/h} = x \text{ component}$ | $23\bar{0} \text{ km/h} = y \text{ component}$ |
| Thus, $x$ component $= -23\bar{0}$ km/h | $y$ component $= +23\bar{0}$ km/h |

*Wind:* See figure at left.

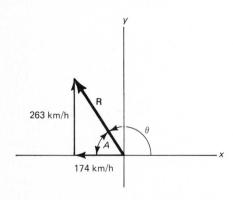

To find:

| x component | y component |
|---|---|
| $\cos A = \dfrac{\text{side adjacent to } A}{\text{hypotenuse}}$ | $\sin A = \dfrac{\text{side opposite } A}{\text{hypotenuse}}$ |
| $\cos 30.0° = \dfrac{x \text{ component}}{65.0 \text{ km/h}}$ | $\sin 30.0° = \dfrac{y \text{ component}}{65.0 \text{ km/h}}$ |
| $(0.8660)(65.0 \text{ km/h}) = x \text{ component}$ | $(0.5000)(65.0 \text{ km/h}) = y \text{ component}$ |
| $56.3 \text{ km/h} = x \text{ component}$ | $32.5 \text{ km/h} = y \text{ component}$ |
| Thus, $x$ component $= +56.3$ km/h | $y$ component $= +32.5$ km/h |

To find **R**:

|  | x component | y component |  |
|---|---|---|---|
| plane: | $-23\bar{0}$ km/h | $+23\bar{0}$ km/h |  |
| wind: | $+\ 56$ km/h | $+\ 33$ km/h | (need same precision |
| sum: | $-174$ km/h | $+263$ km/h | to add measurements) |

We find angle $A$ as follows:

$$\tan A = \frac{\text{side opposite } A}{\text{side adjacent to } A}$$

$$\tan A = \frac{263 \text{ km/h}}{174 \text{ km/h}} = 1.512$$

$$A = 56.5°$$

and

$$\theta = 180° - 56.5° = 123.5°$$

The magnitude of **R** is found by using the Pythagorean theorem:

$$c = \sqrt{a^2 + b^2}$$
$$\mathbf{R} = \sqrt{(174 \text{ km/h})^2 + (263 \text{ km/h})^2}$$
$$= 315 \text{ km/h}$$

That is, the velocity of the plane is 315 km/h at 123.5°.

## PROBLEMS

*Make a sketch of each vector in standard position. Use the scale 1.0 cm = $1\bar{0}$ m.*

1. $\mathbf{v} = 2\bar{0}$ m at 25°
2. $\mathbf{w} = 25$ m at 125°
3. $\mathbf{u} = 25$ m at 245°
4. $\mathbf{s} = 2\bar{0}$ m at 345°

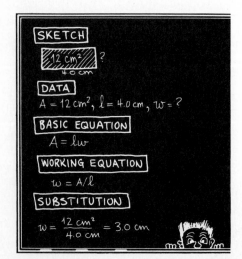

**5.** **t** = 15 m at 105°

**6.** **r** = 35 m at 291°

**7.** **m** = 3$\overline{0}$ m at 405°

**8.** **n** = 25 m at 525°

*Find the x and y components of each vector shown.*

**9.**

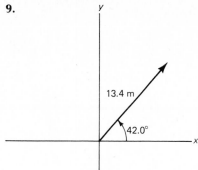

**10.**

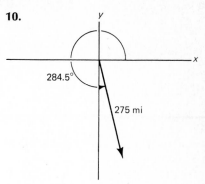

**11.**

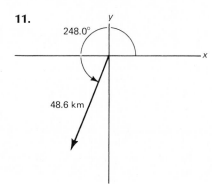

**12.**

**13.**

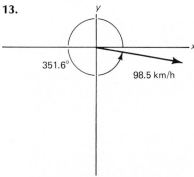

**14.**

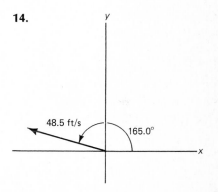

*Find the x and y components of each vector given in standard position.*

**15.** 38.9 m at 10.5°

**16.** 478 ft at 195.0°

**17.** 9.60 km at 310.0°

**18.** 5430 mi at 153.7°

**19.** 29.5 m/s at 101.5°

**20.** 154 mi/h at 273.2°

*In Problems 21–32, find each resultant vector* **R**. *Give* **R** *in standard position.*

**21.**

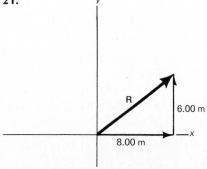

**22.**

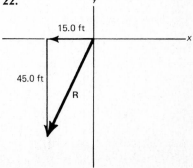

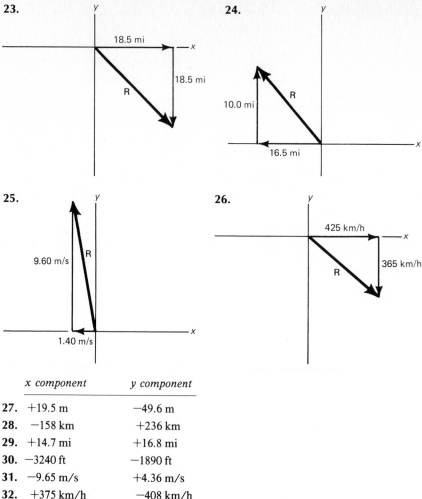

| | x component | y component |
|---|---|---|
| **27.** | +19.5 m | −49.6 m |
| **28.** | −158 km | +236 km |
| **29.** | +14.7 mi | +16.8 mi |
| **30.** | −3240 ft | −1890 ft |
| **31.** | −9.65 m/s | +4.36 m/s |
| **32.** | +375 km/h | −408 km/h |

**33.** A plane is flying due north at 325 km/h. Suddenly there is a wind from the south at 45 km/h. What is the plane's velocity with respect to the ground in standard position?

**34.** A plane is flying due west at 275 km/h. Suddenly there is a wind from the west at 8$\overline{0}$ km/h. What is the plane's velocity with respect to the ground in standard position?

**35.** A plane is flying due west at 235 km/h. Suddenly there is a wind from the north at 45.0 km/h. What is the plane's velocity with respect to the ground in standard position?

**36.** A plane is flying due north at 185 mi/h. Suddenly there is a wind from the west at 35.0 mi/h. What is the plane's velocity with respect to the ground in standard position?

**37.** A plane is flying southwest at 155 mi/h. Suddenly there is a wind from the west at 45.0 mi/h. What is the plane's velocity with respect to the ground in standard position?

**38.** A plane is flying southeast at 215 km/h. Suddenly there is a wind from the north at 75.0 km/h. What is the plane's velocity with respect to the ground in standard position?

**39.** A plane is flying at 25.0° north of west at 19$\overline{0}$ km/h. Suddenly there is a wind from 15.0° north of east at 45.0 km/h. What is the plane's velocity with respect to the ground in standard position?

**40.** A plane is flying at 36.0° south of west at 15$\overline{0}$ mi/h. Suddenly there is a wind from 75.0° north of east at 55.0 mi/h. What is the plane's velocity with respect to the ground in standard position?

# CHAPTER
# *6*

## Concurrent Forces

**6-1**
**EQUILIBRIUM
IN ONE DIMENSION**

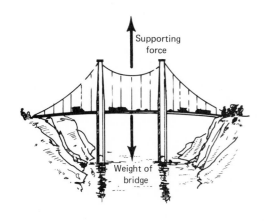

Supporting
force

Weight of
bridge

In Chapter 4 we studied forces that actually caused motion. Forces also tend to cause motion. We discussed a bridge on which the forces balanced each other and did not actually cause motion. The weight was balanced by the supporting force. Because the forces balanced each other, the net force was zero.

There are many important technical applications in which the net force on an object is zero. *An object is said to be in equilibrium when the net force acting on it is zero.* That is, a body is in equilibrium when it remains at rest or is not accelerating (not moving at constant velocity). The study of equilibrium problems is called *statics*.

Let the forces applied to an object act in the same direction or in opposite directions (in one dimension). For the net force to be zero, the forces in one direction must equal the forces in the opposite direction.

We can write the equation for equilibrium in one dimension as

$$\mathbf{F_+} = \mathbf{F_-}$$

where $\mathbf{F_+}$ is the sum of all forces acting in one direction (call it the positive direction) and $\mathbf{F_-}$ is the sum of all the forces acting in the opposite (negative) direction.

## EXAMPLE 1

A cable supports a large crate of weight $100\overline{0}$ N. What is the upward force on the crate if it is in equilibrium?

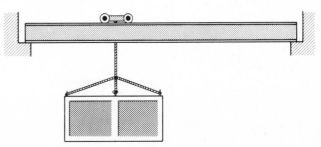

SKETCH: It is helpful to draw a "free-body diagram" of the crate. This is a sketch in which we draw only the object in equilibrium and show the forces which act on it. Note that we call the upward direction positive as indicated by the arrow.

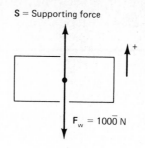

S = Supporting force

$F_w = 100\overline{0}$ N

DATA:

$$F_w = 100\overline{0} \text{ N}$$
$$S = ?$$

BASIC EQUATION:

$$\mathbf{F}_+ = \mathbf{F}_-$$

WORKING EQUATION:

$$S = F_w$$

SUBSTITUTION:

$$S = 100\overline{0} \text{ N}$$

## EXAMPLE 2

Four persons are having a tug-of-war with a rope. Harry and Mary are on the left; Bill and Jill are on the right. Mary pulls with a force of 105 lb, Harry pulls with a force of 255 lb, and Jill pulls with a force of 165 lb. What force must Bill pull to produce equilibrium?

SKETCH:

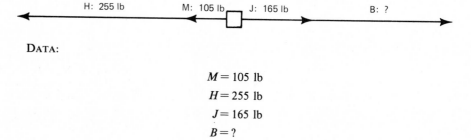

H: 255 lb        M: 105 lb    J: 165 lb        B: ?

DATA:

$$M = 105 \text{ lb}$$
$$H = 255 \text{ lb}$$
$$J = 165 \text{ lb}$$
$$B = ?$$

BASIC EQUATION:

$$M + H = J + B$$

WORKING EQUATION:

$$B = M + H - J$$

SUBSTITUTION:

$$B = 105 \text{ lb} + 255 \text{ lb} - 165 \text{ lb}$$
$$= 195 \text{ lb}$$

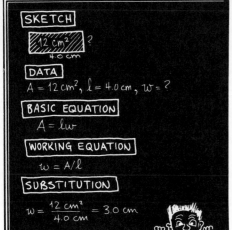

**PROBLEMS**

*Find the force* F *that will produce equilibrium for the free-body diagrams shown. Use the same procedure as in the Examples 1 and 2.*

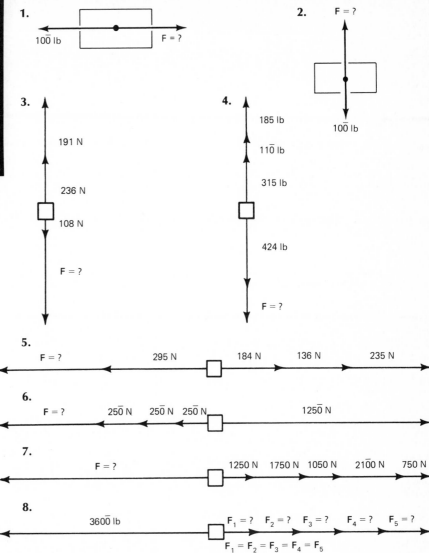

1.  100 lb    F = ?

2.  F = ?    100 lb

3.  191 N    236 N    108 N    F = ?

4.  185 lb    110 lb    315 lb    424 lb    F = ?

5.  F = ?    295 N    184 N    136 N    235 N

6.  F = ?    250 N    250 N    250 N    1250 N

7.  F = ?    1250 N    1750 N    1050 N    2100 N    750 N

8.  3600 lb    $F_1$ = ?    $F_2$ = ?    $F_3$ = ?    $F_4$ = ?    $F_5$ = ?
    $F_1 = F_2 = F_3 = F_4 = F_5$

9.  Five persons are having a tug-of-war. Kurt and Brian are on the left; Amy, Barbara, and Joyce are on the right. Amy pulls with a force of 225 N, Barbara pulls with a force of 495 N, Joyce pulls with a force of 455 N, and Kurt pulls with a force of 605 N. What force must Brian pull to produce equilibrium?

10. A certain wire can support 6450 lb before it breaks. Seven 820-lb weights are suspended from the wire.

    (a) Can the wire support an eighth weight?
    (b) If so, how much more weight can the wire support before it breaks?

11. The frictional force of a loaded pallet in a warehouse is 385 lb. Can three workers, each exerting a force of 135 lb, push it to the side?

12. A long bridge has a weight limit of 7.0 tons. How heavy a load can a 2.5-ton truck carry across?

**6-2**
**CONCURRENT FORCES**
**IN EQUILIBRIUM**

In the construction of buildings and machinery, a technician is often required to solve problems involving the equilibrium of certain parts such as pins and joints. The forces acting on these parts do not always act in

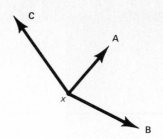

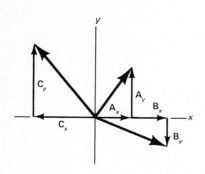

one dimension; instead they may act in several directions as shown. These forces are called *concurrent* because they are all applied at the same point, *x*.

*If an object is to be in equilibrium, the net force acting on it must be zero.* The separate forces, **A, B,** and **C,** are vectors, and the net force is the vector sum of these forces.

We can now apply the component method of adding vectors that we learned in Chapter 5. The *x* and *y* components of the net force can be found by adding the *x* components and the *y* components of the separate forces. The *x* component of the net force is

$$\mathbf{R}_x = \mathbf{A}_x + \mathbf{B}_x + \mathbf{C}_x = \text{sum of } x \text{ components}$$

The *y* component of the net force is

$$\mathbf{R}_y = \mathbf{A}_y + \mathbf{B}_y + \mathbf{C}_y = \text{sum of } y \text{ components}$$

If the net force is zero, the sum of the *x* components and the sum of the *y* components must also be zero. The conditions for equilibrium then are

$$\text{sum of } x \text{ components} = 0$$

$$\text{sum of } y \text{ components} = 0$$

---

Use the following procedure to solve equilibrium problems:

(a) Draw the free-body diagram of the point at which the unknown forces act.

(b) Find the *x* and *y* component of each force.

(c) Substitute the components in the equations

$$\text{sum of } x \text{ components} = 0$$

$$\text{sum of } y \text{ components} = 0$$

(d) Solve for the unknowns. This may involve two simultaneous equations.

---

Tension

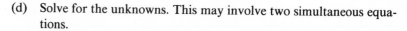

Compression

In many problems we will be interested in finding the tension or compression in part of a structure, such as in a beam or a cable. *Tension* is a stretching force produced by forces pulling outward on the ends of the object. *Compression* is a compressing force produced by forces pushing inward on the ends of an object. A rubber band being stretched is an example of tension. A valve spring whose ends are pushed together is an example of compression.

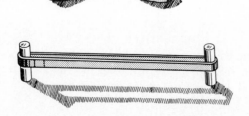

EXAMPLE 1

Find the forces **F** and **F′** necessary to produce equilibrium in the free-body diagram shown.

(a)

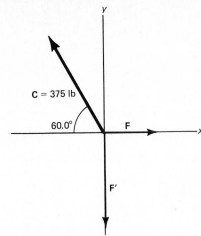

(b) *x components*

$$\mathbf{F}_x = \mathbf{F}$$
$$\mathbf{F}'_x = 0$$
$$\mathbf{C}_x = -(375 \text{ lb})(\cos 60.0°)$$
$$= -188 \text{ lb}$$

*y components*

$$\mathbf{F}_y = 0$$
$$\mathbf{F}'_y = -\mathbf{F}'$$
$$\mathbf{C}_y = (375 \text{ lb})(\sin 60.0°)$$
$$= 325 \text{ lb}$$

(c) Sum of *x* components = 0
$$\mathbf{F} + 0 + (-188 \text{ lb}) = 0$$

(d) $$\mathbf{F} = 188 \text{ lb}$$

Sum of *y* components = 0
$$0 + (-\mathbf{F}') + 325 \text{ lb} = 0$$
$$\mathbf{F}' = 325 \text{ lb}$$

EXAMPLE 2

Find the forces **F** and **F′** necessary to produce equilibrium in the free-body diagram shown.

(a)

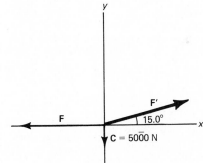

(b) *x components*

$$\mathbf{F}_x = -\mathbf{F}$$
$$\mathbf{F}'_x = \mathbf{F}' \cos 15.0°$$
$$\mathbf{C}_x = 0$$

*y components*

$$\mathbf{F}_y = 0$$
$$\mathbf{F}'_y = \mathbf{F}' \sin 15.0°$$
$$\mathbf{C}_y = -50\overline{0}0 \text{ N}$$

(c)    Sum of *x* components = 0
$$(-\mathbf{F}) + \mathbf{F}' \cos 15.0° + 0 = 0$$

Sum of *y* components = 0
$$0 + \mathbf{F}' \sin 15.0° + (-50\overline{0}0 \text{ N}) = 0$$

(d) *Note:* Let's solve for **F′** in the right-hand equation first. Then substitute this value in the left-hand equation to solve for **F**.

$$\mathbf{F}' = \frac{50\overline{0}0 \text{ N}}{\sin 15.0°}$$
$$= 19{,}300 \text{ N}$$

$$\mathbf{F} = \mathbf{F}' \cos 15.0°$$
$$= (19{,}300 \text{ N})(\cos 15.0°)$$
$$= 18{,}600 \text{ N}$$

EXAMPLE 3

The crane in the figure is supporting a beam that weighs 60$\overline{0}$0 N. Find the tension in the horizontal supporting cable and the compression in the boom.

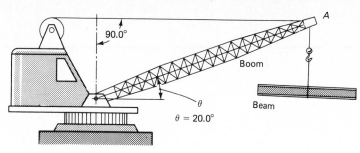

$\theta = 20.0°$

(a) We want to draw the free-body diagram showing the forces *on* the point labeled *A*.

**T** is the force exerted on *A* by the horizontal supporting cable.
**C** is the force exerted by the boom on *A*.
**W** is the force (weight of the beam) pulling straight down on *A*.
**R** is the sum of forces **W** and **T**, which is equal in magnitude but opposite in direction to force **C**. (**R** = − **C**.)

(b) *x components*

$\mathbf{C}_x = \mathbf{C} \cos 20.0°$
$\mathbf{T}_x = -\mathbf{T}$
$\mathbf{W}_x = 0$

*y components*

$\mathbf{C}_y = \mathbf{C} \sin 20.0°$
$\mathbf{T}_y = 0$
$\mathbf{W}_y = -60\overline{0}0$ N

(c) Sum of *x* components = 0
$\mathbf{C} \cos 20.0° + (-\mathbf{T}) = 0$

Sum of *y* components = 0
$\mathbf{C} \sin 20.0° + (-60\overline{0}0 \text{ N}) = 0$

(d) $\mathbf{T} = \mathbf{C} \cos 20.0°$

$\mathbf{C} = \dfrac{60\overline{0}0 \text{ N}}{\sin 20.0°}$
$= 17{,}500 \text{N}$

$\mathbf{T} = (17{,}500 \text{ N})(\cos 20.0°)$
$= 16{,}400$ N

EXAMPLE 4

A homeowner pushes a 40.0-lb lawn mower at a constant velocity. If the frictional force on the mower is 20.0 lb, what force must the person exert on the handle? Also, find the normal (perpendicular to ground) force.

This is an equilibrium problem because the mower is not accelerating and the net force is zero.

(a) Free-body diagram of the mower.

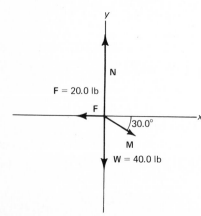

**M** is the force exerted on the mower by the person; this force is directed down along the handle.
**W** is the weight of the mower directed straight down.
**N** is the force exerted upward on the mower by the ground, which keeps the mower from falling through the ground.
**F** is the frictional force that opposes the motion.

(b) *x components*

$\mathbf{N}_x = 0$
$\mathbf{W}_x = 0$
$\mathbf{F}_x = -20.0$ lb
$\mathbf{M}_x = \mathbf{M} \cos 30.0°$

*y components*

$\mathbf{N}_y = \mathbf{N}$
$\mathbf{W}_y = -40.0$ lb
$\mathbf{F}_y = 0$
$\mathbf{M}_y = -\mathbf{M} \sin 30.0°$

(c) Sum of *x* components = 0
$0 + 0 + (-20.0 \text{ lb}) + \mathbf{M} \cos 30.0° = 0$

Sum of *y* components = 0
$\mathbf{N} + (-40.0 \text{ lb}) + 0$
$\qquad + (-\mathbf{M} \sin 30.0°) = 0$
$\mathbf{N} = \mathbf{M} \sin 30.0° + 40.0$ lb

(d) $\mathbf{M} = \dfrac{20.0 \text{ lb}}{\cos 30.0°}$
$= 23.1$ lb

$\mathbf{N} = (23.1 \text{ lb})(\sin 30.0°) + 40.0$ lb
$= 51.6$ lb

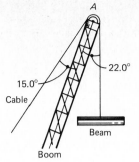

A

22.0°

15.0°
Cable

Beam

Boom

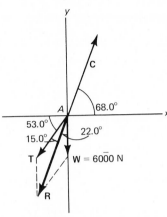

y

C

A 68.0°
53.0°
15.0° 22.0°

x

T W = 60̄0̄0 N

R

# EXAMPLE 5

The crane shown is supporting a beam that weighs 60̄0̄0 N. Find the tension in the supporting cable and the compression in the boom.

(a) Draw the free-body diagram showing the forces *on* point *A*.

**W** is the weight of the beam, which pulls straight down.

**T** is the force exerted on *A* by the supporting cable.

**C** is the force exerted by the boom on *A*.

**R** is the sum of forces **W** and **T**, which is equal in magnitude but opposite in direction to force **C**. **(R = − C.)**

(b)

| *x* components | *y* components |
|---|---|
| $C_x = C \cos 68.0°$ | $C_y = C \sin 68.0°$ |
| $T_x = -T \cos 53.0°$ | $T_y = -T \sin 53.0°$ |
| $W_x = 0$ | $W_y = -60̄0̄0$ N |

(c)

Sum of *x* components = 0      Sum of *y* components = 0

$C \cos 68.0° +$          $C \sin 68.0° +$

    $(-T \cos 53.0°) + 0 = 0$       $(-T \sin 53.0°) + (-60̄0̄0$ N$) = 0$

(d) *Note:* Let us solve the left equation for **C**. Substitute this quantity in the other equation. Then solve the other equation for **T**.

$$C = \frac{T \cos 53.0°}{\cos 68.0°} \qquad \left(\frac{T \cos 53.0°}{\cos 68.0°}\right)(\sin 68.0°) - T \sin 53.0° = 60̄0̄0 \text{ N}$$

$$1.490T - 0.799T = 60̄0̄0 \text{ N}$$

$$0.691T = 60̄0̄0 \text{ N}$$

$$T = \frac{60̄0̄0 \text{ N}}{0.691}$$

$$= 8680 \text{ N}$$

$$C = \frac{(8680 \text{ N})(\cos 53.0°)}{\cos 68.0°}$$

$$= 13{,}900 \text{ N}$$

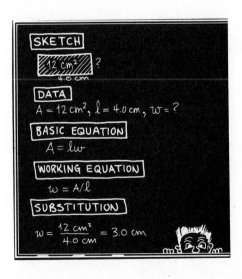

SKETCH

12 cm² ?
4.0 cm

DATA
$A = 12 \text{ cm}^2$, $\ell = 4.0 \text{ cm}$, $w = ?$

BASIC EQUATION
$A = \ell w$

WORKING EQUATION
$w = A/\ell$

SUBSTITUTION
$w = \frac{12 \text{ cm}^2}{4.0 \text{ cm}} = 3.0 \text{ cm}$

# PROBLEMS

*Find forces $F_1$ and $F_2$ necessary to produce equilibrium in the following free-body diagrams.*

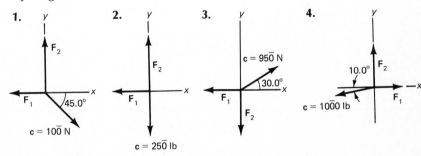

**1.**

y

$F_2$

$F_1$ 45.0°
x

c = 10̄0 N

**2.**

y

$F_2$

$F_1$
x

c = 25̄0 lb

**3.**

y

c = 95̄0 N

30.0°

$F_1$
x

$F_2$

**4.**

y

10.0° $F_2$

$F_1$
x

c = 10̄0̄0 lb

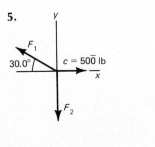

**5.**

y

$F_1$

30.0°
c = 50̄0 lb
x

$F_2$

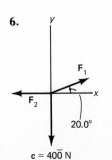

**6.**

y

$F_1$

$F_2$
x

20.0°

c = 40̄0 N

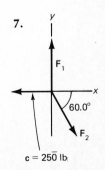

**7.**

y

$F_1$

x

60.0°

$F_2$

c = 25̄0 lb

**101**

8. The rope shown is attached to two buildings and supports a 50$\overline{0}$-lb sign. Find the tension in the two ropes, $T_1$ and $T_2$. (*Hint:* Draw the free-body diagram of the point labeled $A$.)

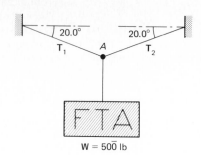

9. If the angle between the horizontal and the ropes in Problem 8 is changed to 10.0°, what are the tensions in the two ropes, $T_1$ and $T_2$?

10. Find the tension in the horizontal supporting cable and the compression in the boom of the crane shown, which supports an 89$\overline{0}$0-N beam.

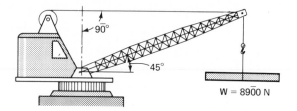

11. Find the tension in the horizontal supporting cable and the compression in the boom of the crane shown, which supports a 15$\overline{0}$0-lb beam.

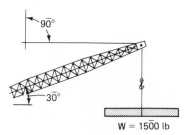

12. The frictional force of the mower is 2$\overline{0}$ lb. What force must the man exert along the handle to push it at a constant velocity?

13. An automobile that weighs 16,200 N is parked on a 20.0° hill as shown. What braking force is necessary to keep the auto from rolling downward? Neglect frictional forces. (*Hint:* When you draw the free-body diagram of the auto, tilt the $x$ and $y$ axes as shown. $B$ is the braking force directed up the hill and along the $x$-axis.)

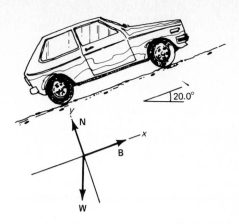

14. Find the tension in the cable and the compression in the support of the sign shown.

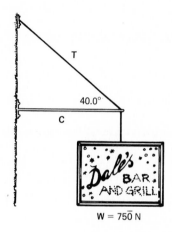

W = 75$\overline{0}$ N

15. The crane in the figure is supporting a load of 1850 lb. Find the tension in the supporting cable and the compression in the boom.

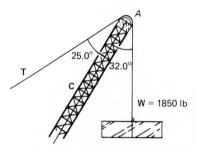

W = 1850 lb

16. The crane shown is supporting a load of 11,500 N. Find the tension in the supporting cable and the compression in the boom.

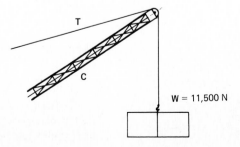

W = 11,500 N

# CHAPTER 7

## Work and Energy

**7-1**
**WORK**
What is *work?* Are we doing work when we try to lift a large crate but it will not budge? Because we have exerted a force and feel tired, we would probably say that we had been working. But is it proper to say that we have done work *on* the crate?

Is work done on the crate if we push the crate across the floor?

Work does have a technical meaning. When a stake is driven into the ground, work is done *by* the moving sledgehammer, and work is done *on* the stake.

When a bulldozer pushes a boulder, work is done *by* the bulldozer, and work is done *on* the boulder.

The examples above show a limited meaning of work: that work is done when a force acts through a distance. The meaning of work used by scientists and technicians is even narrower: *Work is the product of the force in the direction of the motion and the displacement.*

$$W = Fs$$

where: $W$ = work
$F$ = force applied
$s$ = displacement *in the direction of the force*

Now, let us apply our technical definition of work to trying unsuccessfully to lift the crate. We applied a force by lifting on the crate. Have we done work? No work has been done. We were unable to move the crate.

Therefore, the displacement was zero, and the product of the force and the displacement must also be zero. Therefore, no work was done.

By studying the equation for work, we can ourselves determine the correct units for work. In the metric system, force is expressed in newtons and the displacement in metres:

$$\text{work} = \text{force} \times \text{displacement} = \text{newton} \times \text{metre} = \text{N m}$$

This unit (N m) has a special name in honor of an English physicist, James P. Joule. It is the joule (J) [pronounced jo͞ol].

$$1 \text{ N m} = 1 \text{ joule} = 1 \text{ J}$$

In the English system, force is expressed in pounds (lb) and the displacement in feet (ft):

$$\text{work} = \text{force} \times \text{displacement} = \text{pounds} \times \text{feet} = \text{ft lb}$$

The English unit of work is called the foot-pound.

Work is not a vector quantity because it has no particular direction. It is a scalar and has only magnitude.

### EXAMPLE 1

A worker lifting 50.0 lb of bricks to a height of 5.00 ft does $25\overline{0}$ ft lb of work.

$$W = Fs$$
$$W = (50.0 \text{ lb})(5.00 \text{ ft})$$
$$= 25\overline{0} \text{ ft lb}$$

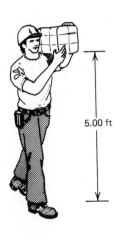

5.00 ft

### EXAMPLE 2

A worker pushing a 60.0-lb pallet (portable platform used in warehouses) a distance of 30.0 ft by exerting a constant force of 10.0 lb does $30\overline{0}$ ft lb of work.

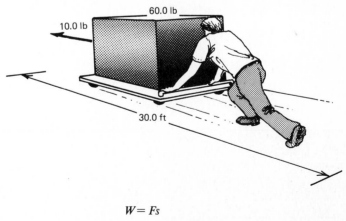

60.0 lb

10.0 lb

30.0 ft

$$W = Fs$$
$$W = (10.0 \text{ lb})(30.0 \text{ ft})$$
$$= 30\overline{0} \text{ ft lb}$$

Note in Example 2 that the pallet weighs 60.0 lb. (Recall that the weight of an object is the measure of its gravitational attraction to the earth and is represented by a vertical vector pointing down to the center of the earth.) There is no motion in the direction *this* force is exerted. Therefore, the weight of the box is not the force used to determine the work being done in the problem.

Work is being done by the worker pushing the pallet. He is exerting a force of 10.0 lb and there is a resulting displacement in the direction the force is applied. The work done is the product of this force (10.0 lb) and the displacement (30.0 ft) in the direction the force is applied.

### EXAMPLE 3

A person pulls a sled along level ground for a distance of 15.0 m by exerting a constant force of $20\overline{0}$ N at an angle of 30.0° with the ground. How much work does he do?

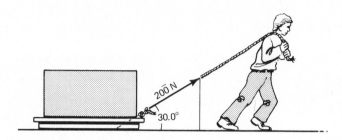

Note that in this example the applied force ($20\overline{0}$ N) is *not* in the direction of the displacement. Therefore, we must determine the component of the applied force in the direction of the motion.

Using the figure at left, we can see that we must determine the magnitude of component force vector *a*. We may determine *a* by applying the trigonometric ratio

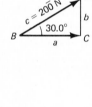

$$\cos B = \frac{\text{side adjacent to } B}{\text{hypotenuse}} = \frac{a}{c}$$

Since $B = 30.0°$, $\cos 30.0° = 0.8660$ and $c = 20\overline{0}$ N, we obtain

$$\cos B = \frac{a}{c}$$

$$\cos 30.0° = \frac{a}{20\overline{0} \text{ N}}$$

$$0.8660 = \frac{a}{20\overline{0} \text{ N}}$$

$$a = (0.8660)(20\overline{0} \text{ N})$$

$$= 173 \text{ N}$$

The work done may now be found by the following:

$$W = Fs$$
$$W = (173 \text{ N})(15.0 \text{ m})$$
$$= 26\overline{0}0 \text{ N m}$$
$$= 26\overline{0}0 \text{ J}$$

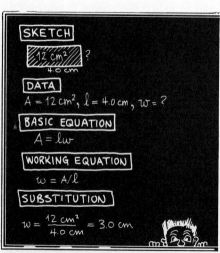

### PROBLEMS

1. Given: $F = 10.0$ lb
   $s = 3.43$ ft
   $W = ?$

2. Given: $W = 697$ ft lb
   $s = 976$ ft
   $F = ?$

3. Given: $s = 384$ m
   $W = 171$ J
   $F = ?$

4. Given: $F = ma$
   $m = 16.0$ kg
   $a = 9.80$ m/s²
   $s = 13.0$ m
   $W = ?$

5. How much work is required for a mechanical hoist to lift a $36\overline{0}0$-lb automobile to a height of 6.00 ft for repairs?

6. A hay wagon is used to move bales from the field to the barn. The tractor pulling the wagon exerts a constant force of 350 lb. The distance from field to barn is $\frac{1}{2}$ mi. How much work (ft lb) is done in moving one load of hay to the barn?

7. A worker lifts 75 concrete blocks a distance of 1.50 m to the bed of a truck. Each block has a mass of 4.00 kg. How much work must he do to lift all the blocks to the truck bed?

8. The work required to lift eleven 94.0-lb bags of cement from the ground to the back of a truck is 4340 ft lb. What is the distance from the ground to the bed of the truck?

9. A worker carries bricks to a mason 20.0 m away. If the amount of work required to carry one brick is 6.00 J, what force must the worker exert on each brick?

10. How much work is done lifting a $20\overline{0}$-kg wrecking ball 6.50 m above the ground?

11. A gardener pushes a mower a distance of $90\overline{0}$ m in mowing a yard. The handle of the mower makes an angle of 40.0° with the ground. The gardener exerts a force of 35.0 N along the handle of the mower. How much work does the gardener do in mowing the lawn?

12. The handle of a vegetable wagon makes an angle of 25.0° with the horizontal. If the peddler exerts a force of 35.0 lb along the handle, how much work does the peddler do in pulling the cart 1.00 mi?

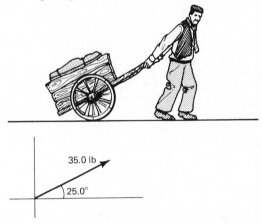

35.0 N

40.0°

35.0 lb

25.0°

**7-2**
**POWER**

Many situations arise where we are concerned not only with the amount of work required for a certain task but also the time to do the job. Power also has a special meaning in physics.

> Power is the *rate* of doing work:
>
> $$P = \frac{W}{t}$$

where:  $P =$ power
$\quad\quad\quad W =$ work
$\quad\quad\quad\; t =$ time

The units of power are familiar to most of us. In the metric system the unit of power is the *watt*. Since the unit is inconveniently small, power is usually expressed in kilowatts and megawatts.

$$P = \frac{W}{t} = \frac{Fs}{t} = \frac{N\,m}{s} = \frac{J}{s} = watt$$

and

1000 watts (W) = 1 kilowatt (kW)

1,000,000 watts = 1 megawatt (MW)

In the English system the unit of power used is either ft lb/s or horsepower.

$$P = \frac{W}{t} = \frac{Fs}{t} = \frac{ft\ lb}{s}$$

Horsepower (hp) is a unit derived by James Watt (who also designed the first practical steam engine) to approximate the power delivered by a workhorse.

$$1 \text{ horsepower (hp)} = 55\overline{0} \text{ ft lb/s} = 33,\overline{0}00 \text{ ft lb/min}$$

## EXAMPLE 1

A freight elevator with operator weighs $50\overline{0}$ lb. If it is raised to a height of 50.0 ft in 10.0 s, how much power is developed?

SKETCH:

DATA:

$$F = 50\overline{0} \text{ lb}$$
$$s = 50.0 \text{ ft}$$
$$t = 10.0 \text{ s}$$
$$P = ?$$

BASIC EQUATION:

$$P = \frac{W}{t}$$

WORKING EQUATION:

$$P = \frac{Fs}{t}$$

SUBSTITUTION:

$$P = \frac{(50\overline{0} \text{ lb})(50.0 \text{ ft})}{10.0 \text{ s}}$$
$$= 25\overline{0}0 \text{ ft lb/s}$$

We may find the horsepower developed by the elevator by using the conversion factor

$$1.00 \text{ hp} = 55\overline{0} \text{ ft lb/s}$$

THEREFORE:

$$25\overline{0}0 \frac{\text{ft lb}}{\text{s}} \left( \frac{1.00 \text{ hp}}{55\overline{0} \text{ ft lb/s}} \right) = 4.55 \text{ hp}$$

## EXAMPLE 2

The power expended in lifting a $55\overline{0}$-lb girder to the top of a building $10\overline{0}$ ft high is 10.0 hp. How much time is required to raise the girder?

SKETCH:

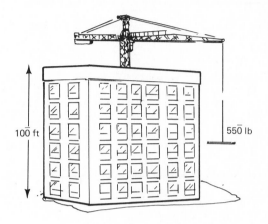

DATA:

$$F = 55\overline{0} \text{ lb}$$
$$s = 10\overline{0} \text{ ft}$$
$$P = 10.0 \text{ hp}$$
$$t = ?$$

BASIC EQUATION:

$$P = \frac{W}{t}$$

WORKING EQUATION:

$$t = \frac{W}{P} = \frac{Fs}{P}$$

SUBSTITUTION:

$$t = \frac{(55\overline{0} \text{ lb})(10\overline{0} \text{ ft})}{10.0 \text{ hp}}$$

$$= \frac{(55\overline{0} \text{ lb})(10\overline{0} \text{ ft})}{10.0 \text{ hp}} \times \frac{1 \text{ hp}}{55\overline{0} \dfrac{\text{ft lb}}{\text{s}}}$$

$$= 10.0 \text{ s}$$

$$\boxed{\frac{\text{lb ft}}{\text{hp}} \cdot \frac{\text{hp}}{\dfrac{\text{ft lb}}{\text{s}}} = \frac{\text{lb ft}}{\text{hp}} \cdot \left( \text{hp} \div \frac{\text{ft lb}}{\text{s}} \right) = \frac{\text{lb ft}}{\text{hp}} \cdot \left( \text{hp} \cdot \frac{\text{s}}{\text{ft lb}} \right) = \text{s}}$$

(*Note:* We must use a conversion factor to obtain time units.)

EXAMPLE 3

The mass of a large steel wrecking ball is $20\overline{0}0$ kg. What power is used to raise it to a height of 40.0 m if the work is done in 20.0 s?

DATA:

$$m = 20\overline{0}0 \text{ kg}$$
$$s = 40.0 \text{ m}$$
$$t = 20.0 \text{ s}$$
$$P = ?$$

BASIC EQUATION:

$$P = \frac{W}{t} = \frac{Fs}{t}$$

WORKING EQUATION:

$$P = \frac{Fs}{t}$$

Note that we cannot directly substitute into the working equation because our data are given in terms of *mass* and we must find *force* to substitute in $P = Fs/t$. Recall that the force is the weight of the ball.

$$F = mg = (20\overline{0}0 \text{ kg})(9.80 \text{ m/s}^2) = 19{,}600 \frac{\text{kg m}}{\text{s}^2} = 19{,}600 \text{ N}$$

SUBSTITUTION:

$$P = \frac{(19{,}600 \text{ N})(40.0 \text{ m})}{20.0 \text{ s}}$$
$$= 39{,}200 \text{ N m/s}$$
$$= 39{,}200 \text{ W} \quad \text{or} \quad 39.2 \text{ kW}$$

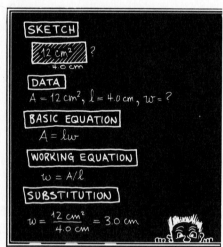

## PROBLEMS

1. Given: $W = 132$ ft lb
   $t = 7.00$ s
   $P = ?$

2. Given: $P = 231$ ft lb/s
   $t = 14.3$ s
   $W = ?$

3. Given: $W = 55.0$ J
   $t = 11.0$ s
   $P = ?$

4. Given: $P = 75.0$ W
   $W = 40.0$ J
   $t = ?$

5. The work required to lift a crate is $31\overline{0}$ ft lb. If it can be lifted in 25.0 s, what power is developed?

6. A $36\overline{0}0$-lb automobile is pushed by its unhappy driver a quarter of a mile (0.250 mi) when it runs out of gas. To keep the car rolling, the driver must exert a constant force of 175 lb.
   (a) How much work does he do?
   (b) If it takes him 15.0 minutes, how much power does he develop?
   (c) Expressed in horsepower, how much power does he develop?

7. An electric golf cart develops 1.25 kW of power while moving at a constant speed.
   (a) Express its power in horsepower.
   (b) If the cart travels $20\overline{0}$ m in 35.0 s, what force is exerted by the cart?

8. How many seconds would it take a 7.00-hp motor to raise a 475-lb boiler to a platform in a factory 38.0 ft high?

9. How long would it take a $95\overline{0}$-W motor to raise a $36\overline{0}$-kg mass to a height of 16.0 m?

10. A $15\overline{0}0$-lb casting is raised 22.0 ft in 2.50 min. Find the required horsepower.

11. What is the rating in kW of a 2.00-hp motor?

12. A wattmeter shows that a motor is drawing $22\overline{0}0$ W. What horsepower is being delivered?

**13.** A 525-kg steel beam is raised 30.0 m in 25.0 s. How many kilowatts of pow[er] are needed?

**14.** How long would it take a 4.50-kW motor to raise a 175-kg boiler to a platform 15.0 m above the floor?

**15.** An escalator is needed to carry 75 passengers per minute a vertical distance of 8.0 m. Assume that the mass of each passenger is $7\overline{0}$ kg.
(a) What is the power (in kW) of the motor needed?
(b) Express this power in horsepower.
(c) What is the power (in kW) of the motor needed if 35% of the power is lost to friction and other losses?

**16.** A pump is needed to lift $75\overline{0}$ L of water per minute a distance of 25.0 m. What power (in kW) must the pump be able to deliver? (1 L of water has a mass of 1 kg.)

**7-3
ENERGY**

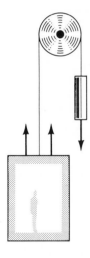

Physicists and technicians define mechanical energy as the ability to do work. Mechanical energy exists in two forms. One type is energy of motion, the kind of energy an automobile possesses as it moves along the highway. This type is called *kinetic energy.*

The other type is energy due to an object's condition or position—stored energy. This type is called *potential energy.* It may be due to internal characteristics of a substance, like gasoline, or its position. The counterweight on an elevator illustrates potential energy of position.

Note that when it is in the raised position (and the elevator is down), the weight has energy available to do work because of the pull of gravity on it. This is called gravitational potential energy. A compressed spring or a stretched rubber band also has mechanical potential energy.

Since energy is defined as the ability to do work, it is not very surprising that energy is expressed in the same units as work—the joule and the foot-pound.

We are now discussing only two kinds of energy—kinetic and potential. Keep in mind that energy exists in many forms—chemical, atomic, electrical, sound, and heat. These forms and the conversion of energy from one form to another will be studied later.

Let us take a closer look at kinetic energy. The formula for kinetic energy follows:

$$\boxed{\text{Kinetic energy (KE)} = \tfrac{1}{2}mv^2}$$

where:  $m$ = mass of moving object
$v$ = velocity of moving object

A pile driver shows the relation of energy of motion to useful work. The energy of the driver is kinetic energy just before it hits. When the driver strikes the pile, work is done on the pile and it is forced into the ground. The depth it goes down is determined by the force applied to it. The force applied is determined by the energy of the driver. If all the kinetic energy of the driver is converted to useful work, then $\tfrac{1}{2}mv^2 = Fs$. Consider the following example.

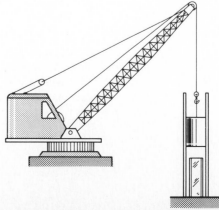

The driver has a mass of $10,\overline{0}00$ kg and upon striking the pile has a velocity of 10.0 m/s.

(a) What is the kinetic energy of the driver when it strikes the pile?

(b) If the pile is driven 20.0 cm into the ground, what force is applied to the pile?

SKETCH:

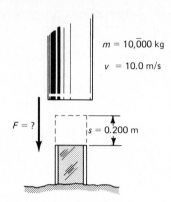

$m = 10,\overline{0}00$ kg

$v = 10.0$ m/s

$F = ?$

$s = 0.200$ m

DATA:

$$m = 1.00 \times 10^4 \text{ kg}$$

$$v = 10.0 \text{ m/s}$$

$$s = 20.0 \text{ cm} = 0.200 \text{ m}$$

$$F = ?$$

(a) BASIC EQUATION:

$$KE = \tfrac{1}{2} mv^2$$

WORKING EQUATION: Same

SUBSTITUTION:

$$KE = \tfrac{1}{2}(1.00 \times 10^4 \text{ kg})(10.0 \text{ m/s})^2$$

$$= 5.00 \times 10^5 \frac{\text{kg m}^2}{\text{s}^2} \times \frac{1 \text{ J s}^2}{\text{kg m}^2}$$

$$= 5.00 \times 10^5 \text{ J}$$

(*Note:* Conversion factor is needed for desired units.)

(b) BASIC EQUATIONS:

$$KE = W = Fs$$

WORKING EQUATION:

$$F = \frac{KE}{s} \qquad \text{[KE from part (a)]}$$

SUBSTITUTION:

$$F = \frac{5.00 \times 10^5 \text{ J}}{0.200 \text{ m}} \times \frac{\text{N m}}{1 \text{ J}} \qquad (\textit{Note: } \text{conversion factor})$$

$$= 2.50 \times 10^6 \text{ N}$$

Potential energy will now be considered more fully. Internal potential energy is determined by the nature or condition of the substance, for example gasoline, a compressed spring, or a stretched rubber band. Gravitational potential energy is determined by the position of the object relative to a particular reference level. The formula for potential energy is:

Potential energy (PE) $= mgh$

where:  $m =$ mass
$g = 9.80$ m/s² or 32.2 ft/s²
$h =$ height above reference level

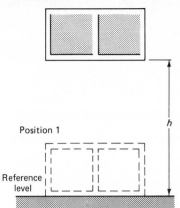

Position 2

Position 1

Reference
level

$h$

## EXAMPLE 2

In position 1 (see the figure), the crate is at rest on the floor. It has no ability to do work as it is in its lowest position. To raise the crate to position 2, work must be done to lift it. In the raised position, however, it now has stored ability to do work (by falling to the floor!). Its PE (potential energy) can be calculated by multiplying the mass of the crate times acceleration of gravity *(g)* times height above reference level *(h)*. Note that we could calculate the potential energy of the crate with respect to any level we may choose. In this example we have chosen the floor as the zero or lowest reference level.

## EXAMPLE 3

A wrecking ball of mass $20\bar{0}$ kg is poised 4.00 m above a concrete platform whose top is 2.00 m above the ground.
(a) With respect to the platform, what is the potential energy of the ball?
(b) With respect to the ground, what is the potential energy of the ball?

SKETCH:

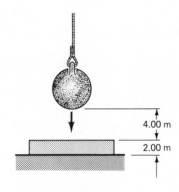

4.00 m

2.00 m

DATA:

$$m = 20\bar{0} \text{ kg}$$
$$h_1 = 4.00 \text{ m}$$
$$h_2 = 2.00 \text{ m}$$
$$\text{PE} = ?$$

BASIC EQUATION:

$$\text{PE} = mgh$$

WORKING EQUATION:  Same

(a) SUBSTITUTION:

$$\text{PE} = (20\bar{0} \text{ kg})\left(9.80 \frac{\text{m}}{\text{s}^2}\right)(4.00 \text{ m})$$

$$= 7840 \frac{\text{kg m}^2}{\text{s}^2} \times \frac{1 \text{ J s}^2}{\text{kg m}^2}$$

$$= 7840 \text{ J}$$

(b) SUBSTITUTION:

$$\text{PE} = (20\bar{0} \text{ kg})\left(9.80 \frac{\text{m}}{\text{s}^2}\right)(6.00 \text{ m})$$

$$= 11,800 \frac{\text{kg m}^2}{\text{s}^2} \times \frac{1 \text{ J s}^2}{\text{kg m}^2}$$

$$= 11,800 \text{ J}$$

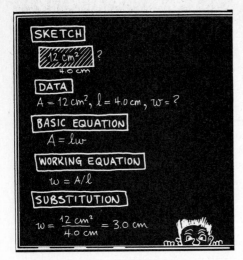

SKETCH

DATA

$A = 12 \text{ cm}^2, l = 4.0 \text{ cm}, w = ?$

BASIC EQUATION

$A = lw$

WORKING EQUATION

$w = A/l$

SUBSTITUTION

$w = \dfrac{12 \text{ cm}^2}{4.0 \text{ cm}} = 3.0 \text{ cm}$

## PROBLEMS

1. Given: $m = 11.4$ slugs
$g = 32.2 \text{ ft/s}^2$
$h = 22.0 \text{ ft}$
$PE = ?$

2. Given: $m = 3.50 \text{ kg}$
$g = 9.80 \text{ m/s}^2$
$h = 15.0 \text{ m}$
$PE = ?$

3. Given: $m = 4.70 \text{ kg}$
$v = 9.60 \text{ m/s}$
$KE = ?$

4. Given: $PE = 93.6 \text{ J}$
$g = 9.80 \text{ m/s}^2$
$m = 2.30 \text{ kg}$
$h = ?$

5. A truck is going along a highway with a velocity of 55.0 mi/h.
   (a) What is its velocity in ft/s?
   The mass of the truck is $95\overline{0}$ slugs.
   (b) What is the kinetic energy of the truck?

6. A bullet travels at 415 m/s. If it has a mass of 12.0 g, what is its kinetic energy? (*Hint:* Convert 12.0 g to kg.)

7. A bicycle and rider together have a mass of 74.0 slugs. If the kinetic energy is 742 ft lb, what is the velocity?

8. A crate of mass 475 kg is raised to a height of 17.0 m from the floor. What potential energy has it acquired with respect to the floor?

9. A shop manual weighing 3.00 lb is on a shelf 4.40 ft above a tabletop which is itself 2.70 ft above the floor.
   (a) What is the potential energy of the book with respect to the table top?
   (b) What is the potential energy of the book with respect to the floor?

10. The potential energy possessed by a girder after being lifted to the top of a new building is $5.17 \times 10^5$ ft lb. If the mass of the girder is 173 slugs, how high is the girder?

    Water is pumped at the rate of $25\overline{0}$ m³/min from a lake into a tank that is 65.0 m above the lake.

11. What power (in kW) must be delivered by the pump?

12. What horsepower rating does this pump motor have?

13. What is the increase in potential energy of the water each minute?

7-4

CONSERVATION
OF MECHANICAL ENERGY

Now let's consider kinetic and potential energy together. Can you see how the two might be related? They are, in fact, related by an important principle:

> *LAW OF CONSERVATION OF MECHANICAL ENERGY*
>
> The sum of the kinetic energy and potential energy in a system is constant.

A pile driver shows this energy conservation. When the driver is at its highest position, the potential energy is maximum and the kinetic energy is zero. Its potential energy is

$$PE = mgh$$

and its kinetic energy is

$$KE = \tfrac{1}{2}mv^2 = \tfrac{1}{2}m(0)^2 = 0$$

When the driver hits the top of the pile, it has its maximum kinetic energy and the potential energy is

$$PE = mgh = mg(0) = 0$$

**114**

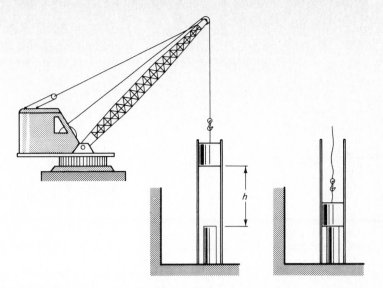

Since the total energy in the system must remain constant, the maximum potential energy must equal the maximum kinetic energy.

$$PE_{max} = KE_{max}$$

$$mgh = \tfrac{1}{2}mv^2$$

Solving for the velocity of the driver just before it hits the pile, we get

$$\boxed{v = \sqrt{2gh}}$$

EXAMPLE

A pile driver falls freely from a height of 3.50 m above a pile. What is its velocity just as it hits the pile?

DATA:

$$h = 3.50 \text{ m}$$
$$g = 9.80 \text{ m/s}^2$$
$$v = ?$$

BASIC EQUATION:

$$v = \sqrt{2gh}$$

WORKING EQUATION:   Same

SUBSTITUTION:

$$v = \sqrt{2(9.80 \text{ m/s}^2)(3.50 \text{ m})}$$
$$= 8.28 \text{ m/s}$$

The conservation of mechanical energy can also be illustrated by considering a swinging pendulum bob where there is no resistance involved. Pull the bob over to the right side so that the string makes an angle of 65° with the vertical. At this point, the bob contains its maximum potential energy and its minimum kinetic energy (zero). Note that a larger maximum potential energy is possible when an initial deflection of greater than 65° is made.

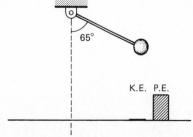

An instant later, the bob has lost some of its potential or stored energy, but it has gained in kinetic energy due to its motion.

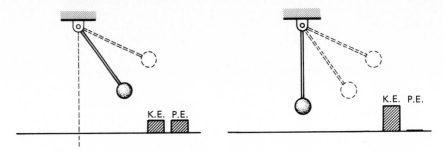

At the bottom of its arc of swing, its potential energy is zero and its kinetic energy is maximum (its velocity is maximum).

The kinetic energy of the bob then causes the bob to swing upward to the left. As it completes its swing, its kinetic energy is decreasing and its potential energy is increasing. That is, its kinetic energy is changing to potential energy.

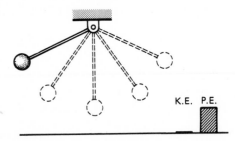

According to the law of conservation of mechanical energy, the sum of the kinetic energy and the potential energy of the bob at any instant is a constant. Assuming no resistant forces, such as friction or air resistance, the bob would swing uniformly "forever."

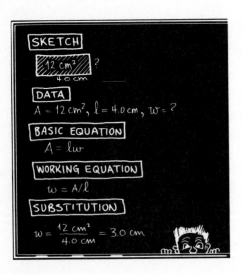

## PROBLEMS

1. A pile driver falls through a distance of 12.0 ft before hitting a pile. What is the velocity of the driver just before it hits the pile?

2. A sky diver jumps out of a plane at a height of $50\overline{0}0$ ft. If her parachute does not open until she reaches $10\overline{0}0$ ft, what is her velocity if air resistance can be neglected?

3. A piece of shattered glass falls from the 82nd floor of a skyscraper, $27\overline{0}$ m above the ground. What is the velocity of the glass when it hits the ground, if air resistance can be neglected?

4. A 2.00-kg projectile is fired vertically with an initial velocity of 98.0 m/s. Find its kinetic energy, its potential energy, and the sum of its kinetic and potential energies at each of the following times.

   (a) the instant of its being fired
   (b) $t = 1.00$ s
   (c) $t = 2.00$ s
   (d) $t = 5.00$ s
   (e) $t = 10.00$ s
   (f) $t = 12.00$ s
   (g) $t = 15.0$ s
   (h) $t = 20.0$ s

# CHAPTER

## Simple Machines

**8-1**
MACHINES
AND ENERGY TRANSFER

Machines are used to transfer energy from one place to another. By using a pulley system one person can easily lift an engine from an automobile. Pliers allow a person to cut a wire with the strength of his or her hand.

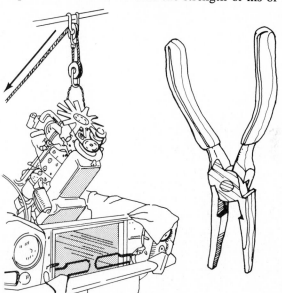

Machines are sometimes used to multiply force. By pushing with a small force, we can use a machine to jack up an automobile.

Machines are sometimes used to multiply speed. The gears on a bicycle are used to multiply speed.

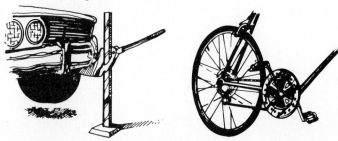

117

Machines are used to change direction. For example, when we use a single fixed pulley on a flag pole to raise a flag, the only advantage we get is the change in direction. (We pull the rope down, and the flag goes up.)

There are six basic or simple machines:

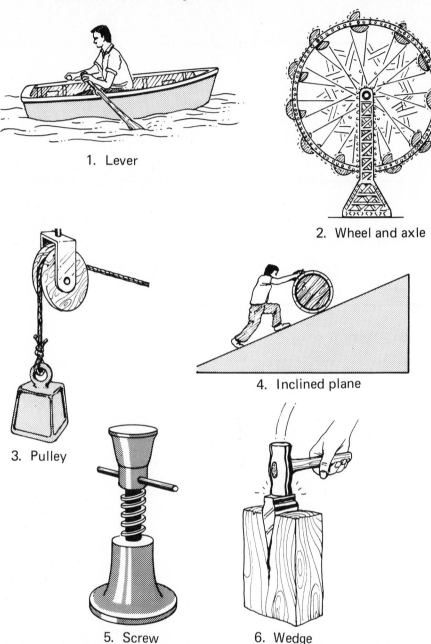

1. Lever

2. Wheel and axle

3. Pulley

4. Inclined plane

5. Screw

6. Wedge

All other machines—no matter how complex—are combinations of two or more of these simple machines.

In every machine we are concerned with two forces—effort and resistance. The *effort* is the force applied *to* the machine. The *resistance* is the force overcome *by* the machine.

The man in the figure on page 119 applies $3\overline{0}$ lb on the jack handle to produce a lifting force of $6\overline{0}0$ lb on the camper's bumper. The effort force is $3\overline{0}$ lb. The resistance force is $6\overline{0}0$ lb.

---

*LAW OF SIMPLE MACHINES*

resistance force $\times$ resistance distance $=$ effort force $\times$ effort distance

---

Jack

**8-2**
MECHANICAL ADVANTAGE
AND EFFICIENCY

The *mechanical advantage* is the ratio of the resistance force to the effort force. By formula:

$$\text{MA} = \frac{\text{resistance force}}{\text{effort force}}$$

The MA of the jack in the preceding example is found as follows:

$$\text{MA} = \frac{\text{resistance force}}{\text{effort force}} = \frac{6\overline{0}0 \, \text{lb}}{3\overline{0} \, \text{lb}} = \frac{2\overline{0}}{1}$$

This MA means that, for each pound applied by the mechanic, he lifts $2\overline{0}$ pounds.

Each time a machine is used, part of the energy or effort applied to the machine is lost due to friction.

The *efficiency* of a machine is the ratio of the work output to the work input. By formula:

$$\text{efficiency} = \frac{\text{work output}}{\text{work input}} \times 100\% = \frac{F_{\text{output}} \times s_{\text{output}}}{F_{\text{input}} \times s_{\text{input}}} \times 100\%$$

**8-3**
THE LEVER

A *lever* consists of a rigid bar free to turn on a pivot called a fulcrum.
The mechanical advantage (MA) is the ratio of the effort arm $(d_E)$ to the resistance arm $(d_R)$.

$$\text{MA}_{\text{lever}} = \frac{\text{effort arm}}{\text{resistance arm}} = \frac{d_E}{d_R}$$

The *effort arm* is the distance from the effort to the fulcrum. The *resistance arm* is the distance from the fulcrum to the resistance.

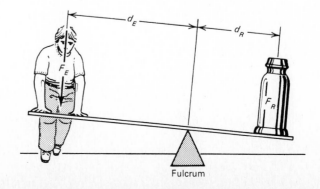

Fulcrum

There are three types or classes of levers to study:

*First class:* The fulcrum is between the resistance force $(F_R)$ and the effort force $(F_E)$.

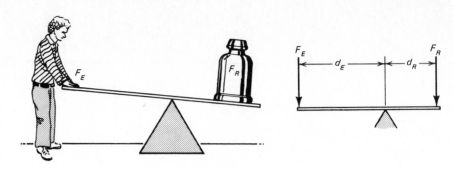

*Second class:* The resistance force $(F_R)$ is between the fulcrum and the effort force $(F_E)$.

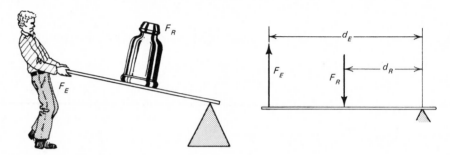

*Third class:* The effort force $(F_E)$ is between the fulcrum and the resistance force $(F_R)$.

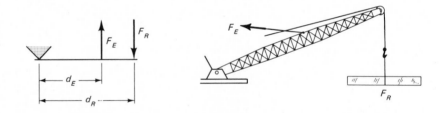

*The law of simple machines as applied to levers* (basic equation):

$$F_R \cdot d_R = F_E \cdot d_E$$

## EXAMPLE 1

A bar is used to raise an $18\overline{0}0$-lb stone. The pivot is placed 9.00 in. from the stone. The worker pushes 108 in. from the pivot. What is the mechanical advantage? What force does he exert?

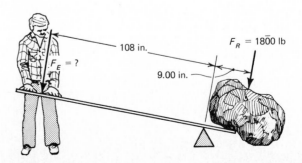

$$MA = \frac{d_E}{d_R} = \frac{108 \text{ in.}}{9.00 \text{ in.}} = \frac{12.0}{1}$$

To find the force:

DATA:

$$d_E = 108 \text{ in.}$$
$$d_R = 9.00 \text{ in.}$$
$$F_R = 18\overline{0}0 \text{ lb}$$
$$F_E = ?$$

BASIC EQUATION:

$$F_R \cdot d_R = F_E \cdot d_E$$

WORKING EQUATION:

$$F_E = \frac{F_R \cdot d_R}{d_E}$$

SUBSTITUTION:

$$F_E = \frac{(18\overline{0}0 \text{ lb})(9.00 \text{ in.})}{108 \text{ in.}}$$
$$= 15\overline{0} \text{ lb}$$

## EXAMPLE 2

A wheelbarrow 2.00 m long has a $90\overline{0}$ N load 0.500 m from the axle. What is the MA? What force is needed to lift the wheelbarrow?

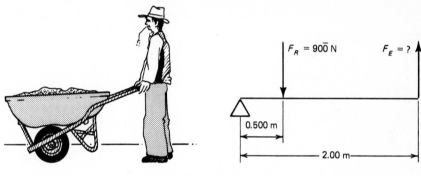

$$MA = \frac{d_E}{d_R} = \frac{2.00 \text{ m}}{0.500 \text{ m}} = 4.00$$

To find the force:

DATA:

$$d_E = 2.00 \text{ m}$$
$$d_R = 0.500 \text{ m}$$
$$F_R = 90\overline{0} \text{ N}$$
$$F_E = ?$$

BASIC EQUATION:

$$F_R \cdot d_R = F_E \cdot d_E$$

WORKING EQUATION:

$$F_E = \frac{F_R \cdot d_R}{d_E}$$

SUBSTITUTION:

$$F_E = \frac{(90\overline{0} \text{ N})(0.500 \text{ m})}{2.00 \text{ m}}$$
$$= 225 \text{ N}$$

The MA of a pair of pliers is 6.0/1. A force of 8.0 lb is exerted on the handle. What force is exerted on a wire in the pliers?

MA = 6.0/1 means that, for each pound of force applied on the handle, 6.0 lb are exerted on the wire. Therefore, if a force of 8.0 lb is applied on the handle, a force of (6.0)(8.0 lb) or 48 lb is exerted on the wire.

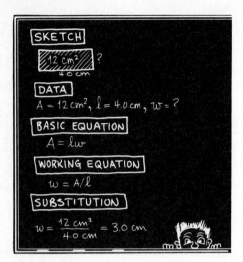

SKETCH

DATA
$A = 12 \text{ cm}^2, \ell = 4.0 \text{ cm}, w = ?$

BASIC EQUATION
$A = \ell w$

WORKING EQUATION
$w = A/\ell$

SUBSTITUTION
$w = \dfrac{12 \text{ cm}^2}{4.0 \text{ cm}} = 3.0 \text{ cm}$

## PROBLEMS

*Given* $F_R \cdot d_R = F_E \cdot d_E$, *find the missing quantities.*

| | $F_R$ | $F_E$ | $d_R$ | $d_E$ |
|---|---|---|---|---|
| **1.** | 20.0 lb | 5.00 lb | 3.70 in. | ____ in. |
| **2.** | ____ N | 176 N | 49.2 cm | 76.3 cm |
| **3.** | 37.0 N | 12.0 N | ____ cm | 112 cm |
| **4.** | 23.4 lb | 9.80 lb | ____ in. | 53.9 in. |
| **5.** | 119 N | ____ N | 29.7 cm | 67.4 cm |

*Given* $MA = F_R/F_E$, *find the missing quantities.*

| | MA | $F_R$ | $F_E$ |
|---|---|---|---|
| **6.** | ____ | 20.0 N | 5.00 N |
| **7.** | ____ | 23.4 lb | 9.80 lb |
| **8.** | 7.00 | 119 N | ____ N |
| **9.** | 4.00 | ____ lb | 12.2 lb |
| **10.** | ____ | 37.0 N | 12.0 N |

*Given* $MA = d_E/d_R$, *find the missing quantities.*

| | MA | $d_R$ | $d_E$ |
|---|---|---|---|
| **11.** | ____ | 49.2 cm | 76.3 cm |
| **12.** | 7.00 | 29.7 in. | ____ in. |
| **13.** | ____ | 29.7 cm | 67.4 cm |
| **14.** | 4.00 | ____ cm | 67.4 cm |
| **15.** | 3.00 | 13.7 in. | ____ in. |

**16.** A pole is used to lift a car that fell off a jack. The pivot is 2.00 ft from the car. Two people together exert 275 lb of force 8.00 ft from the pivot. What force is applied to the car?

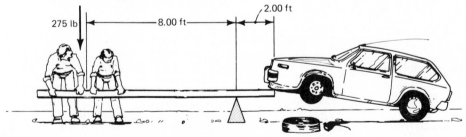

**17.** Calculate the MA of Problem 16.

**18.** A bar is used to lift a $10\overline{0}$-kg block of concrete. The pivot is 1.00 m from the block. If the worker pushes down on the other end of the bar a distance of 2.50 m from the pivot, what force (in N) must he apply?

**19.** Calculate the MA of Problem 18.

**20.** A wheelbarrow 6.00 ft long is to be used to haul a $18\overline{0}$-lb load. How far from the wheel is the load placed so that a person can lift the load with a force of 45.0 lb?

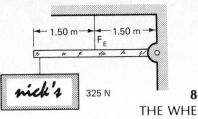

1.50 m — 1.50 m

$F_E$

nick's        325 N

**21.** Calculate the MA of Problem 20.

**22.** Find the force, $F_E$, pulling up on the beam holding the sign shown.

**23.** Calculate the MA of Problem 22.

**8-4**
**THE WHEEL**
**AND AXLE**

This simple machine consists of a large wheel attached to an axle so that both turn together.

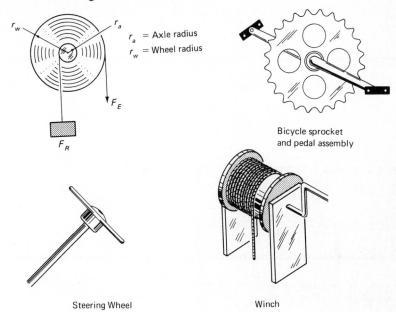

$r_a$ = Axle radius
$r_w$ = Wheel radius

$F_E$

$F_R$

Bicycle sprocket
and pedal assembly

Steering Wheel        Winch

Other examples include a doorknob, a screwdriver with a thick handle, and a water faucet.

*The law of simple machines as applied to the wheel and axle* (basic equation):

$$F_R \cdot r_a = F_E \cdot r_w$$

EXAMPLE 1

The winch shown has a handle that turns in a radius of 9.00 in. The radius of the drum or axle is 3.00 in. Find the force required to lift a bucket weighing 90.0 lb.

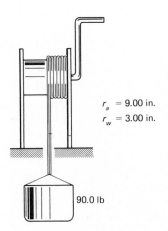

$r_a$ = 9.00 in.
$r_w$ = 3.00 in.

90.0 lb

DATA:

$F_R = 90.0$ lb
$r_w = 9.00$ in.
$r_a = 3.00$ in.
$F_E = ?$

BASIC EQUATION:

$$F_R \cdot r_a = F_E \cdot r_w$$

WORKING EQUATION:

$$F_E = \frac{F_R \cdot r_a}{r_w}$$

SUBSTITUTION:

$$F_E = \frac{(90.0 \text{ lb})(3.00 \text{ in.})}{(9.00 \text{ in.})}$$

$$= 30.0 \text{ lb}$$

123

The mechanical advantage (MA) of the wheel and axle is the ratio of the radius of the wheel to the radius of the axle.

$$\text{MA} = \frac{\text{radius of wheel}}{\text{radius of axle}} = \frac{r_w}{r_a}$$

## EXAMPLE 2

Calculate the MA of the winch in Example 1.

$$\text{MA} = \frac{r_w}{r_a}$$

$$= \frac{9.00 \text{ in.}}{3.00 \text{ in.}}$$

$$= \frac{3.00}{1}$$

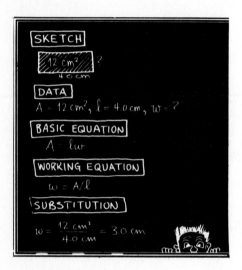

## PROBLEMS

Given $F_R \cdot r_a = F_E \cdot r_w$, find the missing quantities.

| | $F_R$ | $F_E$ | $r_a$ | $r_w$ |
|---|---|---|---|---|
| 1. | 20.0 lb | 5.30 lb | 3.70 in. | ___ in. |
| 2. | 37$\overline{0}$ N | 12$\overline{0}$ N | ___ m | 1.12 m |
| 3. | ___ N | 175 N | 49.2 cm | 76.3 cm |
| 4. | 23.4 lb | 9.80 lb | ___ in. | 53.9 in. |
| 5. | 1190 N | ___ N | 29.7 cm | 67.4 cm |

Given $\text{MA} = r_w/r_a$, find the missing quantities.

| | MA | $r_w$ | $r_a$ |
|---|---|---|---|
| 6. | 7.00 | 119 mm | ___ mm |
| 7. | 4.00 | ___ in. | 12.2 in. |
| 8. | ___ | 49.2 cm | 31.7 cm |
| 9. | 3.00 | 61.3 cm | ___ cm |
| 10. | ___ | 67.4 mm | 29.7 mm |

11. The radius of the axle of a winch is 3.00 in. The length of the handle (radius of wheel) is 1.50 ft. What weight will be lifted by an effort of 73.0 lb?
12. Calculate the MA of Problem 11.
13. A wheel having a radius of 70.0 cm is attached to an axle that has a radius of 20.0 cm. What force must be applied to the rim of the wheel to raise a weight of 15$\overline{0}$0 N?
14. Calculate the MA of Problem 13.
15. What weight can be lifted in Problem 13 if a force of 575 N is applied?
16. The diameter of the wheel of a wheel and axle is 10.0 m. If a force of 475 N is to be raised by applying a force of 142 N, find the diameter of the axle.
17. Calculate the MA of Problem 16.

**8-5**
**THE PULLEY**

A *pulley* is a grooved wheel that turns readily on an axle and is supported in a frame. It can be fastened to a fixed object or it may be fastened to the resistance that is to be moved. If the pulley is fastened to a fixed object, it is called a *fixed pulley*. If the pulley is fastened to the resistance to be moved, it is called a *movable pulley*.

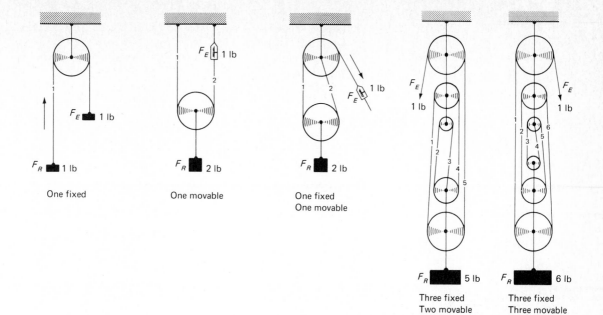

One fixed

One movable

One fixed
One movable

$F_E$ 1 lb    $F_R$ 5 lb

Three fixed
Two movable

$F_E$ 1 lb    $F_R$ 6 lb

Three fixed
Three movable

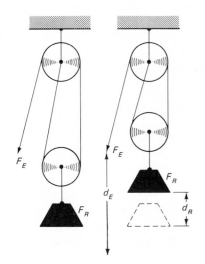

*The law of simple machines as applied to pulleys* (basic equation):

$$F_R \cdot d_R = F_E \cdot d_E$$

From this equation,

$$\boxed{\frac{F_R}{F_E} = \frac{d_E}{d_R} = \mathbf{MA}}$$

However, when one continuous cord is used, this ratio reduces to be the number of strands holding the resistance in the pulley system. Therefore,

$$\boxed{MA = \textit{number of strands holding the resistance}}$$

*Note:* The mechanical advantage of the pulley does not depend on the diameter of the pulley.

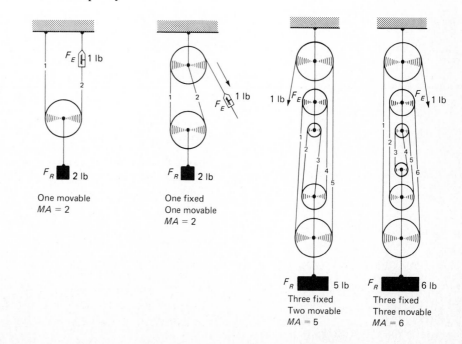

One fixed
$MA = 1$

One movable
$MA = 2$

One fixed
One movable
$MA = 2$

Three fixed
Two movable
$MA = 5$

Three fixed
Three movable
$MA = 6$

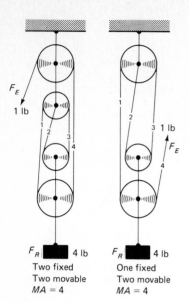

$F_E$
1 lb

$F_R$    4 lb
Two fixed
Two movable
$MA = 4$

3 1 lb
$F_E$

$F_R$    4 lb
One fixed
Two movable
$MA = 4$

## EXAMPLE 1

Draw two different sets of pulleys, each with an MA of 4. See figures at left.

## EXAMPLE 2

What effort will lift a resistance of $25\bar{0}$ N in the pulley system in Example 1?

DATA:

$$MA = 4$$
$$F_R = 25\bar{0} \text{ N}$$
$$F_E = ?$$

BASIC EQUATION:

$$MA = \frac{F_R}{F_E}$$

WORKING EQUATION:

$$F_E = \frac{F_R}{MA}$$

SUBSTITUTION:

$$F_E = \frac{25\bar{0} \text{ N}}{4}$$
$$= 62.5 \text{ N}$$

## EXAMPLE 3

If the resistance moves 7.00 ft, what is the effort distance of the pulley system in Example 1?

DATA:

$$MA = 4$$
$$d_R = 7.00 \text{ ft}$$
$$d_E = ?$$

BASIC EQUATION:

$$MA = \frac{d_E}{d_R}$$

WORKING EQUATION:

$$d_E = d_R(MA)$$

SUBSTITUTION:

$$d_E = (7.00 \text{ ft})(4)$$
$$= 28.0 \text{ ft}$$

## EXAMPLE 4

The pulley system shown is used to raise a $65\bar{0}$-lb object 25.0 ft. What is the mechanical advantage? What force is exerted?

$$MA = \text{number of strands holding up the resistance}$$
$$= 5$$

To find the force exerted:

DATA:

$$MA = 5$$
$$F_R = 65\bar{0} \text{ lb}$$
$$F_E = ?$$

$F_E = ?$

$F_R$    $65\bar{0}$ lb
Three fixed
Two movable
$MA = ?$

126

BASIC EQUATION:

$$MA = \frac{F_R}{F_E}$$

WORKING EQUATION:

$$F_E = \frac{F_E}{MA}$$

SUBSTITUTION:

$$F_E = \frac{65\overline{0} \text{ lb}}{5}$$

$$= 13\overline{0} \text{ lb}$$

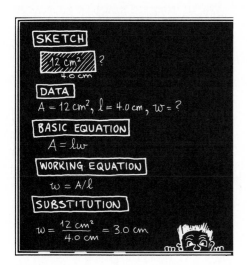

SKETCH

12 cm² ?
4.0 cm

DATA
$A = 12 \text{ cm}^2, \ell = 4.0 \text{ cm}, w = ?$

BASIC EQUATION
$A = \ell w$

WORKING EQUATION
$w = A/\ell$

SUBSTITUTION
$w = \dfrac{12 \text{ cm}^2}{4.0 \text{ cm}} = 3.0 \text{ cm}$

## PROBLEMS

*What is the mechanical advantage of each system?*

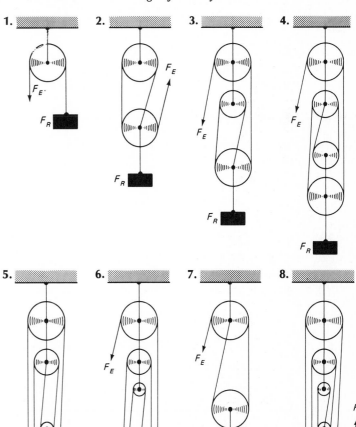

*In Problems 9 through 14, draw each pulley system.*

**9.** One fixed and two movable.

**10.** Two fixed and two movable with a MA of 5.

**11.** Three fixed and three movable with a MA of 6.

**12.** Four fixed and three movable.

13. Four fixed and four movable with a MA of 8.
14. Three fixed and three movable with a MA of 7.
15. What is the MA of a single movable pulley?
16. What effort will lift a $25\bar{0}$-lb weight by using a single movable pulley?
17. If the weight is moved 15.0 ft in Problem 16, how many feet of rope must be pulled up by the person exerting the effort?
18. Draw a system consisting of two fixed pulleys and two movable pulleys with a mechanical advantage of 4. If a force of 97.0 N is to be exerted, what weight can be raised?
19. If the weight in Problem 18 is raised 20.5 m, what length of rope is pulled?
20. Draw a system of one fixed and two movable pulleys. What is the mechanical advantage of this system?
21. A $40\bar{0}$-lb weight is to be lifted 30.0 ft. Using the system in Problem 20, find the effort force and effort distance.
22. If an effort force of 65.0 N and an effort distance of 13.0 m is to be applied, find the weight of the resistance and the distance it is moved using the pulley system in Problem 20.

## 8-6
## THE INCLINED PLANE

Gangplanks, chutes, and ramps are all examples of the inclined plane. Inclined planes are used to raise objects that are too heavy to lift vertically.

The work done in raising a resistance using the inclined plane equals the resistance times the height. This must also equal the work input, which can be found by multiplying the effort times the length of the plane.

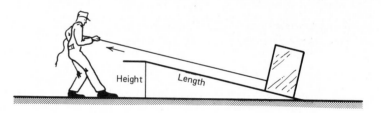

$$F_R \cdot d_R = F_E \cdot d_E \qquad \text{(law of machines)}$$

$$\boxed{F_R \cdot \text{height} = F_E \cdot \text{length}}$$

From the preceding equation:

$$\boxed{\frac{F_R}{F_E} = \frac{\text{length of plane}}{\text{height of plane}} = \text{MA}}$$

## EXAMPLE 1

A worker is pushing a box weighing 375 lb up a ramp 17.0 ft long onto a platform 4.50 ft above the ground. What is the mechanical advantage? What effort is applied?

$$\text{MA} = \frac{\text{length of plane}}{\text{height of plane}} = \frac{17.0 \text{ ft}}{4.50 \text{ ft}} = 3.78$$

To find the effort force:

DATA:

$$F_R = 375 \text{ lb}$$
$$\text{MA} = 3.78$$
$$F_E = ?$$

BASIC EQUATION:

$$MA = \frac{F_R}{F_E}$$

WORKING EQUATION:

$$F_E = \frac{F_R}{MA}$$

SUBSTITUTION:

$$F_E = \frac{375 \text{ lb}}{3.78}$$

$$= 99.2 \text{ lb}$$

## EXAMPLE 2

What is the shortest ramp that can be used to push a $60\overline{0}$-lb resistance onto a platform 3.50 ft high by exerting a force of 72.0 lb?

DATA:

$$F_R = 60\overline{0} \text{ lb}$$
$$F_E = 72.0 \text{ lb}$$
$$\text{height} = 3.50 \text{ lb}$$
$$\text{length} = ?$$

BASIC EQUATION:

$$F_R \cdot \text{height} = F_E \cdot \text{length}$$

WORKING EQUATION:

$$\text{length} = \frac{F_R \cdot \text{height}}{F_E}$$

SUBSTITUTION:

$$\text{length} = \frac{(60\overline{0} \text{ lb})(3.50 \text{ ft})}{72.0 \text{ lb}}$$

$$= 29.2 \text{ ft}$$

## EXAMPLE 3

An inclined plane is 13.0 m long and 7.00 m high. What is the mechanical advantage and what weight can be raised by exerting a force of $22\overline{0}$ N?

$$MA = \frac{\text{length of plane}}{\text{height of plane}} = \frac{13.0 \text{ m}}{7.00 \text{ m}} = 1.86$$

To find the weight of the resistance:

DATA:

$$MA = 1.86$$
$$F_E = 22\overline{0} \text{ N}$$
$$F_R = ?$$

BASIC EQUATION:

$$MA = \frac{F_R}{F_E}$$

WORKING EQUATION:

$$F_R = (F_E)(MA)$$

SUBSTITUTION:

$$F_R = (22\overline{0} \text{ N})(1.86)$$

$$= 409 \text{ N}$$

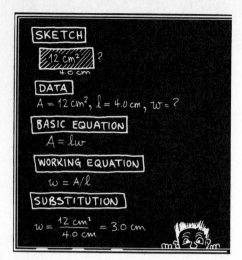

SKETCH

$12 \text{ cm}^2$ ?

4.0 cm

DATA

$A = 12 \text{ cm}^2, \; l = 4.0 \text{ cm}, \; w = ?$

BASIC EQUATION

$A = lw$

WORKING EQUATION

$w = A/l$

SUBSTITUTION

$w = \dfrac{12 \text{ cm}^2}{4.0 \text{ cm}} = 3.0 \text{ cm}$

## PROBLEMS

Given $F_R \cdot height = F_E \cdot length$, find the missing quantities.

| | $F_R$ | $F_E$ | Height of plane | Length of plane |
|---|---|---|---|---|
| 1. | 20.0 lb | 5.30 lb | 3.40 ft | ____ ft |
| 2. | 2340 N | 98$\overline{0}$0 N | ____ m | 3.79 m |
| 3. | 119 lb | ____ lb | 13.2 in. | 74.0 in. |
| 4. | ____ N | 1760 N | 0.821 m | 3.79 m |
| 5. | 37$\overline{0}$0 N | 12$\overline{0}$0 N | ____ cm | 112 cm |

Given $MA = \dfrac{length\ of\ plane}{height\ of\ plane}$, find the missing quantities.

| | MA | Length of plane | Height of plane |
|---|---|---|---|
| 6. | 9.00 | 3.40 ft | ____ ft |
| 7. | ____ | 3.79 m | 0.821 m |
| 8. | 1.30 | ____ ft | 9.72 ft |
| 9. | ____ | 74.0 cm | 13.2 cm |
| 10. | 17.4 | ____ in. | 13.4 in. |

11. An inclined plane is 10.0 m long and 2.50 m high. What is the mechanical advantage?

12. A resistance of 727 N is to be pushed up the plane in Problem 11. What effort is needed?

13. An effort of 20$\overline{0}$ N is applied to push an 815-N resistance up the inclined plane in Problem 11. Is the effort enough?

14. A safe is to be loaded onto a truck whose bed is 5.50 ft off the ground. The safe weighs 538 lb. If the effort to be applied is 14$\overline{0}$ lb, what length of ramp is needed to raise the safe?

15. What is the MA of the inclined plane in Problem 14?

16. Another safe weighing only 257 lb is to be loaded onto the same truck as in Problem 14. If the ramp is 21.1 ft long, what effort is needed?

17. A resistance of 325 N is to be raised by using a ramp 5.76 m long and by applying a force of 75.0 N. How high can it be raised?

18. What is the MA of the ramp in Problem 17?

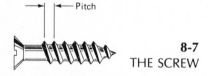

Pitch

Wood screw

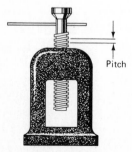

Pitch

Jack screw

**8-7**
**THE SCREW**

A *screw* is an inclined plane wrapped around a cylinder. The jackscrew and wood screw are examples of this simple machine.

The distance a beam rises or the distance the wood screw advances into a piece of wood in one revolution is called the *pitch* of the screw. Therefore, the pitch of a screw is actually the distance between two successive threads.

From the law of machines:

$$F_R \cdot d_R = F_E \cdot d_E$$

However,

$$d_R = pitch$$
$$d_E = circumference\ of\ the\ handle\ of\ the\ screwdriver$$

or

$$d_E = 2\pi r$$

where $r$ is the radius of the handle of the screwdriver. Therefore,

$$F_R \cdot \text{pitch} = F_E \cdot 2\pi r$$

so

$$\frac{F_R}{F_E} = \frac{2\pi r}{\text{pitch}} = \text{MA}$$

In the case of a jackscrew, $r$ is the length of the handle turning the screw and not the radius of the screw.

## EXAMPLE 1

Find the mechanical advantage of a jackscrew having a pitch of 0.125 in. and a handle radius of 12.0 in.

DATA:

$$\text{pitch} = 0.125 \text{ in.}$$
$$r = 12.0 \text{ in.}$$
$$\text{MA} = ?$$

BASIC EQUATION:

$$\text{MA} = \frac{2\pi r}{\text{pitch}}$$

WORKING EQUATION:   Same

SUBSTITUTION:

$$\text{MA} = \frac{2\pi(12.0 \text{ in.})}{0.125 \text{ in.}}$$
$$= 603$$

## EXAMPLE 2

What resistance can be lifted using the jackscrew in Example 1 if an effort of 203 N is exerted?

DATA:

$$\text{MA} = 603$$
$$F_E = 203 \text{ N}$$
$$F_R = ?$$

BASIC EQUATION:

$$\text{MA} = \frac{F_R}{F_E}$$

WORKING EQUATION:

$$F_R = (F_E)(\text{MA})$$

SUBSTITUTION:

$$F_R = (203 \text{ N})(603)$$
$$= 122,000 \text{ N}$$

## EXAMPLE 3

A 19,400-N pickup is to be raised using a jackscrew having a pitch of 5.00 mm and a handle radius of 255 mm. What force must be applied?

$$\text{pitch} = 5.00 \text{ mm}$$
$$r = 255 \text{ mm}$$
$$F_R = 19{,}400 \text{ N}$$
$$F_E = ?$$

BASIC EQUATION:

$$F_R \cdot \text{pitch} = F_E \cdot 2\pi r$$

WORKING EQUATION:

$$F_E = \frac{F_R(\text{pitch})}{2\pi r}$$

SUBSTITUTION:

$$F_E = \frac{(19{,}400 \text{ N})(5.00 \text{ mm})}{2\pi(255 \text{ mm})}$$
$$= 60.5 \text{ N}$$

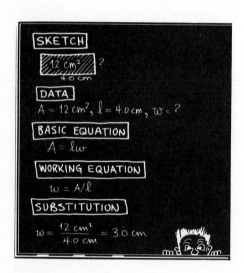

## PROBLEMS

*Given $F_R \cdot \text{pitch} = F_E \cdot 2\pi r$, find the missing quantities.*

| | $F_R$ | $F_E$ | Pitch | $r$ |
|---|---|---|---|---|
| 1. | 20.7 N | 5.30 N | 3.70 mm | ____ mm |
| 2. | ____ lb | 17.6 lb | 0.130 in. | 24.5 in. |
| 3. | 234 N | 9.80 N | ____ mm | 53.9 mm |
| 4. | 1190 N | ____ | 2.97 mm | 67.4 mm |
| 5. | $37\overline{0}$ lb | 12.0 lb | ____ in. | 11.2 in. |

*Given $\text{MA} = \dfrac{2\pi r}{\text{pitch}}$, find the missing quantities.*

| | MA | $r$ | Pitch |
|---|---|---|---|
| 6. | 7.00 | 34.0 mm | ____ mm |
| 7. | ____ | 3.79 in. | 0.812 in. |
| 8. | 9.00 | ____ in. | 0.970 in. |
| 9. | ____ | 7.40 mm | 1.32 mm |
| 10. | 13.0 | ____ mm | 2.10 mm |

11. A 3650-lb car is to be raised using a jackscrew having eight threads to the inch and a handle 15.0 in. long. What effort must be applied?

12. What is the MA in Problem 11?

13. The mechanical advantage of a jackscrew is 97.0. If the handle is 34.5 cm long, what is the pitch?

14. How much weight can be raised by applying an effort of 405 N to the jackscrew in Problem 13?

15. The radius of the handle of a screwdriver represents the effort distance of a wood screw. If a wood screw whose pitch 0.125 in. is advanced into a piece of wood using a screwdriver whose handle is 1.50 in. in diameter, what is the mechanical advantage of the screw?

16. What is the resistance of the wood if 15.0 lb of effort is applied on the wood screw in Problem 15?

17. What is the resistance of the wood if 15.0 lb of effort is applied to the wood screw in Problem 15 using a screwdriver whose handle is 0.500 in. in diameter?

18. The handle of a jackscrew is 60.0 cm long. If the mechanical advantage is 78.0, what is the pitch?

19. How much weight can be raised by applying a force of $43\overline{0}$ N to the jackscrew handle in Problem 18?

**8-8**
**THE WEDGE**

A *wedge* is an inclined plane in which the plane is moved instead of the resistance.

Finding the mechanical advantage of a wedge is not practical because of the large amount of friction. A narrow wedge is easier to drive than a thick wedge. Therefore, the mechanical advantage depends on the ratio of its length to its thickness.

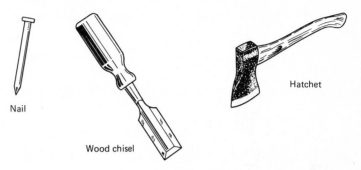

Nail

Wood chisel

Hatchet

**8-9**
**COMPOUND MACHINES**

A *compound machine* is a combination of simple machines. In most all compound machines, the *total mechanical advantage is the product of the mechanical advantages of each simple machine.*

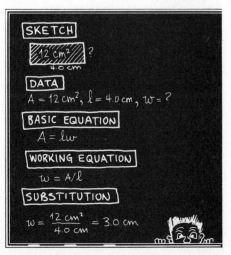

SKETCH

12 cm² ?
4.0 cm

DATA
$A = 12\ cm^2,\ \ell = 4.0\ cm,\ w = ?$

BASIC EQUATION
$A = \ell w$

WORKING EQUATION
$w = A/\ell$

SUBSTITUTION
$w = \dfrac{12\ cm^2}{4.0\ cm} = 3.0\ cm$

**PROBLEMS**

1. The box shown being pulled up an inclined plane using the indicated pulley system (called a block and tackle) weighs $40\overline{0}$ lb. If the inclined plane is 21.0 ft long and the height of the platform is 7.00 ft, find the mechanical advantage of this compound machine.

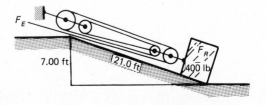

$F_E$

7.00 ft          21.0 ft          $F_R$          400 lb

2. What effort force must be exerted to move the box to the platform in Problem 1?

3. Find the mechanical advantage of the compound machine shown. The radius of the wheel is 1.00 ft and the radius of the axle is 0.500 ft.

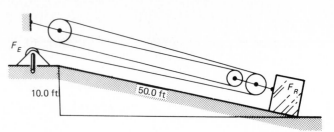

4. If an effort of $30\overline{0}$ lb is exerted, what weight can be moved up the plane using the compound machine in Problem 3?

5. What effort is required to move a load of 1.50 tons up the plane using the compound machine in Problem 3? (1 ton = $200\overline{0}$ lb.)

6. Find the mechanical advantage of the compound machine in Problem 1 if the inclined plane is 8.00 m long and 2.00 m high.

7. What effort force is needed to move a box of weight $250\overline{0}$ N to the platform in Problem 6?

8. Find the mechanical advantage of the compound machine in Problem 3 if the radius of the wheel is 40.0 cm, the radius of the axle is 12.0 cm, the length of the inclined plane is 12.0 m, and the height of the inclined plane is 50.0 cm.

9. If an effort of $45\overline{0}$ N is exerted in Problem 8, what weight can be moved up the inclined plane?

10. What effort force (in newtons) is needed to move 2.50 metric tons up the plane in Problem 8?

## 8-10
### THE EFFECT
### OF FRICTION
### ON SIMPLE MACHINES

Energy is lost in every machine by overcoming friction. This lost energy decreases the efficiency of the machine. That is, more work must be put into a machine than what you get out of the machine. This lost energy becomes heat energy, which results in machine wear or even burning out of certain parts of the machine.

Throughout this chapter we have been discussing simple machines, mechanical advantages of simple machines, resistance forces, effort forces, resistance distances, and effort distances.

In the previous calculations, the effect of friction was not considered. Therefore, when we found the effort to lift a resistance of $25\overline{0}$ N to be 62.5 N (using the pulley system shown), actually an effort of 70.0 N was needed due to the friction of the pulleys on their axles.

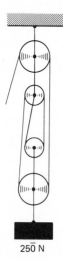

$25\overline{0}$ N

# CHAPTER
# *9*

## Rotational Motion

**9-1**
**MEASUREMENT OF ROTATIONAL MOTION**

Until now we have considered only motion in a straight line, called *rectilinear motion*. Technicians are faced with many problems, however, that deal with motion along a curved path or objects that are rotating about an axis. Although these kinds of motion are similar, we must distinguish between them.

Motion along a curved path is called *curvilinear motion*. A satellite in orbit around the earth is an example of this kind of motion.

*Rotary motion* occurs when the body itself is spinning. Examples of this kind of motion are the earth spinning on its axis, any turning wheel, a turning driveshaft, and the turning shaft of an electric motor.

We can see a wheel turn, but to gather useful information, we need a system of measurement.

There are three basic systems of defining angle measurement. One unit of measurement in rotary motion is the number of *rotations*—how many times the object goes around. We usually need to know not only the number of rotations but also the time for each rotation. The unit of rotation (most often used in industry) is the *revolution* (rev). A second system of measurement divides the circle of rotation into 360 degrees (360° = 1 rev).

The *radian* (rad) is a third angular measure used by scientists, which is approximately 57.3° or exactly $\left(\frac{360}{2\pi}\right)^{\circ}$. A radian is defined as that angle with its vertex at the center of a circle whose sides cut off an arc on the circle equal to its radius (see figure at left):
where $s = r$ and $\theta$ (Greek lowercase letter theta, used as a variable describing the angle) = 1 rad.

Stated as a formula,

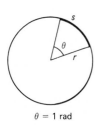

$\theta = 1$ rad

$$\theta = \frac{s}{r}$$

where: $\theta$ = angle determined by $s$ and $r$
$s$ = length of the arc of circle cut off
$r$ = radius of the circle

Technically, an angle measured in radians has no units since the radian is really a ratio (comparison) of the length of an arc on the circle to the

135

length of its radius: $\theta = s/r$. As a matter of convenience, though, "rad" is often used in practice to show radian measurement and therefore often appears in formulas.

A useful relationship is that there are $2\pi$ rad in one revolution. Therefore,

$$1 \text{ rev} = 360° = 2\pi \text{ rad}$$

You will find it necessary to refer to this box when a conversion between systems of measurement is required.

## EXAMPLE 1

Convert the angle $10\pi$ rad

(a) to rev
(b) to degrees

Using 1 rev $= 360° = 2\pi$ rad, form the following conversion factors, so that the old units are in the denominator (are canceled out) and the new units are in the numerator.

$$\text{(a)} \quad \theta = (10\pi \text{ rad})\left(\frac{1 \text{ rev}}{2\pi \text{ rad}}\right)$$

$$= 5 \text{ rev}$$

$$\text{(b)} \quad \theta = (10\pi \text{ rad})\left(\frac{360°}{2\pi \text{ rad}}\right)$$

$$= 1800°$$

*Angular displacement* is the angle through which any point on a rotating body moves. Note that on any rotating body, all points on that body move through the same angle in any given amount of time—even though each may travel different linear distances.

Flywheel

Point $A$ on the flywheel shown in the figure travels much farther than point $B$ (along a curved line), but during one revolution both travel through the same angle (have equal angular displacements).

In the automobile industry, technicians are concerned with the *rate* of rotary motion. Recall that in the linear system, velocity is the rate of motion (displacement/time). Similarly, we have what is called *angular velocity* in the rotary system (angular displacement/time). Angular velocity (designated $\omega$, the Greek lowercase letter omega) is usually measured in rev/min (rpm) for relatively slow rotations (e.g., automobile engines) and rev/s or rad/s for high-speed instruments. It is defined as the rate of angular displacement.

$$\omega = \text{angular velocity} = \frac{\text{number of revolutions}}{\text{time}} = \frac{\text{angular displacement}}{\text{time}}$$

As a formula,

$$\omega = \frac{\theta}{t}$$

where:   $\omega$ = angular velocity
$\theta$ = angle (in radians)
$t$ = time (in seconds)

EXAMPLE 2

A motorcycle wheel turns $36\overline{0}0$ times while being ridden for 6.40 min. What is its angular velocity in rev/min?

SKETCH: None needed

DATA:

$$t = 6.40 \text{ min}$$

$$\text{number of revolutions} = 36\overline{0}0 \text{ rev}$$

$$\omega = ?$$

BASIC EQUATION:

$$\omega = \frac{\text{number of revolutions}}{\text{time}}$$

WORKING EQUATION: Same

SUBSTITUTION:

$$\omega = \frac{36\overline{0}0 \text{ rev}}{6.40 \text{ min}}$$

$$= 563 \text{ rev/min or } 563 \text{ rpm}$$

Formulas for linear velocity of a rotating point on a circle and angular velocity may be related as follows:
We know:

$$(1) \ \theta = \frac{s}{r} \qquad (2) \ v = \frac{s}{t} \qquad (3) \ \omega = \frac{\theta}{t}$$

Therefore, combining and substituting $s/r$ for $\theta$ in (3),

$$\omega = \frac{s/r}{t}$$

$$\omega(r) = \frac{(s/r)(r)}{t} \qquad \text{Multiply both sides by } r.$$

$$\omega r = \frac{s}{t}$$

and recalling that $v = s/t$,

$$\omega r = v$$

Thus,

$$\boxed{v = \omega r}$$

where: $\omega$ = angular velocity
$r$ = radius
$v$ = linear velocity of a point on the circle

EXAMPLE 3

A wheel of 1.00 m radius turns at $10\overline{0}0$ rpm.

(a) Express the angular speed in rad/s.

(b) Find the angular displacement in 2.00 s.

(c) Find the linear velocity of a point on the rim of the wheel.

SKETCH:

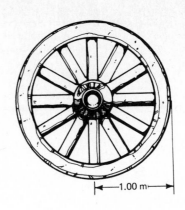

|←——1.00 m——→|

(a) DATA:

$$\omega = 10\overline{0}0 \text{ rpm} \qquad \text{(change it to rad/s)}$$

$$\omega = 10\overline{0}0 \frac{\text{rev}}{\text{min}} \times \frac{2\pi \text{ rad}}{1 \text{ rev}} \times \frac{1 \text{ min}}{60 \text{ s}} = 105 \text{ rad/s}$$

(b) DATA:

$$\omega = 105 \text{ rad/s}$$
$$t = 2.00 \text{ s}$$
$$\theta = ?$$

BASIC EQUATION:

$$\omega = \frac{\theta}{t}$$

WORKING EQUATION:

$$\theta = \omega t$$

SUBSTITUTION:

$$\theta = (105 \text{ rad/s})(2.00 \text{ s})$$
$$= 21\overline{0} \text{ rad}$$

(c) DATA:

$$\omega = 105 \text{ rad/s}$$
$$r = 1.00 \text{ m}$$
$$v = ?$$

BASIC EQUATION:

$$v = \omega r$$

WORKING EQUATION:  Same

SUBSTITUTION:

$$v = (105 \text{ rad/s})(1.00 \text{ m})$$
$$= 105 \text{ m/s}$$

A device called a *stroboscope* or strobe light may be used to measure or check the speed of rotation of a shaft or other machinery part. It can be used to "slow down" repeating motion to be observed more conveniently. The light flashes rapidly and the rate of flash can be adjusted to coincide with the rotation of a point or points on the rotating object. Knowing the rate of flashing will also then reveal the rate of rotation. A slight variation in the rate of rotation and flash will cause the observed point to appear to move either forward or backward as the stagecoach wheels in western movies sometimes appear to do.

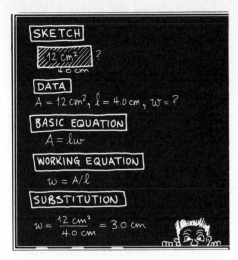

## PROBLEMS

1. Convert $6\frac{1}{2}$ revolutions
   (a) to radians
   (b) to degrees

2. Convert 2880°
   (a) to revolutions
   (b) to radians

3. Convert $25\pi$ rad
   (a) to revolutions
   (b) to degrees

4. Convert $60\bar{0}$ rpm to rad/s.

In Problems 5 through 14, find the angular velocity.

5. Number of revolutions = 525
   $t = 3.42$ min
   $\omega$ in rpm = ?

6. Number of revolutions = 7360
   $t = 37.0$ s
   $\omega$ in rev/s = ?

7. A motor turns at a rate of 11.0 rev/s. What is its angular velocity in rpm?

8. Number of revolutions = 4.00
   $t = 3.00$ s
   $\omega$ in rad/s = ?

9. Number of revolutions = 6370
   $t = 4.00$ min
   $\omega$ in rev/s = ?

10. Number of revolutions = $30\bar{0}$
    $t = 5.00$ min
    $\omega$ in rpm = ?

11. Number of revolutions = 2540
    $t = 18$ s
    $\omega$ in rev/s = ?

12. A motor turns at a rate of 15 rev/s. What is its angular velocity in rpm?

13. Number of revolutions = 6.25
    $t = 5.05$ s
    $\omega$ in rad/s = ?

14. Number of revolutions = 5680
    $t = 3.07$ min
    $\omega$ in rev/s = ?

15. Convert $60\bar{0}$ rad/s to rpm.

16. A flywheel is rotating at 1050 rpm.
    (a) How long does it take to complete one revolution?
    (b) How many revolutions does it complete in 5.00 s?

17. A rotating wheel completes one revolution in 0.150 s. Find its angular velocity
    (a) in rev/s
    (b) in rpm
    (c) in rad/s

18. If a wheel of radius 0.240 m turns at a rate of 4.00 rev/s, find the angular displacement in 13.0 s.

19. A shaft of 0.085 m radius rotates 7.00 rad/s. What is its angular displacement in radians in 1.20 s?

20. A wheel has an angular velocity of 47.0 rpm. If its radius is 0.270 m, what is the linear velocity of a point on the rim of the wheel?

21. A pendulum of length 1.50 m swings through an arc of 5.0°. Find the length of the arc through which the pendulum swings.

22. A flywheel of radius 25.0 cm is rotating at 655 rpm.
    (a) Express its angular speed in rad/s.
    (b) Find its angular displacement (in rad) in 3.00 min.
    (c) Find the linear distance traveled by a point on the rim in one complete revolution.
    (d) Find the linear distance traveled by a point on the rim in 3.00 min.
    (e) Find the linear velocity of a point on the rim.

23. An airplane propeller whose blades are 2.00 m long is rotating at $220\bar{0}$ rpm.
    (a) Express the angular speed in rad/s.
    (b) Find the angular displacement in 4.00 s.
    (c) Find the linear velocity of a point on the end of a blade.

24. An automobile is traveling at 60.0 km/h. Its tires have a radius of 33.0 cm.
    (a) Find the angular speed of the tires (in rad/s).
    (b) Find the angular displacement in 30.0 s.
    (c) Find the linear distance traveled by a point on the tread in 30.0 s.

25. Find the angular speed in rad/s of the following hands on a clock:
    (a) second
    (b) minute
    (c) hour

26. A belt is wrapped around a pulley that is 30.0 cm in diameter. If the pulley rotates at $25\overline{0}$ rpm, what is the linear speed (in m/s) of the belt? (Assume no belt slippage on the pulley.)

27. The earth rotates on its axis at an angular velocity of 1 rev/24 h. Find the linear velocity (in km/h) of a point on the equator where the radius is $64\overline{0}0$ km.

**9-2**

**TORQUE**

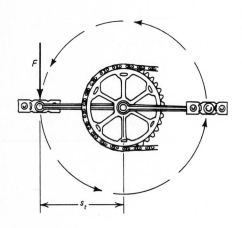

We have already discussed force in the linear system and defined it as a push or a pull. In rotational systems we have a "twist" which we call *torque*. Torque is the tendency to produce change in rotary motion. It is, in rotational systems, similar to force in the linear system. You may already have studied torque in connection with automobile engines. We shall consider the simpler example of pedaling a bicycle.

In pedaling we apply a torque to the sprocket causing it to rotate. The torque developed depends on two factors:

1. The size of the force applied.
2. How far from the axle (shaft) center point the force is applied.

We can express torque with an equation:

$$\boxed{T = Fs_t}$$

where:  $T$ = torque (lb ft or N m)
$F$ = applied force (lb or N)
$s_t$ = length of torque arm (ft or m)

Note that $s_t$, the length of the torque arm, is different from $s$ in the equation defining work ($W = Fs$). Recall that $s$ in the work equation is the linear distance over which the force acts.

In all torque problems we are concerned with motion about a point or axis of rotation. The torque arm is the *perpendicular* distance from the point of rotation to the direction of the applied force.

In torque problems, $s_t$ is always perpendicular to the force. Note in the figure that $s_t$ is the distance from the axle to the pedal.

The units of torque look similar to those of work, but do not forget the difference between $s$ and $s_t$.

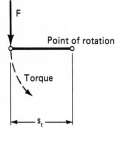

## EXAMPLE

A force of 10.0 lb is applied to a bicycle pedal as shown. If the length of the pedal arm is 0.850 ft, what torque is applied to the shaft?

SKETCH:

10.0 lb

0.850 ft

DATA:

$F = 10.0$ lb
$s_t = 0.850$ ft
$T = ?$

BASIC EQUATION:

$$T = Fs_t$$

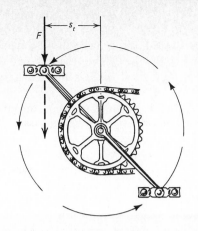

WORKING EQUATION: Same

SUBSTITUTION:

$$T = (10.0 \text{ lb})(0.850 \text{ ft})$$
$$= 8.50 \text{ lb ft}$$

If the force is not exerted tangent to the circle made by the pedal, the length of the torque arm is *not* the length of the pedal arm. $s_t$ is measured as the perpendicular distance to the force.

Since $s_t$ is therefore shorter, the product $F \cdot s_t$ is smaller and the turning effect, the torque, is less in the pedal position illustrated. Maximum torque is produced when the pedals are horizontal and the force applied is straight down.

## PROBLEMS

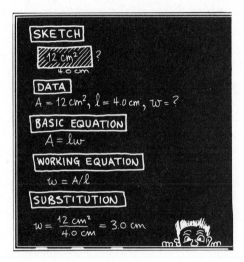

1. Given: $F = 16.0$ lb
   $s_t = 6.00$ ft
   $T = ?$

2. Given: $F = 10\bar{0}$ N
   $s_t = 0.420$ m
   $T = ?$

3. Given: $T = 35.7$ lb ft
   $s_t = 0.0240$ ft
   $F = ?$

4. Given: $T = 60.0$ N m
   $F = 30.0$ N
   $s_t = ?$

5. Given: $T = 65.4$ N m
   $s_t = 35.0$ cm
   $F = ?$

6. Given: $F = 63\bar{0}$ N
   $s_t = 0.740$ m
   $T = ?$

7. If the torque on a shaft of radius 0.0237 m is 38.0 N m, what force is applied to the shaft?

rad. = 0.0237 m

8. If the force applied to a torque wrench 1.50 ft long is 56.2 lb, what torque is indicated by the wrench?

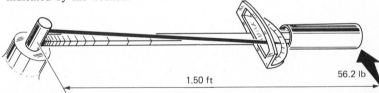

1.50 ft          56.2 lb

9. If a motorcycle head bolt is torqued to 20.0 lb ft, how long a shaft do we need on a wrench on which we can exert a maximum force of 35.0 lb?

10. If a force of 24.0 lb is applied to a shaft of radius 0.140 ft, what is the torque on the shaft?

11. If a torque of 175 lb ft is needed to free a large rusted-on nut and the length of the wrench is 1.10 ft, what force must be applied to free it?

12. A torque wrench reads 14.5 N m. If its length is 25.0 cm, what force is being applied to the handle?

**9-3**
**MOTION**
**IN A CURVED PATH**

Newton's laws of motion apply to motion along a curved path as well as in a straight line. Recall that a moving body tends to continue in a straight line because of inertia.

If we are to cause the body to move in a circle, we must constantly apply a force perpendicular to the line of motion of the body. The simplest example is a rock being swung in a circle on the end of a string. By Newton's

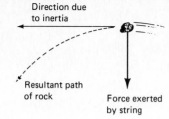

Direction due
to inertia

Resultant path
of rock

Force exerted
by string

first law, the rock tends to go in a straight line but the string exerts a constant force on the rock perpendicular to this line of travel. The resulting path of the rock is a circle. The force of the string on the rock is the *centripetal* (toward the center) *force.*

If the string should break, however, there would no longer be a centripetal force acting on the rock, and it would fly off tangent to the circle.

The equation for calculating the centripetal force on any body moving along a curved path is

$$F = \frac{mv^2}{r}$$

where:  $F$ = centripetal force
$m$ = mass of the body
$v$ = velocity of the body
$r$ = radius of curvature of the path of the body

What then is the difference between centripetal and centrifugal forces? The term "centrifugal force" is widely misused to mean centripetal force. Recall Newton's third law of motion that for every action there is an equal and opposite reaction. *Centrifugal force* is the reaction force. It is present only when there is a centripetal force. It is equal in magnitude to the centripetal force but is in the opposite direction. It is exerted *by* the rock *on* the string.

EXAMPLE

An automobile of mass 113 slugs rounds a curve of radius 75.0 ft with a velocity of 44.0 ft/s (30.0 mi/h). What centripetal force is exerted on the automobile while rounding the curve?

SKETCH:

$r = 75.0$ ft

44.0 ft/s

DATA:

$m = 113$ slugs
$v = 44.0$ ft/s
$r = 75.0$ ft
$F = ?$

BASIC EQUATION:

$$F = \frac{mv^2}{r}$$

WORKING EQUATION:   Same

SUBSTITUTION:

$$F = \frac{(113 \text{ slugs})(44.0 \text{ ft/s})^2}{75.0 \text{ ft}}$$

$$= 2920 \text{ slug ft/s}^2 \quad \left( \text{Recall: } 1 \text{ lb} = \frac{1 \text{ slug ft}}{\text{s}^2} \right)$$

$$= 2920 \text{ lb}$$

142

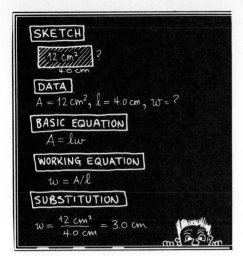

**PROBLEMS**

**1.** Given: $m = 64.0$ kg
$v = 34.0$ m/s
$r = 17.0$ m
$F$ in N $= ?$

**2.** Given: $m = 11.3$ slugs
$v = 3.00$ ft/s
$r = 3.24$ ft
$F$ in lb $= ?$

**3.** Given: $F = 25\overline{0}0$ lb
$v = 47.6$ ft/s
$r = 72.0$ ft
$m$ in slugs $= ?$

**4.** Given: $F = 587$ N
$v = 0.780$ m/s
$m = 67.0$ kg
$r$ in m $= ?$

**5.** Given: $F = 602$ N
$m = 63.0$ kg
$r = 3.20$ m
$v$ in m/s $= ?$

**6.** Given: $m = 37.5$ kg
$v = 17.0$ m/s
$r = 3.75$ m
$F$ in N $= ?$

**7.** Given: $F = 75.0$ N
$v = 1.20$ m/s
$m = 100$ kg
$r$ in m $= ?$

**8.** Given: $F = 80.0$ N
$m = 43.0$ kg
$r = 17.5$ m
$v$ in m/s $= ?$

**9.** An automobile of mass 117 slugs follows a curve of radius 79.0 ft with a velocity of 49.3 ft/s. What centripetal force is exerted on the automobile while it is rounding the curve?

**10.** What is the centripetal force exerted on a 7.12-kg mass moving at a speed of 2.98 m/s in a circle of radius 2.72 m?

**11.** The centripetal force on a car of mass 97.0 slugs rounding a curve is 3250 lb. If its velocity is 40.1 ft/s, what is the radius of the curve?

**12.** The centripetal force on a runner is 17.0 lb. If the runner weighs 175 lb and his velocity is 14.0 miles per hour, what is the radius of the curve?

**13.** An automobile whose mass is 1650 kg is driven around a circular curve of radius $15\overline{0}$ m at 80.0 km/h. What is the centripetal force of the road on the automobile?

**9-4**
**POWER**
**IN ROTARY SYSTEMS**

Probably the most important aspect of rotational motion to the technician is the power developed in rotary systems. Recall that torque in rotary systems was discussed in Section 9-2. Power, however, must be considered whenever an engine or motor is used to turn a shaft. Some common examples are the use of winches and automobile drive trains.

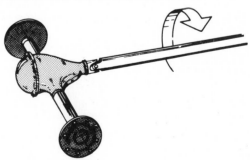

Earlier we learned that

$$\text{power} = \frac{\text{force} \times \text{displacement}}{\text{time}} = \frac{\text{work}}{\text{time}}$$

in the linear system. In the rotary system:

$$\text{power} = \frac{(\text{torque})(\text{angular displacement})}{\text{time}}$$
$$= (\text{torque})(\text{angular velocity})$$
$$= T\omega$$

**143**

Recall that angular displacement is the angle through which the shaft is turned.

In the English system we measure angular displacement by multiplying the number of revolutions by $2\pi$:

$$\text{angular displacement} = \text{number of revolutions} \times 2\pi$$

For the rotary system

$$\text{power} = \frac{\text{torque} \times 2\pi \text{ revolutions}}{\text{time}}$$

When time is in minutes

$$\text{power} = \text{torque} \times 2\pi \times \frac{\text{rev}}{\text{min}} \times \frac{1 \text{ min}}{60 \text{ s}}$$

$$\boxed{\text{power in } \frac{\text{ft lb}}{\text{s}} = \text{torque in ft lb} \times \frac{\text{number of revolutions}}{\text{min}} \times 0.105 \frac{\text{min}}{\text{rev s}}}$$

Another common unit of power is the horsepower (hp). The conversion factor between $\frac{\text{ft lb}}{\text{s}}$ and hp is

$$\boxed{\text{power in hp} = \text{power in } \frac{\text{ft lb}}{\text{s}} \times \frac{\text{hp}}{55\overline{0} \text{ ft lb/s}}}$$

## EXAMPLE 1

What power (in ft lb/s) is developed by an electric motor with torque of 5.70 ft lb and speed of 425 rpm?

SKETCH: None needed

DATA:

$$T = 5.70 \text{ ft lb}$$
$$\omega = 425 \text{ rpm}$$
$$P = ?$$

BASIC EQUATION:

$$P = \text{torque} \times \frac{\text{rev}}{\text{min}} \times 0.105 \frac{\text{min}}{\text{rev s}}$$

WORKING EQUATION: Same

SUBSTITUTION:

$$P = (5.70 \text{ ft lb}) \left( 425 \frac{\cancel{\text{rev}}}{\cancel{\text{min}}} \right) \left( 0.105 \frac{\cancel{\text{min}}}{\cancel{\text{rev}} \text{ s}} \right)$$
$$= 254 \text{ ft lb/s}$$

## EXAMPLE 2

What horsepower is developed by a racing engine with torque of 545 ft lb at $65\overline{0}0$ rpm? First, find power in ft lb/s and then convert to hp.

SKETCH: None needed

DATA:

$$T = 545 \text{ ft lb}$$
$$\omega = 65\bar{0}0 \text{ rpm}$$
$$P = ?$$

BASIC EQUATION:

$$\text{Power} = \text{torque} \times \frac{\text{rev}}{\text{min}} \times 0.105 \frac{\text{min}}{\text{rev s}}$$

WORKING EQUATION:  Same

SUBSTITUTION:

$$P = (545 \text{ ft lb})\left(65\bar{0}0 \frac{\text{rev}}{\text{min}}\right)\left(0.105 \frac{\text{min}}{\text{rev s}}\right)$$

$$= 372{,}000 \frac{\text{ft lb}}{\text{s}} \times \frac{1 \text{ hp}}{55\bar{0} \frac{\text{ft lb}}{\text{s}}}$$

$$= 676 \text{ hp}$$

In the metric system, angular displacement must be expressed in radians. (Recall: 1 rev = $2\pi$ radians.) The power formula is then

$$\boxed{\begin{aligned} \text{power} &= \text{torque} \times \frac{\text{angular displacement}}{\text{time}} \\ &= \text{torque} \times \text{angular velocity} \end{aligned}}$$

Substituting symbols and units:
In watts (W):

$$\boxed{\begin{aligned} \text{power} &= T\omega \\ &= (\text{N m})\left(\frac{1}{\text{s}}\right) \\ &= \frac{\text{N m}}{\text{s}} \\ &= \text{W} \end{aligned}}$$

To find the power in kilowatts (kW), multiply the number of watts by the conversion factor:

$$\frac{1 \text{ kW}}{1000 \text{ W}}$$

*Note:* In problem solving, the radian unit is not used, and $\omega$ is expressed with the unit 1/s.

## EXAMPLE 3

How many watts of power are developed by a mechanic tightening bolts using 50.0 N m of torque at a rate of 2.10 rad/s? How many kW?

SKETCH:  None needed

DATA:

$$T = 50.0 \text{ N m}$$
$$\omega = 2.10/\text{s}$$
$$P = ?$$

BASIC EQUATION:

$$P = T\omega$$

WORKING EQUATION:   Same

SUBSTITUTION:

$$P = (50.0 \text{ N m})(2.10/s)$$
$$= 105 \text{ N m/s}$$
$$= 105 \text{ W}$$

To find the power in kW:

$$105 \text{ W} \times \frac{1 \text{ kW}}{1000 \text{ W}} = 0.105 \text{ kW}$$

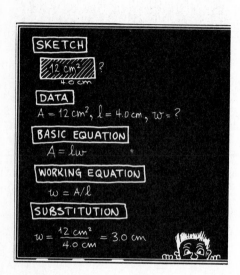

## PROBLEMS

1. Given:   $T = 372$ ft lb
   $\omega = 264$ rpm
   $P = ?$ hp

2. Given:   $T = 39.4$ N m
   $\omega = 6.70/s$
   $P = ?$ W

3. Given:   $T = 125$ ft lb
   $\omega = 555$ rpm
   $P = ? \dfrac{\text{ft lb}}{\text{s}}$

4. Given:   $T = 650$ N m
   $\omega = 45.0/s$
   $P = ?$ W

5. Given:   $P = 8950$ W
   $\omega = 4.80/s$
   $T = ?$

6. Given:   $P = 650$ W
   $T = 540$ N m
   $\omega = ?$

7. What horsepower is developed by an engine with torque of $40\overline{0}$ ft lb at $45\overline{0}0$ rpm?

8. What torque must be applied to develop 175 (ft lb)/s of power in a motor if $\omega = 394$ rpm?

9. What is the angular velocity of a motor developing 649 W of power with torque of 131 N m?

10. A high-speed industrial drill develops 0.500 hp at $16\overline{0}0$ rpm. What torque is applied to the drill bit?

11. An engine has torque of 505 ft lb at 3250 rev/min. What power in (ft lb)/s does it develop?

12. What is the angular velocity of a motor developing $35\overline{0}$ (ft lb)/s of power with a torque of 96.0 ft lb?

13. What power (in hp) is developed by an engine with torque of 524 ft lb
    (a) at $30\overline{0}0$ rpm?
    (b) at $60\overline{0}0$ rpm?

14. What is the angular velocity of a motor developing $65\overline{0}$ W of power with a torque of $13\overline{0}$ N m?

15. A drill develops 0.50 kW of power at $18\overline{0}0$ rpm. What torque is applied to the drill bit?

16. What power is developed by an engine whose torque of 750 N m is applied at $45\overline{0}0$ rpm?

17. A tangential force of 150 N is applied to a flywheel whose diameter is 0.45 m to maintain a constant angular velocity of 175 rpm. How much work is done per minute?

# CHAPTER
# *10*

# *Gears and Pulleys*

**10-1**
**TRANSFERRING ROTATIONAL MOTION**

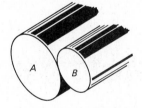

Suppose that we have two discs touching each other as indicated in the figure. Disc $A$ is driven by a motor and turns disc $B$ by making use of the friction between them.

The relationship between the diameters of the two discs and their number of revolutions is

$$D \cdot N = d \cdot n$$

where:   $D =$ diameter of the driver disc
$d =$ diameter of the driven disc
$N =$ number of revolutions of the driver disc
$n =$ number of revolutions of the driven disc

However, using two discs to transfer rotational motion is not very efficient due to the slippage that usually occurs between discs. The most common ways to prevent disc slippage are the placing of teeth on the edge of the disc and the connecting of discs with a belt. Therefore, instead of using discs, we use gears or belt-driven pulleys to transfer this motion. The teeth on the gears eliminate the slippage, and the belt connecting the pulleys helps reduce the slippage.

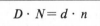
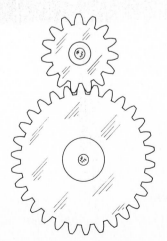
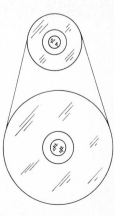

147

We can change the equation $D \cdot N = d \cdot n$ to the form $D/d = n/N$ by dividing both sides by $dN$. The left side indicates the ratio of the diameters of the discs. If the ratio is 2, this means that the larger disc must have a diameter two times the diameter of the smaller disc. The same ratio would apply to gears and pulleys. The ratio of the diameters of the gears must be 2 to 1, and the ratio of the diameters of the pulleys must be 2 to 1. In fact, the ratio of the number of teeth on the gears must be 2 to 1.

The right side of the equation indicates the ratio of the number of revolutions of the two discs. If the ratio is 2, this means that the smaller disc makes two revolutions while the larger disc makes one revolution. The same would be true for gears and for pulleys connected by a belt.

**10-2
GEARS**

Gears are used to transfer rotational motion from one gear to another. The gear that causes the motion is called the *driver gear*. The gear to which the motion is transferred is called the *driven gear*.

There are many different sizes, shapes, and types of gears. A few examples are shown in the following figures.

Some of these gears and more are shown in the transmission (first photograph).

Photograph by R. T. Gladin.

Photograph courtesy Illinois Gear, Wallace-Murray Corporation.

Photograph courtesy Illinois Gear, Wallace-Murray Corporation.

Photograph courtesy Illinois Gear, Wallace-Murray Corporation.

For any type of gear, we use one basic formula:

$$T \cdot N = t \cdot n$$

where: $T$ = number of teeth on the driver
$N$ = number of revolutions of the driver
$t$ = number of teeth on the driven
$n$ = number of revolutions of the driven

## EXAMPLE 1

A driver gear has 30 teeth. How many revolutions does the driven gear with 20 teeth make while the driver makes 1 revolution?

DATA:

$$T = 30 \text{ teeth}$$
$$N = 1 \text{ revolution}$$
$$t = 20 \text{ teeth}$$
$$n = ?$$

BASIC EQUATION:

$$T \cdot N = t \cdot n$$

WORKING EQUATION:

$$n = \frac{T \cdot N}{t}$$

SUBSTITUTION:

$$n = \frac{(30 \text{ teeth})(1 \text{ rev})}{20 \text{ teeth}}$$
$$= 1.5 \text{ rev}$$

149

A driven gear of 70 teeth makes 63.0 revolutions per minute (rpm). The driver gear makes 90.0 rpm. What is the number of teeth required for the driver gear?

DATA:

$$N = 90.0 \text{ rpm}$$
$$t = 70 \text{ teeth}$$
$$n = 63.0 \text{ rpm}$$
$$T = ?$$

BASIC EQUATION:

$$T \cdot N = t \cdot n$$

WORKING EQUATION:

$$T = \frac{t \cdot n}{N}$$

SUBSTITUTION:

$$T = \frac{(70 \text{ teeth})(63.0 \text{ rpm})}{90.0 \text{ rpm}}$$
$$= 49 \text{ teeth}$$

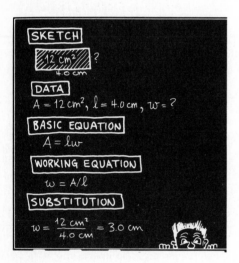

## PROBLEMS

*Fill in the blanks using the equation $T \cdot N = t \cdot n$.*

| | $T$ | $N$ | $t$ | $n$ |
|---|---|---|---|---|
| **1.** | 17 | $16\bar{0}$ | 37 | ____ |
| **2.** | 36 | ____ | 14 | $37\bar{0}$ |
| **3.** | 60 | $16\bar{0}0$ | ____ | $48\bar{0}$ |
| **4.** | ____ | $10\bar{0}$ | 20 | $15\bar{0}$ |
| **5.** | 12 | ____ | 44 | 234 |
| **6.** | 73 | 169 | 99 | ____ |
| **7.** | 39 | 23.4 | ____ | 70.2 |
| **8.** | ____ | 39.0 | 23 | 156 |
| **9.** | 19 | ____ | 34 | 93.4 |
| **10.** | 7 | 324 | 19 | ____ |

**11.** A driver gear has 38 teeth and makes 85.0 rpm. What is the rpm of the driven gear with 72 teeth?

**12.** A motor turning at 1250 rpm is fitted with a gear having 56 teeth. Find the speed of the driven gear if it has 66 teeth.

**13.** A gear running at $25\bar{0}$ rpm meshes with another revolving at $10\bar{0}$ rpm. If the smaller gear has 30 teeth, how many teeth does the larger gear have?

*Fill in the blanks:*

| | Number of teeth | | rpm | |
|---|---|---|---|---|
| | Driver | Driven | Driver | Driven |
| **14.** | 58 | ____ | $23\bar{0}$ | 145 |
| **15.** | ____ | 150 | $24\bar{0}$ | $12\bar{0}$ |
| **16.** | 190 | 240 | ____ | 162 |
| **17.** | 70 | ____ | $42\bar{0}$ | $70\bar{0}$ |
| **18.** | 80 | 65 | $26\bar{0}$ | ____ |
| **19.** | ____ | 80 | $48\bar{0}$ | $78\bar{0}$ |

**20.** A driver gear with 40 teeth makes 154 rpm. How many teeth must the driver have if it makes $22\bar{0}$ rpm?

**21.** Two gears have a speed ratio of 4.6 to 1. If the smaller gear has 15 teeth, what must be the number of teeth on the larger gear?

**22.** What size gear should be mated with a 15-tooth pinion to achieve a speed reduction of 10 to 3?

**10-3
GEAR TRAINS**

When two gears mesh as shown,* they turn in opposite directions. If gear *A* turns clockwise, gear *B* turns counterclockwise. If gear *A* turns counterclockwise, gear *B* turns clockwise.

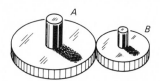

If a third gear is inserted between the two, as shown, then gears *A* and *B* are rotating in the same direction. This third gear is called an *idler,* and such an arrangement of gears is called a *gear train.*

---

*RULE*

*When the number of shafts in a gear train is odd (such as 1, 3, 5, . . .), the first gear and the last gear rotate in the same direction. When the number of shafts is even, the gears rotate in opposite directions.*

---

When a very complicated gear train is considered, the relationship between revolutions and number of teeth is still present. This relationship is: The number of revolutions of the first driver times the product of the numbers of teeth of all the driver gears equals the number of revolutions of the final driven gear times the product of the number of teeth on all the driven gears. That is,

$$NT_1\,T_2\,T_3\,T_4\,\ldots = nt_1\,t_2\,t_3\,t_4\,\ldots$$

where:  $N$ = number of revolutions of first driver
$T_1$ = teeth on first driver
$T_2$ = teeth on second driver
$T_3$ = teeth on third driver
$T_4$ = teeth on fourth driver
$n$ = number of revolutions of last driven
$t_1$ = teeth on first driven
$t_2$ = teeth on second driven
$t_3$ = teeth on third driven
$t_4$ = teeth on fourth driven

---

* Although gears have teeth, in technical work they are usually shown as cylinders.

EXAMPLE 1

Determine the relative motion of gears $A$ and $B$ in the figures.

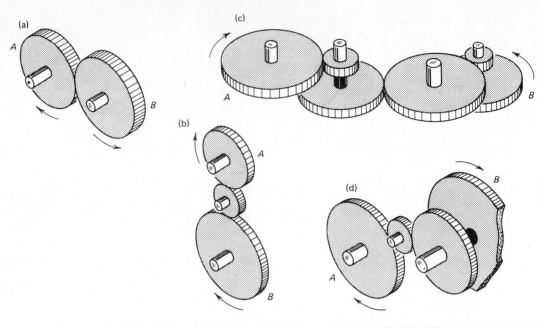

EXAMPLE 2

Find the number of revolutions per minute of gear $D$ in the following figure if gear $A$ rotates at 20.0 revolutions per minute.

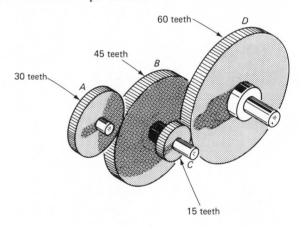

Gears $A$ and $C$ are drivers and gears $B$ and $D$ are driven.

DATA:

$$N = 20.0 \text{ rpm}$$
$$T_1 = 30 \text{ teeth}$$
$$T_2 = 15 \text{ teeth}$$
$$t_1 = 45 \text{ teeth}$$
$$t_2 = 60 \text{ teeth}$$
$$n = ?$$

BASIC EQUATION:

$$NT_1 T_2 = n t_1 t_2$$

WORKING EQUATION:

$$n = \frac{NT_1 T_2}{t_1 t_2}$$

$$n = \frac{(20.0 \text{ rpm})(30 \text{ teeth})(15 \text{ teeth})}{(45 \text{ teeth})(60 \text{ teeth})}$$

$$= 3.33 \text{ rpm}$$

## EXAMPLE 3

Find the revolutions per minute of gear *D* in the train shown.

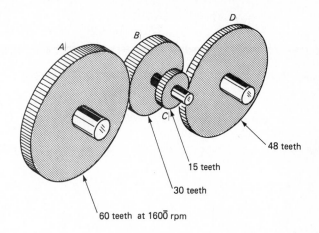

60 teeth at $160\overline{0}$ rpm
30 teeth
15 teeth
48 teeth

Gears *A* and *C* are drivers and gears *B* and *D* are driven.

Data:

$$N = 160\overline{0} \text{ rpm}$$
$$T_1 = 60 \text{ teeth}$$
$$T_2 = 15 \text{ teeth}$$
$$t_1 = 30 \text{ teeth}$$
$$t_2 = 48 \text{ teeth}$$
$$n = ?$$

Basic Equation:

$$N T_1 T_2 = n t_1 t_2$$

Working Equation:

$$n = \frac{N T_1 T_2}{t_1 t_2}$$

Substitution:

$$n = \frac{(160\overline{0} \text{ rpm})(60 \text{ teeth})(15 \text{ teeth})}{(30 \text{ teeth})(48 \text{ teeth})}$$

$$= 100\overline{0} \text{ rpm}$$

## EXAMPLE 4

In the gear train shown, find the speed in rpm of gear *A*.

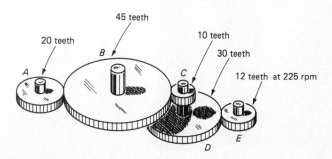

20 teeth
45 teeth
10 teeth
30 teeth
12 teeth at 225 rpm

DATA:

$$T_1 = 20 \text{ teeth}$$
$$T_2 = 45 \text{ teeth}$$
$$T_3 = 30 \text{ teeth}$$
$$t_1 = 45 \text{ teeth}$$
$$t_2 = 10 \text{ teeth}$$
$$t_3 = 12 \text{ teeth}$$
$$n = 225 \text{ rpm}$$
$$N = ?$$

Gear $B$ is both a driver and a driven gear.

BASIC EQUATION:

$$NT_1 T_2 T_3 = n t_1 t_2 t_3$$

WORKING EQUATION:

$$N = \frac{n t_1 t_2 t_3}{T_1 T_2 T_3}$$

SUBSTITUTION:

$$N = \frac{(225 \text{ rpm})(45 \text{ teeth})(10 \text{ teeth})(12 \text{ teeth})}{(20 \text{ teeth})(45 \text{ teeth})(30 \text{ teeth})}$$
$$= 45.0 \text{ rpm}$$

In a gear train, when a gear is both a driver gear and a driven gear, it may be left out of the computation.

## EXAMPLE 5

The problem in Example 4 could have been worked as follows since gear $B$ is both a driver and a driven.

BASIC EQUATION:

$$NT_1 T_3 = n t_2 t_3$$

WORKING EQUATION:

$$N = \frac{n t_2 t_3}{T_1 T_3}$$

SUBSTITUTION:

$$N = \frac{(225 \text{ rpm})(10 \text{ teeth})(12 \text{ teeth})}{(20 \text{ teeth})(30 \text{ teeth})}$$
$$= 45.0 \text{ rpm}$$

## PROBLEMS

*If gear A turns in a clockwise motion, determine the motion of gear B in each gear train.*

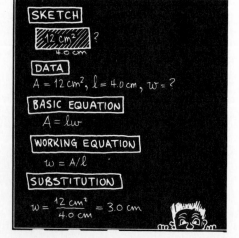

**1.**

**2.**

**3.**

**4.**

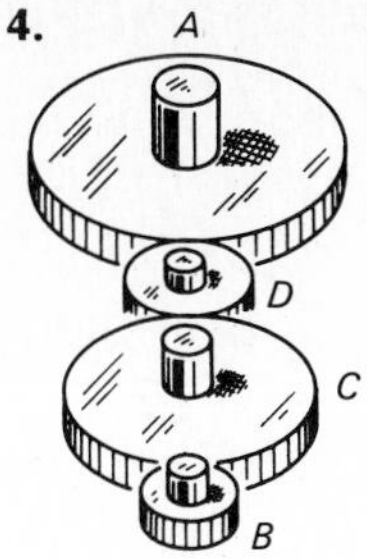

A
D
C
B

**5.**

C
D
A
B

**6.**

A
C
B
D

**7.**

E
A
C
B
D

**8.**

A
D
E
B
F
C

**9.**

F
B
E
D
A
C

**10.**

C
D
A
B
E

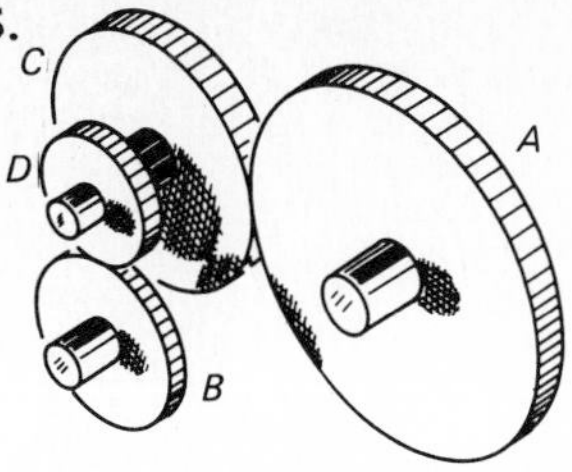

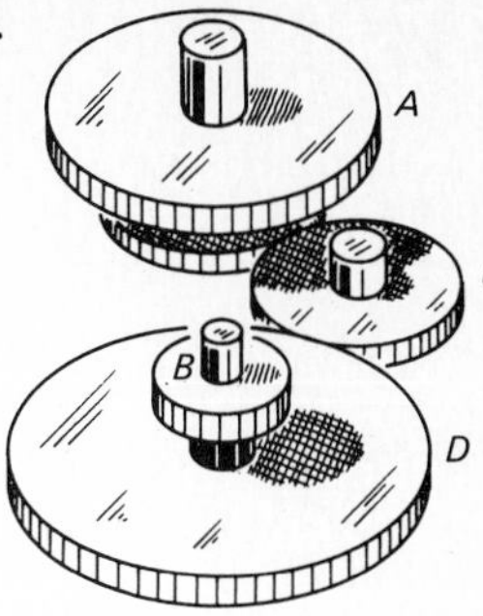

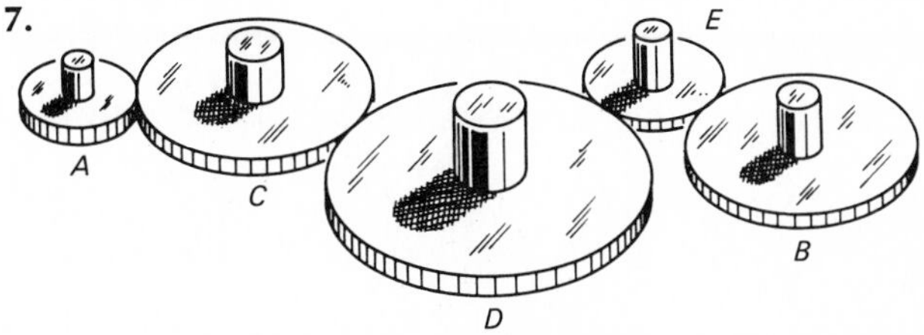

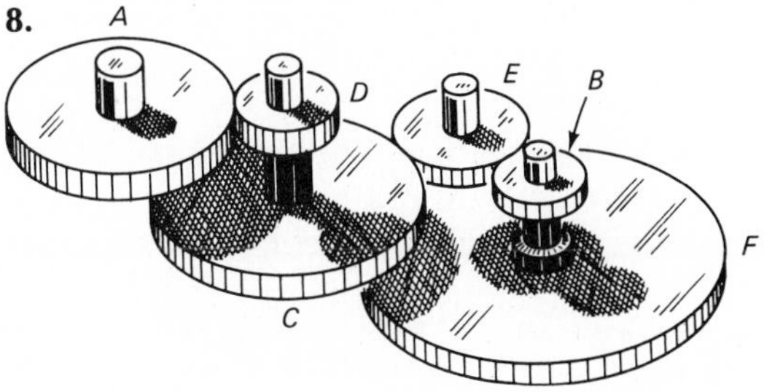

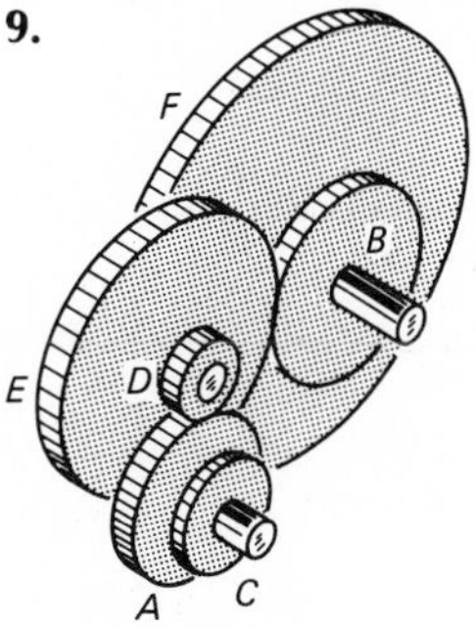

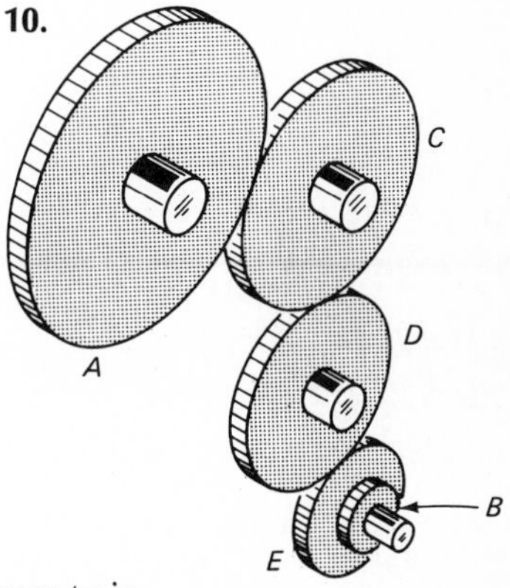

**11–15.** Find the speed in rpm of gear *D* in each gear train.

**11.**

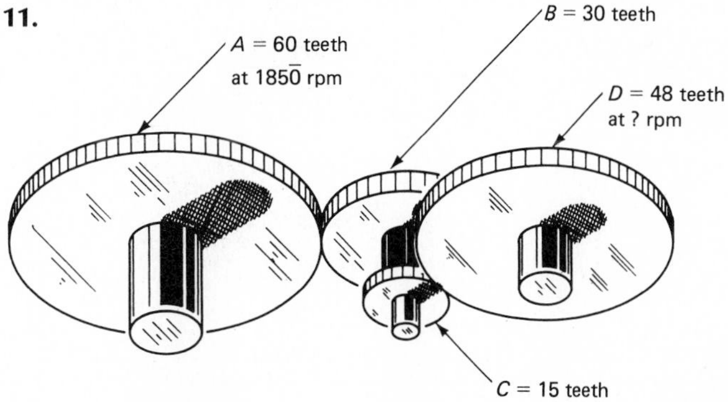

B = 30 teeth
A = 60 teeth
at 1850 rpm
D = 48 teeth
at ? rpm
C = 15 teeth

**12.**

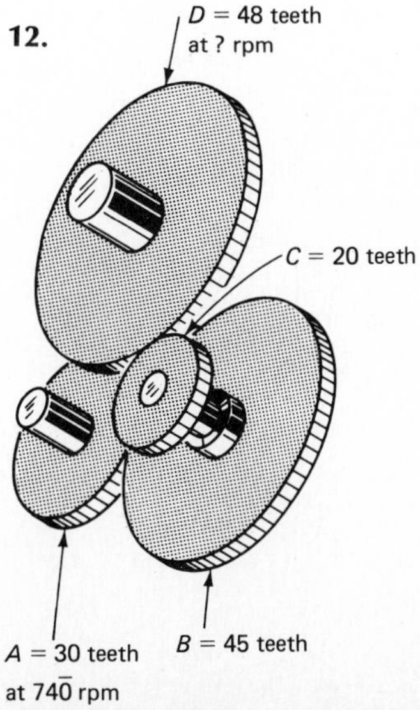

D = 48 teeth
at ? rpm
C = 20 teeth
A = 30 teeth
at 740 rpm
B = 45 teeth

**13.**

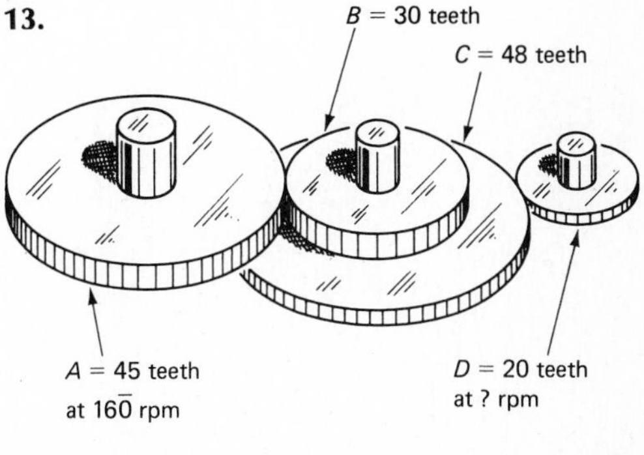

B = 30 teeth
C = 48 teeth
A = 45 teeth
at 160 rpm
D = 20 teeth
at ? rpm

**14.** A = 20 teeth
at 250 rpm

D = 12 teeth
at ? rpm

C = 10 teeth

B = 30 teeth

E = 45 teeth

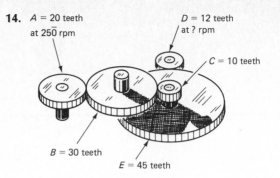
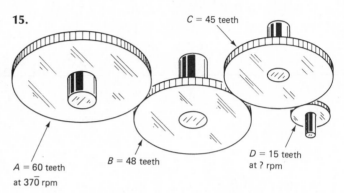

**15.**

C = 45 teeth

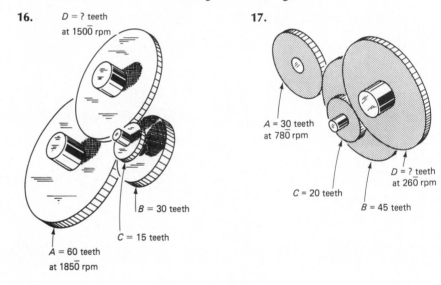

A = 60 teeth
at 370 rpm

B = 48 teeth

D = 15 teeth
at ? rpm

**16–20.** Find the number of teeth for gear *D* in each gear train.

**16.** D = ? teeth
at 1500 rpm

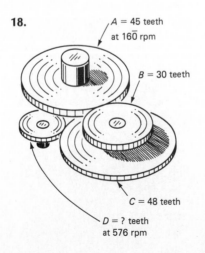

B = 30 teeth

C = 15 teeth

A = 60 teeth
at 1850 rpm

**17.**

A = 30 teeth
at 780 rpm

C = 20 teeth

D = ? teeth
at 260 rpm

B = 45 teeth

**18.** A = 45 teeth
at 160 rpm

B = 30 teeth

C = 48 teeth

D = ? teeth
at 576 rpm

**19.** D = ? teeth
at 1125 rpm

E = 45 teeth

C = 10 teeth

B = 30 teeth

A = 20 teeth
at 250 rpm

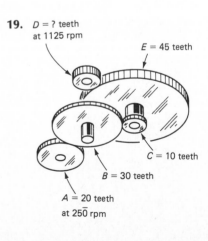

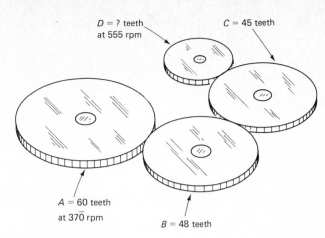

D = ? teeth
at 555 rpm

C = 45 teeth

A = 60 teeth
at 37$\overline{0}$ rpm

B = 48 teeth

**10-4**

**PULLEYS CONNECTED
WITH A BELT**

Pulleys connected with a belt are used to transfer rotational motion from one shaft to another.

Two pulleys connected with a belt have a relationship similar to gears. Assuming no slippage, when two pulleys are connected

$$D \cdot N = d \cdot n$$

where:  $D =$ diameter of the driver pulley
$N =$ number of revolutions per minute of the driver pulley
$d =$ diameter of the driven pulley
$n =$ number of revolutions per minute of the driven pulley

The preceding equation may be generalized in the same manner as for gear trains to get

$$ND_1D_2D_3\cdots = nd_1d_2d_3\cdots$$

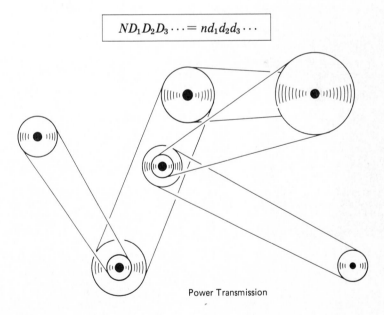

Power Transmission

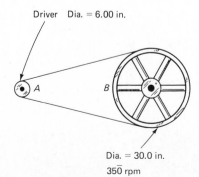

Driver  Dia. = 6.00 in.

A

B

Dia. = 30.0 in.
35$\overline{0}$ rpm

## EXAMPLE

Find the speed in rpm of pulley *A* shown.

DATA:

$$D = 6.00 \text{ in.}$$
$$d = 30.0 \text{ in.}$$
$$n = 35\overline{0} \text{ rpm}$$
$$N = ?$$

BASIC EQUATION:

$$D \cdot N = d \cdot n$$

WORKING EQUATION:

$$N = \frac{dn}{D}$$

SUBSTITUTION:

$$N = \frac{(30.0 \text{ in.})(35\bar{0} \text{ rpm})}{6.00 \text{ in.}}$$

$$= 1750 \text{ rpm}$$

When two pulleys are connected with an open-type belt, the pulleys turn in the same direction. When two pulleys are connected with a cross-type belt, the pulleys turn in opposite directions. This is illustrated in the following figure.

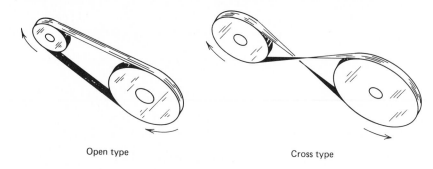

Open type                    Cross type

## PROBLEMS

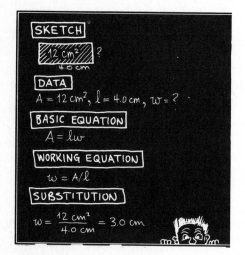

SKETCH

12 cm² ?
4.0 cm

DATA
$A = 12 \text{ cm}^2$, $l = 4.0 \text{ cm}$, $w = ?$

BASIC EQUATION
$A = lw$

WORKING EQUATION
$w = A/l$

SUBSTITUTION
$w = \dfrac{12 \text{ cm}^2}{4.0 \text{ cm}} = 3.0 \text{ cm}$

*Find the missing quantities using* $D \cdot N = d \cdot n$.

| | D | N | d | n |
|---|---|---|---|---|
| **1.** | 18.0 | 150̄0̄ | 14.0 | ___ |
| **2.** | 8.00 | ___ | 9.00 | 972 |
| **3.** | 12.0 | 180̄0̄ | 6.00 | ___ |
| **4.** | ___ | 225̄0̄ | 9.00 | 1125 |
| **5.** | 49.0 | 186̄0̄ | ___ | 62̄0̄ |

6. The diameter of a driving pulley is 6.50 in. and revolves at 165̄0̄ rpm. At what speed will the driven pulley revolve if it is 26.0 in. in diameter?

7. The diameter of a driving pulley is 25.0 cm and makes 12̄0̄ rpm. At what speed will the driven pulley turn if it is 42.0 cm in diameter?

8. The diameter of a driving pulley is 18.0 in. and makes 60̄0̄ rpm. What is the diameter of the driven pulley if it rotates at 36̄0̄ rpm?

9. A driving pulley rotates at 44̄0̄ rpm. The diameter of the driven pulley is 15.0 in. and makes 68̄0̄ rpm. What is the diameter of the driving pulley?

10. The radius of a driving pulley is 4.00 in. and rotates at 12̄0̄ rpm. The radius of the driven pulley is 6.00 in. Find the rpm of the driven pulley.

*Determine the direction of pulley B in each pulley system.*

**11.**                              **12.**

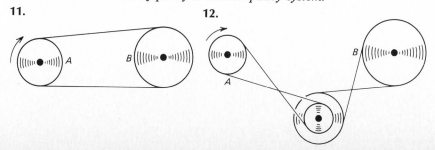

**13.**

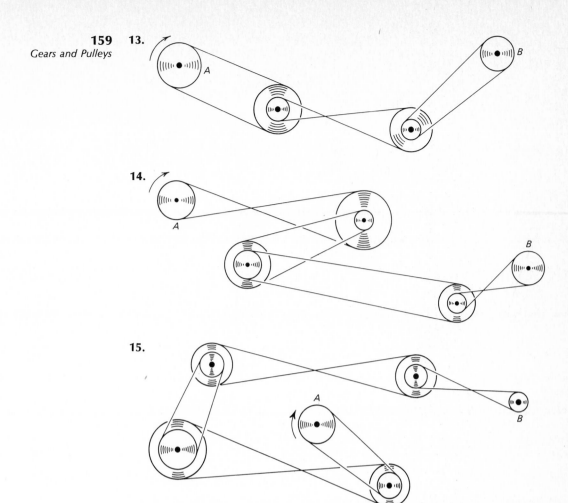

**14.**

**15.**

**16.** What size pulley should be placed on a countershaft turning $15\overline{0}$ rpm to drive a grinder with a 4.00-in. pulley which is to turn at $120\overline{0}$ rpm?

# CHAPTER
# 11

## Nonconcurrent Forces

**11-1**
PARALLEL FORCE
PROBLEMS

A painter stands 2.00 ft from one end of a 6.00-ft plank that is supported at each end by a ladder. How much of the painter's weight must each ladder support?

Problems of this kind are often faced in the construction industry, particularly in the design of bridges and buildings. Using some things we learned about torques and equilibrium, we can now solve problems of this type.

Let's look at the painter problem described above. A force diagram below shows the forces and distances involved. The arrow pointing down represents the weight of the person. The arrows pointing up represent the forces exerted by each of the ladders in holding up the plank and painter. (For now, we will neglect the weight of the plank.)

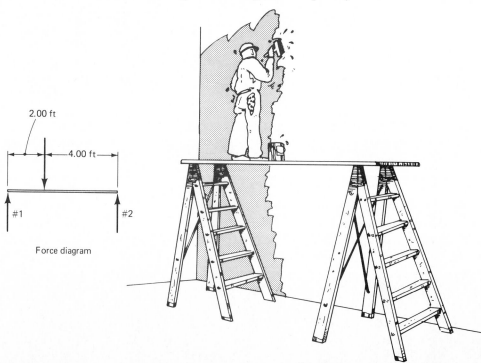

2.00 ft

4.00 ft

#1    #2

Force diagram

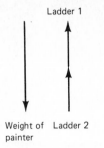

Ladder 1

Weight of    Ladder 2
painter

We have a condition of equilibrium. The plank and painter are not moving. The sum of the forces exerted by the ladders is equal to the weight of the painter. Since these forces are vectors and are parallel, we can show that their sum is zero.

Using engineering notation,

$$\Sigma \mathbf{F} = 0$$

where $\Sigma$ (Greek capital letter sigma) means summation or "the sum of" and $\mathbf{F}$ is force, a vector quantity. So $\Sigma \mathbf{F}$ means "the sum of forces," in this case the sum of parallel forces.

> *FIRST CONDITION OF EQUILIBRIUM:*
>
> *The sum of all parallel forces must be zero.*

If the vector sum was not zero (forces up unequal to forces down), we would have an unbalanced force tending to cause motion.

Now consider this situation: One ladder is strong enough to support the man, and the other is removed. What happens to the painter? The plank, supported only on one end, falls, and the painter has a mess to clean up!

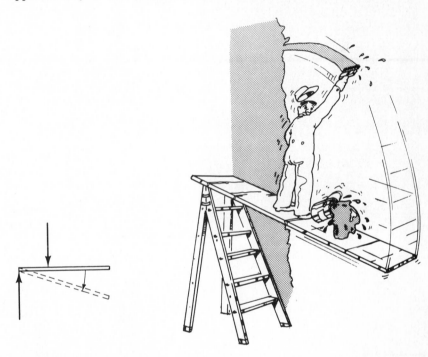

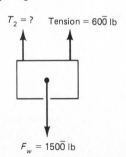

## EXAMPLE 1

A sign of weight $150\overline{0}$ lb is supported by two cables. If one cable has a tension of $60\overline{0}$ lb, what is the tension in the other cable?

**SKETCH:** Draw the free-body diagram.

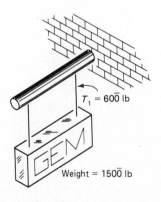

$T_1 = 60\overline{0}$ lb

Weight = $150\overline{0}$ lb

$T_2 = ?$    Tension = $60\overline{0}$ lb

$F_w = 150\overline{0}$ lb

$$F_w = 150\overline{0}\ \text{lb}$$
$$T_1 = 60\overline{0}\ \text{lb}$$
$$T_2 = \ ?$$

**BASIC EQUATION:**

$$F_+ = F_-$$

**WORKING EQUATION:**

$$T_1 + T_2 = F_w$$
$$T_2 = F_w - T_1$$

**SUBSTITUTION:**

$$T_2 = 150\overline{0}\ \text{lb} - 60\overline{0}\ \text{lb}$$
$$= 90\overline{0}\ \text{lb}$$

Note that for purposes of this section, possible rotational motion is not considered.

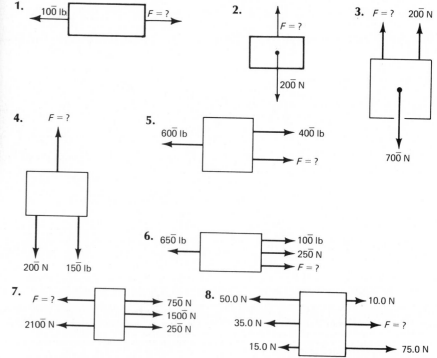

## PROBLEMS

*Find the force F that will produce equilibrium for the free-body diagrams shown. Use the same procedure as in Example 1.*

1. 100 lb ← ☐ → F = ?

2. F = ? ↑ ☐ ↓ 200 N

3. F = ? ↑   200 N ↑ ☐ ↓ 700 N

4. F = ? ↑ ☐ 200 N ↓   150 lb ↓

5. 600 lb ← ☐ → 400 lb, → F = ?

6. 650 lb ← ☐ → 100 lb, → 250 N, → F = ?

7. F = ? ← ☐ → 750 N, → 150\overline{0} N, → 250 N; 210\overline{0} N ←

8. 50.0 N ← ☐ → 10.0 N; 35.0 N ← ☐ → F = ?; 15.0 N ← ☐ → 75.0 N

Not only must the forces balance each other (vector sum = 0), but they must also be positioned so that there is no rotation in the system.

Do you see how torques are important in this type of problem? To avoid rotation, we can have no unbalanced torques.

Sometimes there will be a natural point of rotation, as in our painter problem. We can, however, choose any point as our center of rotation as we consider the torques present. We will soon see that one of any number of points could be selected. What is necessary, though, is that there be no rotation (no unbalanced torques).

Again, using engineering notation,

$$\Sigma \ T_{\text{any point}} = 0$$

where $\Sigma \ T_{\text{any point}}$ is the sum of the torques about any chosen point or,

---

*SECOND CONDITION OF EQUILIBRIUM:*

*The sum of the clockwise torques must equal the sum of the counterclockwise torques.*

$$\Sigma \ T_{\text{clockwise (cw)}} = \Sigma \ T_{\text{counterclockwise (ccw)}}$$

---

To illustrate these principles, we will find how much weight each stepladder must support if our painter weighs $15\overline{0}$ lb.

SKETCH:

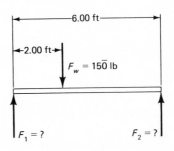

DATA:        $F_w = 15\overline{0}$ lb

plank $= 6.00$ ft

$F_w$ is 2.00 ft from one end

BASIC EQUATIONS:    (1)   $\Sigma \ F = 0$     or

sum of forces $= 0$      (*Note:* $F_w$ is negative here because its direction is opposite that of $F_1$ and $F_2$.)

$F_1 + F_2 - F_w = 0$

or   $F_1 + F_2 = F_w$

$F_1 + F_2 = 15\overline{0}$ lb

(2)   $\Sigma \ T_{\text{clockwise}} = \Sigma \ T_{\text{counterclockwise}}$

We must first select a point of rotation. Picking an end is usually helpful in simplifying the calculations. We will pick the left end in this problem where $F_1$ acts. What are the clockwise torques about this point?

The force due to the weight of the painter tends to cause clockwise motion. The torque arm is 2.00 ft. Therefore, $T = (15\overline{0} \text{ lb})(2.00 \text{ ft})$. This is the only clockwise torque in this problem.

The only counterclockwise torque is $F_2$ times its torque arm, 6.00 ft. $T = (F_2)(6.00 \text{ ft})$.

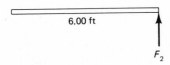

There is no torque involving $F_1$ because its torque arm is zero. Setting $\Sigma \ T_{\text{clockwise}} = \Sigma \ T_{\text{counterclockwise}}$ we have an equation:

$$(15\overline{0} \text{ lb})(2.00 \text{ ft}) = (F_2)(6.00 \text{ ft})$$

Note that by selecting an end as the point of rotation, we were able to set up our equation with just one variable ($F_2$). Solving for $F_2$ gives the working equation.

$$F_2 = \frac{(15\overline{0}\ \text{lb})(2.00\ \text{ft})}{6.00\ \text{ft}}$$

$$F_2 = 50.0\ \text{lb}$$

Since $\Sigma F = F_1 + F_2 = F_w$, we can now substitute for $F_w$ and $F_2$ to find $F_1$.

$$F_1 + 50.0\ \text{lb} = 15\overline{0}\ \text{lb}$$

$$F_1 = 15\overline{0}\ \text{lb} - 50.0\ \text{lb}$$

$$= 10\overline{0}\ \text{lb}$$

---

*OUTLINE OF PROCEDURE TO SOLVE*
*PARALLEL FORCE PROBLEMS*

1. Sketch the problem.
2. Write an equation setting the sums of the opposite forces equal to each other.
3. Pick a point of rotation. Eliminate a variable, if possible (by making its torque arm zero).
4. Write the sum of all clockwise torques.
5. Write the sum of all counterclockwise torques.
6. Set $\Sigma T_{\text{clockwise}} = \Sigma T_{\text{counterclockwise}}$
7. Solve the equation $\Sigma T_{\text{clockwise}} = \Sigma T_{\text{counterclockwise}}$ for the unknown quantity.
8. Substitute the value found in step 7 back into the equation in step 2 to find the other unknown quantity.

---

## EXAMPLE 2

A bricklayer weighing 175 lb stands on an 8.00-ft scaffold 3.00 ft from one end. He has a pile of bricks, which weighs 40.0 lb, 3.00 ft from the other end. How much weight must each end support?

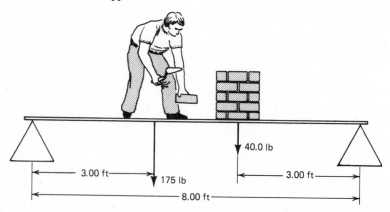

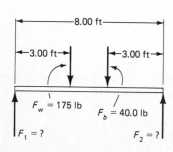

1. SKETCH: (at left)
2. $\Sigma F = F_1 + F_2 = 175\ \text{lb} + 40.0\ \text{lb}$
3. Pick a point of rotation. We should pick either end here to eliminate one of the variables $F_1$ or $F_2$. Let us pick the right end.
4. $\Sigma T_{\text{clockwise}} = (F_1)(8.00\ \text{ft})$
5. $\Sigma T_{\text{counterclockwise}} = (40.0\ \text{lb})(3.00\ \text{ft}) + (175\ \text{lb})(5.00\ \text{ft})$
   Note that there are two counterclockwise torques.

6. $\Sigma\ T_{\text{clockwise}} = \Sigma\ T_{\text{counterclockwise}}$

$F_1(8.00\ \text{ft}) = (40.0\ \text{lb})(3.00\ \text{ft}) + (175\ \text{lb})(5.00\ \text{ft})$

7. $F_1 = \dfrac{(40.0\ \text{lb})(3.00\ \text{ft}) + (175\ \text{lb})(5.00\ \text{ft})}{8.00\ \text{ft}}$

$= \dfrac{12\bar{0}\ \text{lb ft} + 875\ \text{lb ft}}{8.00\ \text{ft}} = \dfrac{995\ \text{lb ft}}{8.00\ \text{ft}} = 124\ \text{lb}$

8. $\hspace{5cm} F_1 + F_2 = 175\ \text{lb} + 40.0\ \text{lb}$

Substituting, we obtain

$$124\ \text{lb} + F_2 = 215\ \text{lb}$$

$$F_2 = 215\ \text{lb} - 124\ \text{lb}$$

$$= 91\ \text{lb}$$

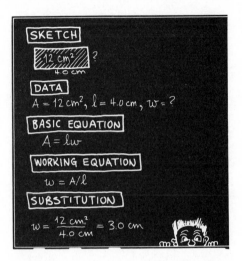

## PROBLEMS

1. A 165-lb painter stands 3.00 ft from one end of an 8.00-ft scaffold. If the scaffold is supported at each end by a stepladder, how much weight must each ladder support?

2. A 50$\bar{0}$0-lb truck is 20.0 ft from one end of a 50.0-ft bridge. A 40$\bar{0}$0-lb car is 40.0 ft from the same end. How much weight must the other end of the bridge support? (Neglect the weight of the bridge.)

3. A 24$\bar{0}$0-kg truck is 6.00 m from one end of a 27.0-m-long bridge. A 15$\bar{0}$0-kg car is 10.0 m from the same end. How much weight must each end of the bridge support?

4. An auto transmission of mass 165 kg is located 1.00 m from one end of a 2.50-m bench. What weight must each end of the bench support?

**11-2**
**CENTER**
**OF GRAVITY**

In Section 11-1 we neglected the weight of the plank in the painter example. In practice, the weight of the plank or bridge is extremely important. An engineer must know the weight of the bridge he or she is designing so as to use methods and materials of sufficient strength to hold up the traffic and not collapse.

An important idea in this kind of problem is center of gravity. *The center of gravity of any object is that point at which all of its weight can be considered to be concentrated.*

An object such as a brick or a uniform rod has its center of gravity at its middle or center. The center of gravity of something like an automobile, however, is not at its center or middle because its weight is not evenly distributed throughout. Its center of gravity is located nearer the heavy engine.

You have probably had the experience of carrying a long board by yourself. If the board was not too heavy, you could carry it yourself by suspending it in about the middle. You didn't have to hold up both ends. You applied the principle of center of gravity and suspended the board at that point.

We shall represent the weight of an object by a vector through its center of gravity. We use a vector to show the weight (force due to gravity) of the object. It is placed through the center of gravity to show that all the weight of the object may be considered concentrated at that point. If the center of gravity is not at the middle of the object, its location will be given.

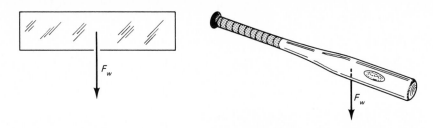

In solving problems, the weight of the plank or bridge is represented like the other forces by a vector at the center of gravity of the object.

## EXAMPLE

A carpenter stands 2.00 ft from one end of a 6.00-ft scaffold which is uniform and weighs 20.0 lb. If the carpenter weighs 165 lb, how much weight must each end support?

1. SKETCH:

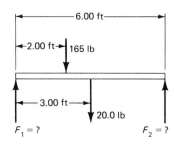

Since the plank is uniform, its center of gravity is at the middle.

2. $\Sigma F = F_1 + F_2 = 165$ lb $+ 20.0$ lb
3. Pick the left end as the point of rotation.
4. $\Sigma T_{\text{clockwise}} = (165 \text{ lb})(2.00 \text{ ft}) + (20.0 \text{ lb})(3.00 \text{ ft})$
5. $\Sigma T_{\text{counterclockwise}} = (F_2)(6.00 \text{ ft})$
6. $(165 \text{ lb})(2.00 \text{ ft}) + (20.0 \text{ lb})(3.00 \text{ ft}) = (F_2)(6.00 \text{ ft})$
7. $F_2 = \dfrac{33\bar{0} \text{ lb ft} + 60.0 \text{ lb ft}}{6.00 \text{ ft}} = \dfrac{39\bar{0} \text{ lb ft}}{6.00 \text{ ft}} = 65.0$ lb
8. $F_1 + 65.0$ lb $= 165$ lb $+ 20.0$ lb

$$F_1 = 165 \text{ lb} + 20.0 \text{ lb} - 65.0 \text{ lb}$$
$$= 12\bar{0} \text{ lb}$$

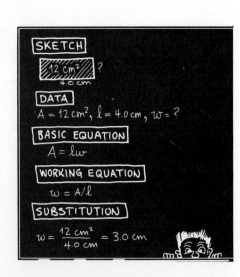

## PROBLEMS

*Solve the following problems using the method outlined in this chapter.*

**1.** Solve for $F_1$:

$$30.0F_1 = (14.0)(18.0) + (25.0)(17.0)$$

**2.** Solve for $F_2$:

$$39.0F_2 + (60.0)(55.0) = (20\bar{0})(40.0) + (52.0)(27.0)$$

**3.** Solve for $F_w$:

$$(12.0)(15.0) + 45.0F_\omega = (21.0)(65.0) + (22.0)(32.0)$$

4. Two workers carry a uniform 15.0-ft plank which weighs 22.0 lb. There is a load of blocks which weighs 165 lb and is located 7.00 ft from the first worker. What force must each worker exert to hold up the plank and load?

5. A wooden beam is 3.30 m long and has its center of gravity 1.30 m from one end. If the beam weighs $2.50 \times 10^4$ N, what force is needed to support each end?

6. An auto engine weighs $65\overline{0}$ lb and is located 4.00 ft from one end of a 10.0-ft workbench. If the bench is uniform and weighs 75.0 lb, what weight must each end of the bench support?

7. An old covered bridge across a country stream weighs $20,\overline{0}00$ lb. A large truck stalls 12.0 ft from one end of the 32.0-ft bridge. What weight must each of the piers support if the truck weighs $22,\overline{0}00$ lb?

8. A window washer's scaffold 12.0 ft long and weighing 75.0 lb is suspended from each end. One washer weighs 155 lb and is 3.00 ft from one end. The other washer is 4.00 ft from the other end. If the force supported by the end near the first washer is $20\overline{0}$ lb, how much does the second washer weigh?

9. A porch swing weighs 29.0 lb. It is 4.40 ft long and has a dog weighing 14.0 lb sleeping on it 1.90 ft from one end. What weight must the support ropes on each end hold up?

10. A wooden plank is 5.00 m long and supports a 75.0-kg block 2.00 m from one end. If the plank is uniform and has a mass of 30.0 kg, how much force is needed to support each end?

11. A bridge has a mass of $1.60 \times 10^4$ kg, is 21.0 m long, and has a $35\overline{0}0$-kg truck 7.00 m from one end. What force must each end of the bridge support?

12. A uniform steel beam is 5.00 m long and weighs $3.60 \times 10^5$ N. What force is needed to lift one end?

13. A wooden pole is 4.00 m long, weighs 315 N, and has its center of gravity 1.50 from one end. What force is needed to lift each end?

14. A bridge has a mass of $2.60 \times 10^4$ kg, is 32.0 m long, and has a $35\overline{0}0$-kg truck 15.0 m from one end. What force must each end of the bridge support?

15. An auto engine of mass 295 kg is located 1.00 m from one end of a 4.00-m workbench. If the uniform bench has a mass of 45.0 kg, what weight must each end of the bench support?

# CHAPTER
# *12*

## *Matter*

**12-1**
INTRODUCTION
TO PROPERTIES
OF MATTER

What are the building blocks of matter? First, matter is anything that occupies space and has mass. Suppose that we take a cube of sugar and divide it into two pieces. Then divide a resulting piece into two pieces. Can we continue this process indefinitely and get smaller and smaller particles of sugar each time? No: at some point the subdivision will result in something different from sugar.

A *molecule* is the smallest particle of a substance that exists in a stable and independent state. Most simple molecules are about $3 \times 10^{-10}$ m in diameter, which is indeed small.

What do we get if we divide the sugar molecule into pieces? We would find that these resulting particles are three simpler kinds of matter—carbon, hydrogen, and oxygen—which are called chemical elements or atoms. An *atom* is the smallest particle of an element that can exist either alone or in combination with other atoms of the same or different elements. Thus, a sugar molecule is made up of carbon, hydrogen, and oxygen atoms.

Models of water and sugar molecules are shown below.

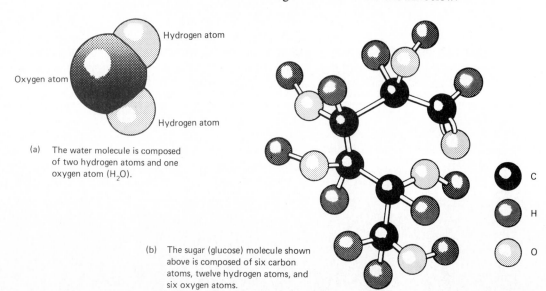

Oxygen atom

Hydrogen atom

Hydrogen atom

(a)  The water molecule is composed of two hydrogen atoms and one oxygen atom ($H_2O$).

C

H

O

(b)  The sugar (glucose) molecule shown above is composed of six carbon atoms, twelve hydrogen atoms, and six oxygen atoms.

Not all atoms are the same size. The hydrogen atom is the smallest. Its diameter is $6 \times 10^{-11}$ m and its mass is $1.67 \times 10^{-27}$ kg. Uranium is one of the heaviest atoms, $3.96 \times 10^{-25}$ kg.

What happens if an atom is subdivided? Several particles of the atom have been discovered. Of these, the three most important particles of the atom are the proton, the electron, and the neutron. We will limit our discussion here to these three. The following table provides some of the basic information known about these three particles.

| Particle | Mass | Diameter | Charge |
|---|---|---|---|
| Proton | $1.672 \times 10^{-27}$ kg | $2-3 \times 10^{-15}$ m | 1 |
| Electron | $9.108 \times 10^{-31}$ kg | $10^{-14}$ (approx.) | $-1$ |
| Neutron | $1.675 \times 10^{-27}$ kg | $2-3 \times 10^{-15}$ m | 0 |

Note that the size of the electron is not accurately known.

Models of the hydrogen atom and the carbon atom are shown next.

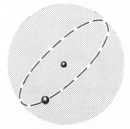

The hydrogen atom is composed of a nucleus which contains one proton. Its one electron moves about or orbits the nucleus.

(a)

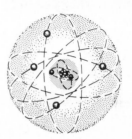

The carbon atom is composed of a nucleus which contains six protons and six neutrons. Its six electrons move about or orbit the nucleus as shown above.

(b)

The nucleus or center part of an atom is made up of protons and neutrons while the electrons orbit (circle) the nucleus. The atoms are held together by strong nuclear forces. The molecules are held together by electrical forces.

In summary, the most basic building blocks of matter are protons, electrons, and neutrons. These particles, formed in various combinations, give us the 105 known atoms or chemical elements. The atoms, formed in various combinations, give us the very long list of known molecules.

Matter exists in three states: solids, liquids, and gases. A solid is a substance that has a definite shape and a definite volume. A liquid is a substance that takes the shape of its container and has a definite volume. A gas is a substance that takes the shape of its container and has the same volume as its container.

The molecules of a solid are fixed in relation to each other. They vibrate in a back-and-forth motion. They are so close that a solid can be compressed only slightly. Solids are usually crystalline substances, meaning that their molecules are arranged in a definite pattern. This is why a solid tends to hold its shape and has a definite volume.

Solid molecules vibrate in fixed positions

Liquid molecules flow over each other

Gas molecules move rapidly in all directions and collide

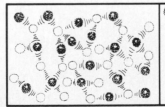

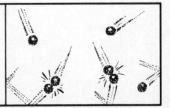

The molecules of a liquid are not fixed in relation to each other. They normally move in a flowing type of motion but yet are so close together that they are practically incompressible, thus having a definite volume. Because the molecules move in a smooth flowing motion and not in any fixed manner, a liquid takes the shape of its container.

The molecules of a gas are not fixed in relation to each other and move rapidly in all directions, colliding with each other. They are much farther apart than molecules in a liquid, and they are extremely far apart when compared to the distance between molecules in solids. The movement of the molecules is limited only by the container. Therefore, a gas takes the shape of its container. Because the molecules are far apart, a gas can easily be compressed and it has the same volume as its container.

**12-2
PROPERTIES
OF SOLIDS**

Solids have a definite shape and a definite volume. Solids have molecules that are usually arranged in a definite pattern. The following properties are common to most solids.

*Cohesion and adhesion*    The molecules of a solid are held together by large internal molecular forces. *Cohesion* is the force of attraction between like molecules. The cohesive forces hold the closely packed molecules of a solid together, which keep its shape and volume from being easily changed.

Cohesion, this force of attraction between like molecules, can also be shown by grinding and polishing the surfaces of two like solids and then sliding their surfaces together. For example, take two pieces of polished plate glass and slide them together. Try to pull them apart. The force of attraction of the like molecules of the two pieces of glass makes it difficult to pull them apart.

*Adhesion* is the force of attraction between different or unlike molecules. Common examples include glue and wood, adhesive tape and skin, and tar and road.

*Tensile strength*    The *tensile strength* of a solid is a measure of its resistance to being pulled apart. That is, the tensile strength of a solid is a measure of its cohesive forces between adjacent molecules. The tensile strength of a rod or wire is found by putting it in a machine that pulls the rod or wire until it breaks. The tensile strength is the ratio

$$\frac{\text{force required to break the rod or wire}}{\text{cross-sectional area of the rod or wire}}$$

*Hardness* The *hardness* of a solid is a measure of the internal resistance of its molecules being forced farther apart or closer together. More commonly, we talk of the hardness of a solid in terms of its difficulty in being scratched. As a matter of classifying the relative hardness of a material, a "scratch test" is used. The given material is scratched in a certain way. Its scratch is then compared with a series of standard scratches of materials that form an arbitrary hardness scale from very soft solids to the hardest known substance, diamond.

The *Brinell method* is a common industrial method used to measure the hardness of a metal. A machine is used to press a 10-mm hardened chrome-steel ball with an equivalent mass of 3000 kg into the metal being tested. The diameter of the resulting impression is used as a measure of the metal's hardness. The Brinell value or number is the ratio

$$\frac{\text{load (in kg)}}{\text{surface area of the impression (in mm}^2)}.$$

This value can also be compared with a scale of the accepted hardnesses of given metals. The larger the Brinell number, the harder the metal.

Steel can be hardened by heating it to a very high temperature, then suddenly cooling it by putting it in water. However, it then becomes brittle. This cooled steel can then be tempered (toughened) by reheating it and allowing it to cool slowly. As the steel cools, it loses hardness and gains toughness. If the steel cools down slowly and completely, we say that it is annealed. Annealed steel is soft and tough but not brittle.

*Ductility* A metal rod that can be drawn through a die to produce a wire is said to have a property called *ductility*. As the rod is pulled through the die, its diameter is decreased and its length is increased as it becomes a wire.

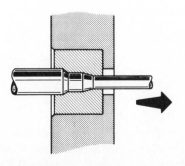

Ductility: A metal being drawn into a wire

*Malleability*   A metal that can be hammered and rolled into sheets is said to have a property called *malleability*. As the metal is hammered or rolled, its shape or thickness is changed. During this process, the atoms slide over each other and change positions. The cohesive forces are relatively strong; thus, the atoms do not become widely separated during their rearrangement and the resulting shape remains relatively stable.

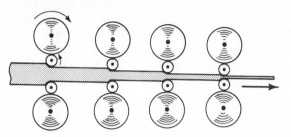

Malleability: a metal being rolled into a sheet

*Elasticity*   An object becomes deformed when outside forces change its shape or size. The object's ability to return to its original size and shape—when the outside forces are removed—is called its *elasticity*. When the solid is being deformed, sometimes these molecules attract each other and sometimes they repel each other. For instance, take a rubber ball and try to pull it apart. You notice that the ball stretches out of shape.

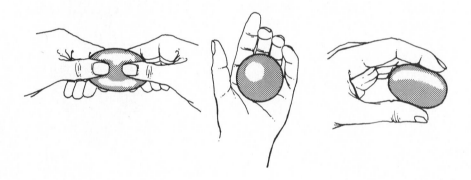

However, when you release the pulling force, the ball returns to its original shape because the molecules, being farther apart than normal, attract each other. If you squeeze the ball together, the ball will again become out of shape. Now release the pressure and the ball will again return to its original shape because the molecules, being too close together, repel each other. Therefore, we can see that when molecules are slightly pulled out of position, they attract each other. When they are pressed too close together, they repel each other.

Most solids have the property of elasticity; however, some are only slightly elastic. For example, wood and Styrofoam are two solids whose elasticity is small.

Not every elastic object returns to its original shape. If too large a deforming force is applied, it will become permanently deformed. Take a door spring and pull it apart far as you can. When you let go, it will probably not return to its exact original shape. When a solid is deformed as such, it is said to have been deformed past its elastic limit. Its molecules have been pulled far enough apart that they have slid past one another beyond which the original molecular forces could return the spring to its original shape. If the deforming force is enough greater, the body breaks apart.

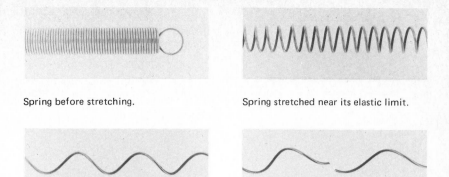

Spring before stretching.

Spring stretched near its elastic limit.

Spring stretched beyond its elastic limit.

Spring stretched much beyond its elastic limit . . . break occurs!

Elasticity properties can be measured in a variety of ways. Whenever an object is deformed by some outside force, the internal molecular forces tend to resist the change in shape and/or volume. Stress is a measure of the tendency of an object to return to its original shape and size after a deforming force has been applied. There are five basic stresses: tension, compression, shear, bending, and twisting. (Shear, bending, and twisting are not addressed here.)

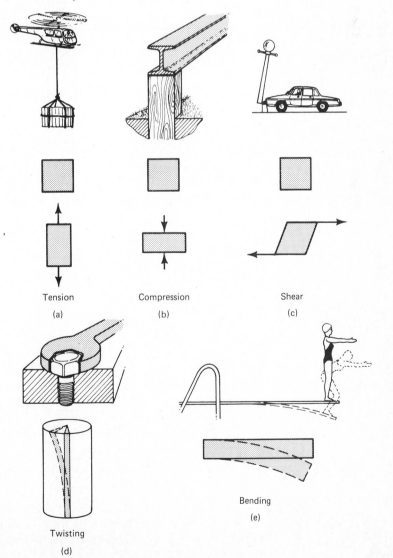

Tension
(a)

Compression
(b)

Shear
(c)

Twisting
(d)

Bending
(e)

*Stress* may be defined as the ratio of the outside applied force, which tends to cause a distortion, to the area over which the force acts. In other words,

$$\text{stress} = \frac{\text{applied force}}{\text{area over which the force acts}}$$

or

$$\boxed{S = \frac{F}{A}}$$

The metric stress unit is usually the pascal (Pa).

$$1 \text{ Pa} = 1 \text{ N/m}^2$$

## EXAMPLE 1

A steel column in a building has a cross-sectional area of 0.250 m² and supports a weight of $1.50 \times 10^5$ N. What is the stress in pascals on the column?

SKETCH:   None needed

DATA:

$$A = 0.250 \text{ m}^2$$
$$F = 1.50 \times 10^5 \text{ N}$$
$$S = ?$$

BASIC EQUATION:

$$S = \frac{F}{A}$$

WORKING EQUATION:   Same

SUBSTITUTION:

$$S = \frac{1.50 \times 10^5 \text{ N}}{0.250 \text{ m}^2}$$
$$= 6.00 \times 10^5 \frac{\text{N}}{\text{m}^2}$$
$$= 6.00 \times 10^5 \text{ Pa}$$
$$= 60\overline{0} \text{ kPa}$$

Whenever a stress is applied to an object, the object will be changed minutely, at least. If you stand on a large steel beam, it will bend—at least slightly. The resulting deformation is called *strain*. That is, strain is the relative amount of deformation of a body that is under stress. Or strain is *change in length per unit of length,* change in volume per unit of volume, and so on. Strain is a direct and necessary consequence of stress.

One of the most basic principles related to the elasticity of solids is Hooke's law. This law is named after Robert Hooke, who first discovered the principle in the seventeenth century while inventing the balance spring for spring-driven clocks.

> *HOOKE'S LAW*
>
> The ratio of the change in length of an object that is stretched or compressed to the force causing this change is constant as long as the elastic limit has not been exceeded.

Or, stated another way: Stress is directly proportional to strain as long as the elastic limit has not been exceeded.

In equation form:

$$\frac{F}{\Delta l} = k$$

where:  $F$ = applied force
$\Delta l$ = change in length
$k$ = elastic constant

(*Note:*  $\Delta$ is often used in mathematics and science to mean "change in.")

## EXAMPLE 2

A force of 5.00 N is applied to a spring whose elastic constant is 0.250 N/cm. Find its change in length.

SKETCH:

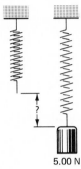

5.00 N

DATA:

$F = 5.00$ N
$k = 0.250$ N/cm
$\Delta l = ?$

BASIC EQUATION:

$$\frac{F}{\Delta l} = k$$

WORKING EQUATION:

$$\Delta l = \frac{F}{k}$$

SUBSTITUTION:

$$\Delta l = \frac{5.00 \text{ N}}{0.250 \text{ N/cm}}$$

$$= 20.0 \text{ cm} \quad \boxed{\frac{N}{N/cm} = N \div \frac{N}{cm} = N \cdot \frac{cm}{N} = cm}$$

## EXAMPLE 3

A force of 3.00 lb stretches a spring 12.0 in. What force is required to stretch the spring 15.0 in.?

SKETCH:

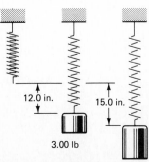

12.0 in.      15.0 in.

3.00 lb

? lb

DATA:

$$F_1 = 3.00 \text{ lb}$$
$$l_1 = 12.0 \text{ in.}$$
$$l_2 = 15.0 \text{ in.}$$
$$F_2 = ?$$

BASIC EQUATION:

$$\frac{F}{\Delta l} = k$$

WORKING EQUATIONS:

$$\frac{F}{\Delta l} = k \quad \text{and} \quad F = k(\Delta l)$$

SUBSTITUTION:

There are two substitutions in this problem, one to find $k$ and one to find the second force $F_2$.

$$\frac{3.00 \text{ lb}}{12.0 \text{ in.}} = k$$
$$0.250 \text{ lb/in.} = k$$

$$F_2 = (0.250 \text{ lb/in.})(15.0 \text{ in.})$$
$$= 3.75 \text{ lb}$$

## EXAMPLE 4

What is the strain on a column with a weight of $5.80 \times 10^6$ N on it if the column is compressed $3.46 \times 10^{-4}$ m under a weight of $6.42 \times 10^5$ N?

SKETCH:  None needed

First find $k$:

DATA:

$$F_2 = 6.42 \times 10^5 \text{ N}$$
$$\Delta l_2 = 3.46 \times 10^{-4} \text{ m}$$
$$k = ?$$

BASIC EQUATION:

$$\frac{F_2}{\Delta l_2} = k$$

WORKING EQUATION:  Same

SUBSTITUTION:

$$k = \frac{6.42 \times 10^5 \text{ N}}{3.46 \times 10^{-4} \text{ m}}$$
$$= 1.86 \times 10^9 \text{ N/m}$$

Then

DATA:

$$k = 1.86 \times 10^9 \text{ N/m}$$
$$F_1 = 5.80 \times 10^6 \text{ N}$$
$$\Delta l_1 = ?$$

BASIC EQUATION:

$$\frac{F_1}{\Delta l_1} = k$$

WORKING EQUATION:

$$\Delta l_1 = \frac{F_1}{k}$$

SUBSTITUTION:

$$\Delta l_1 = \frac{5.80 \times 10^6 \text{ N}}{1.86 \times 10^9 \text{ N/m}}$$

$$= 3.12 \times 10^{-3} \text{ m} \quad \text{or} \quad 3.12 \text{ mm}$$

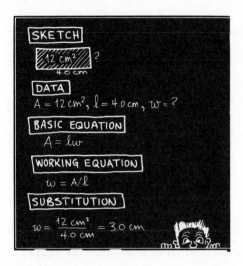

## PROBLEMS

1. A spring is stretched 24.0 in. by a force of 50.0 lb. How far will it stretch if a force of 104 lb is applied?

2. What weight will stretch the spring 9.00 in. in Problem 1?

3. If a 17.0-N force stretches a wire 0.650 cm, what force will stretch a similar piece of wire 1.87 cm?

4. If a force of 21.3 N is applied on a similar piece of wire as in Problem 3, how far will it stretch?

5. A force of 36.0 N stretches a spring 18.0 cm. Find the spring constant (in N/m).

6. A force of 5.00 N is applied to a spring whose spring constant is 0.250 N/cm. Find its change in length (in cm).

7. The vertical steel columns of an office building each support a weight of $1.30 \times 10^5$ N at the second floor, with each being compressed $5.90 \times 10^{-3}$ cm.
   (a) What will be the compression of each column if a weight of $5.50 \times 10^5$ N is supported?
   (b) If the compression for each steel column is 0.0710 cm, what weight is supported by each column?

8. The vertical steel columns of an office building each support a weight of $30,\overline{0}00$ lb at the second floor, with each being compressed 0.00234 in. What will be the compression of each column if a weight of 125,000 lb is supported?

9. If the compression for the steel columns of Problem 8 is 0.0279 in., what weight is supported by each column?

10. A coiled spring is stretched 30.0 cm by a 5.00-N weight. How far will it be stretched by a 15.0-N weight?

11. In Problem 10, what weight will stretch the spring 40.0 cm?

12. A $12,\overline{0}00$-N load is hanging from a steel cable that is 10.0 m long and 16.0 mm in diameter. Find the stress.

13. A rectangular cast-iron column 25.0 cm × 25.0 cm × 5.00 m supports a weight of $6.80 \times 10^6$ N. Find the stress in pascals on the top portion of the column.

**12-3
PROPERTIES
OF LIQUIDS**

As we noted previously, a liquid is a substance that has a definite volume and takes the shape of its container. The molecules move in a flowing motion, yet are so close together that it is very difficult to compress a liquid. In addition, most liquids share the following common properties.

*Cohesion and adhesion*   Cohesion, the force of attraction between like molecules, causes a liquid like molasses to be sticky. Adhesion, the force of attraction between unlike molecules, causes the molasses to also stick to your finger. In the case of water, its adhesive forces are greater than its cohesive forces. Put a plate of glass in a pan of water and pull it out. Some water remains on the glass.

In the case of mercury, the opposite is true. Mercury's cohesive forces are greater than its adhesive forces. If glass is submerged in mercury and then pulled out, virtually no mercury remains on the glass.

We say that a liquid whose adhesive forces are greater than its cohesive forces tends to wet any surface that comes in contact with it. A liquid whose cohesive forces are greater than its adhesive forces tends to leave dry or not wet any surface that comes in contact with it.

*Surface tension*    The ability of the surface of water to support a needle is an example of surface tension. The water's surface acts like a thin, flexible surface film. The surface tension of water can be reduced by adding soap to the water. Soaps are added to laundry water to decrease the surface tension of the water so that the water more easily penetrates the fibers of the clothes being washed.

Water                    Soap added

Surface tension causes a raindrop to hold together. Surface tension causes a small drop of mercury to keep an almost spherical shape. A liquid drop suspended in space would be spherical. A falling raindrop's shape is due to the friction with the air.

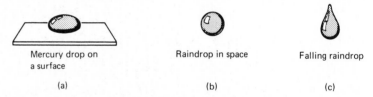

Mercury drop on          Raindrop in space          Falling raindrop
a surface

(a)                              (b)                              (c)

*Viscosity*    In liquids there is friction, which is called *viscosity*. The greater the molecular attraction, the more the friction and the greater the viscosity. For example, it takes more force to move a block of wood through oil than through water. This is because oil is more viscous than water.

If a liquid's temperature is increased, its viscosity decreases. For example, the viscosity of oil in a car engine before it is started on a winter morning at $-10.0°C$ is greater than after the engine has been running for an hour.

One common misunderstanding is that higher viscosity means higher density. One counterexample is oil and water. Oil is more viscous, but water is denser as oil floats on water.

*Capillary action*    Liquids keep the same level in tubes whose ends are submerged in a liquid if the tubes have a large enough diameter.

Cold oil          Hot oil

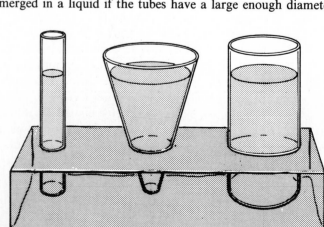

In tubes of different small diameters, water does not stand at the same level. The smaller the diameter, the higher the water rises. If mercury is used, it does not stand at the same level either. But instead of rising up the tube, the mercury level falls or its level is depressed. The smaller the diameter, the lower its level is depressed. This behavior of liquids in small tubes is called *capillary action.*

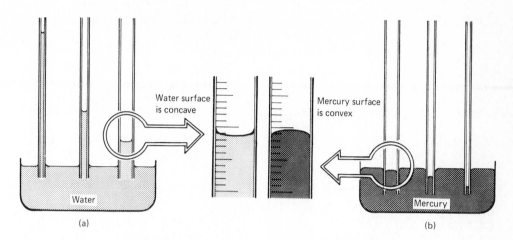

Capillary action is due to both adhesion of the liquid molecules with the tube and the surface tension of the liquid. In the case of water, the adhesive forces are greater than the cohesive molecular forces. Thus, water creeps up the sides of the tube and produces a concave water surface. The surface tension of the water tends to flatten the concave surface. Together, these two forces raise the water up the tubes until it is counterbalanced by the weight of the water column itself.

In the case of mercury, the cohesive molecular forces are greater than the adhesive forces that produce a convex mercury surface. The surface tension of the mercury tends to further hold down the mercury level.

This crescent-shaped surface of a liquid column, whether concave or convex, is called a *meniscus.* In measuring the height of liquid in a tube the measurement must be made to the meniscus: This means measure to the lowest point of a concave meniscus or to the highest point of a convex meniscus.

Experimentally, scientists have found that:

1. Liquids rise in capillary tubes they tend to wet and are depressed in tubes they tend to not wet.
2. Elevation or depression in the tube is inversely proportional to the diameter of the tube.
3. The elevation or depression decreases as the temperature increases.

Capillary action is responsible for the rise of oil (or kerosene) in the wick of an oil lamp. Towels also absorb water because of capillary action.

**12-4
PROPERTIES
OF GASES**

Because of the rapid random movement of its molecules, a gas spreads to completely occupy the volume of its container. This property is called *expansion.*

*Diffusion* of a gas is the process by which molecules of the gas mix with the molecules of a solid, liquid, or another gas. If you remove the cap from a bottle of rubbing alcohol, you soon smell the fumes of the alcohol. The air molecules and the alcohol molecules mix throughout the room because of diffusion.

A balloon inflates due to the pressure of the air molecules on the inside surface of the balloon. This pressure is caused by the bombardment on the walls by the moving molecules. The pressure may be increased by increasing the number of molecules by blowing more air in the balloon. Pressure may also be increased by heating the air molecules already in the balloon. Heat increases the velocity of the molecules.

The behavior of liquids and gases is very similar in many cases. The term *fluid* is used when discussing principles and behaviors common to both liquids and gases.

**12-5
DENSITY**

Density is a property of all three states of matter. *Mass density, $D_m$,* is defined as mass per unit volume. *Weight density, $D_w$,* is defined as weight per unit volume, or,

$$D_m = \frac{M}{V} \qquad D_w = \frac{W}{V}$$

where: $D_m$ = mass density    $D_w$ = weight density
       $M$ = mass            $W$ = weight
       $V$ = volume         $V$ = volume

Although either mass density or weight density can be expressed in both the metric system and the English system, mass density is usually given in the metric units kg/m³ and weight density is usually given in the English units lb/ft³.

| Substance | Mass density (kg/m³) | Weight density (lb/ft³) |
|---|---|---|
| **Solids** | | |
| Copper | 8890 | 555 |
| Iron | 7800 | 490 |
| Lead | 11,300 | 708 |
| Aluminum | 2700 | 169 |
| Ice | 917 | 57 |
| Wood, white pine | 420 | 26 |
| Concrete | 2300 | 145 |
| Cork | 240 | 15 |
| **Liquids** | | |
| Water | 100̄0 | 62.4 |
| Seawater | 1025 | 64.0 |
| Oil | 870 | 54.2 |
| Mercury | 13,600 | 846 |
| Alcohol | 790 | 49.4 |
| Gasoline | 680 | 42.0 |
| **Gases[a]** | at 0°C and 1 atm pressure | at 32°F and 1 atm pressure |
| Air | 1.29 | 0.081 |
| Carbon dioxide | 1.96 | 0.123 |
| Carbon monoxide | 1.25 | 0.078 |
| Helium | 0.178 | 0.011 |
| Hydrogen | 0.0899 | 0.0056 |
| Oxygen | 1.43 | 0.089 |
| Nitrogen | 1.25 | 0.078 |
| Ammonia | 0.760 | 0.047 |
| Propane | 2.02 | 0.126 |

[a] The density of a gas is found by pumping the gas into a container, by measuring its volume and mass or weight, and then by using the appropriate density formula.

The weight density of water is 62.4 lb/ft³, that is, 1 cubic foot of water weighs 62.4 lb. (A suggested project is to make a container 1 cubic foot in volume, pour it full of water, and find the weight of the water. If you fill the container with a gallon container, you will also find that 1 ft³ is approximately $7\frac{1}{2}$ gal.)

The mass density of water is $10\overline{0}0$ kg/m³; that is, 1 cubic metre of water has a mass of $10\overline{0}0$ kg.

In all forms of matter, the density usually decreases as the temperature increases and increases as the temperature decreases.

## EXAMPLE 1

What is the weight density of a block of wood having dimensions 3.00 in. × 4.00 in. × 5.00 in. and a weight of 0.700 lb?

SKETCH:

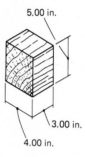

5.00 in.

3.00 in.

4.00 in.

DATA:

$$l = 4.00 \text{ in.}$$
$$w = 3.00 \text{ in.}$$
$$h = 5.00 \text{ in.}$$
$$W = 0.700 \text{ lb}$$
$$D_w = ?$$

BASIC EQUATIONS:

$$V = lwh \quad \text{and} \quad D_w = \frac{W}{V}$$

WORKING EQUATIONS:   Same

SUBSTITUTIONS:

$$V = (4.00 \text{ in.})(3.00 \text{ in.})(5.00 \text{ in.})$$
$$= 60.0 \text{ in}^3$$

$$D_w = \frac{0.700 \text{ lb}}{60.0 \text{ in}^3}$$

$$= 0.0117 \frac{\text{lb}}{\cancel{\text{in}^3}} \times \frac{1728 \cancel{\text{in}^3}}{1 \text{ ft}^3}$$

$$= 20.2 \text{ lb/ft}^3$$

## EXAMPLE 2

Find the mass density of a ball bearing with mass 22.0 g and radius 0.875 cm.

DATA:

$$r = 0.875 \text{ cm}$$
$$M = 22.0 \text{ g}$$
$$D_m = ?$$

BASIC EQUATIONS:

$$V = \tfrac{4}{3}\pi r^3 \quad \text{and} \quad D_m = \frac{M}{V}$$

WORKING EQUATIONS: Same

SUBSTITUTIONS:

$$V = \tfrac{4}{3}\pi(0.875 \text{ cm})^3$$
$$= 2.81 \text{ cm}^3$$

$$D_m = \frac{22.0 \text{ g}}{2.81 \text{ cm}^3}$$
$$= 7.83 \text{ g/cm}^3$$
$$= 7.83 \frac{\text{g}}{\text{cm}^3} \times \frac{10^6 \text{ cm}^3}{1 \text{ m}^3} \times \frac{1 \text{ kg}}{10^3 \text{g}} = 7830 \text{ kg/m}^3$$

## EXAMPLE 3

Find the weight density of a gallon of water weighing 8.34 lb.

DATA:

$$W = 8.34 \text{ lb}$$
$$V = 1 \text{ gal} = 231 \text{ in}^3$$
$$D_w = ?$$

BASIC EQUATION:

$$D = \frac{W}{V}$$

WORKING EQUATION: Same

SUBSTITUTION:

$$D_w = \frac{8.34 \text{ lb}}{231 \text{ in}^3}$$
$$= 0.0361 \frac{\text{lb}}{\text{in}^3} \times \frac{1728 \text{ in}^3}{1 \text{ ft}^3}$$
$$= 62.4 \text{ lb/ft}^3$$

## EXAMPLE 4

Find the weight density of a can of oil (1 quart) weighing 1.90 lb.

DATA:

$$1 \text{ qt} = \tfrac{1}{4} \text{ gal} = \tfrac{1}{4}(231 \text{ in}^3) = 57.8 \text{ in}^3 = V$$
$$W = 1.90 \text{ lb}$$
$$D_w = ?$$

BASIC EQUATION:

$$D_w = \frac{W}{V}$$

WORKING EQUATION: Same

SUBSTITUTION:

$$D_w = \frac{1.90 \text{ lb}}{57.8 \text{ in}^3}$$
$$= 0.0329 \frac{\text{lb}}{\text{in}^3} \times \frac{1728 \text{ in}^3}{1 \text{ ft}^3}$$
$$= 56.9 \text{ lb/ft}^3$$

EXAMPLE 5

A quantity of gasoline weighs 5.50 lb with weight density of 42.0 lb/ft³. What is its volume?

DATA:

$$D_w = 42.0 \, \frac{\text{lb}}{\text{ft}^3}$$

$$W = 5.50 \text{ lb}$$

$$V = ?$$

BASIC EQUATION:

$$D_w = \frac{W}{V}$$

WORKING EQUATION:

$$V = \frac{W}{D_w}$$

SUBSTITUTION:

$$V = \frac{5.50 \cancel{\text{lb}}}{42.0 \cancel{\text{lb}}/\text{ft}^3}$$

$$= 0.131 \text{ ft}^3$$

The density of an irregular solid (rock) cannot be found directly because of the difficulty of finding its volume. However, we could find the amount of water the solid displaces, which is the same as the volume of the irregular solid. Volume of water in beaker = volume of rock.

EXAMPLE 6

A rock of mass 10.8 kg displaces $32\bar{0}0$ cm³ of water. What is the mass density of the rock?

DATA:

$$M = 10.8 \text{ kg}$$

$$V = 32\bar{0}0 \text{ cm}^3$$

$$D_m = ?$$

BASIC EQUATION:

$$D_m = \frac{M}{V}$$

WORKING EQUATION:   Same

SUBSTITUTION:

$$D_m = \frac{10.8 \text{ kg}}{32\overline{0}0 \text{ cm}^3} \times \frac{10^6 \text{ cm}^3}{1 \text{ m}^3}$$

$$= 3380 \text{ kg/m}^3$$

## EXAMPLE 7

A rock displaces 3.00 gal of water and has weight density of 156 lb/ft³. What is its weight?

DATA:

$$D = 156\frac{\text{lb}}{\text{ft}^3}$$

1 gallon water = 231 in³

3 gallon water = 693 in³ = $V$

$W = ?$

BASIC EQUATION:

$$D_w = \frac{W}{V}$$

WORKING EQUATION:

$$W = D_w V$$

SUBSTITUTION:

$$W = 156 \frac{\text{lb}}{\text{ft}^3} \times 693 \text{ in}^3 \times \frac{1 \text{ ft}^3}{1728 \text{ in}^3}$$

$$= 62.6 \text{ lb}$$

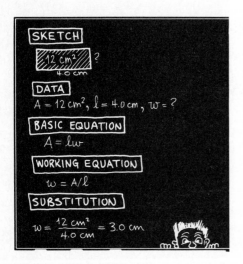

## PROBLEMS

*Express mass density in kg/m³ and weight density in lb/ft³.*

1. What is the mass density of a chunk of rock of mass $21\overline{0}$ g which displaces a volume of 75 cm³ of water?
2. A block of wood is 55.9 in. × 71.1 in. × 25.4 in. and weighs 1810 lb. What is its weight density?
3. If the block of wood in Problem 2 has weight density of 30.0 lb/ft³, what does it weigh?
4. What volume does $13\overline{0}0$ g of mercury occupy?
5. What volume does $13\overline{0}0$ g of cork occupy?
6. What volume does $13\overline{0}0$ g of nitrogen occupy at 0°C and 1 atmosphere pressure?
7. A block of gold 9.00 in. × 8.00 in. × 6.00 in. weighs 301 lb. What is its weight density?
8. If a cylindrical piece of copper is 9.00 in. tall and 1.40 in. in radius, how much does it weigh?
9. A piece of aluminum has a mass of 6.22 kg. If it displaces water that fills a container 12.0 cm × 12.0 cm × 16.0 cm, find the mass density of the aluminum.
10. If 1.00 pint of turpentine weighs 0.907 lb, what is its weight density?
11. Find the mass density of gasoline if 102 g occupy $15\overline{0}$ cm³.
12. How much does 1.00 gal of gasoline weigh?
13. Determine the volume in gallons of $40\overline{0}$ lb of oil.
14. How many ft³ will 573 lb of water occupy?
15. If 20.4 in³ of linseed oil weighs 0.694 lb, what is its weight density?
16. If 108 in³ of ammonia gas weighs 0.00301 lb, what is its weight density?
17. What is the volume of 3.00 kg of propane at 0°C and 1 atm pressure?

**18.** Granite has a mass density of 2650 kg/m³. Find its weight density in lb/ft³.

**19.** What is the mass density of a metal block having dimensions 18.0 cm × 24.0 cm × 8.00 cm with a mass of 9.76 kg?

**20.** Find the mass (in kg) of 1.00 m³ of

    (a) water

    (b) gasoline

    (c) copper

    (d) mercury

    (e) air at 0°C and 1 atm pressure

**21.** What size tank (in litres) is needed for $10\overline{0}0$ kg of

    (a) water?

    (b) gasoline?

    (c) mercury?

# CHAPTER
# *13*

## *Fluids*

In many respects liquids and gases behave in much the same manner. For this reason they are often studied together as fluids. The gas and water piped to your home are fluids having several common characteristics. We will now study some of these characteristics.

If you press your hand against the table, you are applying a force to the table. You are also applying pressure to the table. Is there a difference? The difference is an important one. *Pressure* is the force applied to a unit area. It is the concentration of the force.

$$P = \frac{F}{A}$$

where:   $P$ = pressure [usually in N/m²(Pa) or lb/in² (psi)]
$F$ = force applied (N or lb)
$A$ = area (m² or in²)

Since the SI metric unit for force is the newton (N) and for area is the square metre (m²), the corresponding pressure unit is N/m². This unit is given the special name pascal (Pa), named after Blaise Pascal, a French physicist (1623–1662), who made some important discoveries in the studies of pressure.

$$1 \text{ N/m}^2 = 1 \text{ Pa}$$

Imagine a brick weighing 12.0 N first lying on its side on a table and then standing on one end. The weight of the brick is the same no matter what its position. So the total force (the weight of the brick) on the table is the same in both cases. However, the position of the brick does make a difference on the pressure exerted on the table. In which case is the pressure greater? When standing on end, the brick exerts a greater pressure on the table. The reason is that the area of contact on end is *smaller* than on the side. Using $P = F/A$, let us find the pressure exerted in each case.

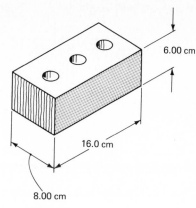
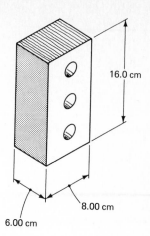

*Case 1:*

$F = 12.0$ N

$A = 8.00$ cm $\times$ 16.0 cm

$A = 128$ cm²

$P = \dfrac{F}{A} = \dfrac{12.0 \text{ N}}{128 \text{ cm}^2} \times \dfrac{10^4 \text{ cm}^2}{1 \text{ m}^2}$

$P = 938$ N/m² $= 938$ Pa

*Case 2:*

$F = 12.0$ N

$A = 6.00$ cm $\times$ 8.00 cm

$A = 48.0$ cm²

$P = \dfrac{F}{A} = \dfrac{12.0 \text{ N}}{48.0 \text{ cm}^2} \times \dfrac{10^4 \text{ cm}^2}{1 \text{ m}^2}$

$P = 25\overline{0}0$ N/m² $= 25\overline{0}0$ Pa

This shows that when the same force is applied to a smaller area, the pressure is greater. From the discussion so far, would you rather a woman step on your foot with a pointed-heel shoe or with a flat-heel shoe? Before you snicker, thinking this question is silly, you should know that the aircraft industry does not think it is. They must design and construct floors light in weight but strong enough to stand the pressure of women wearing pointed-heel shoes. This was a serious problem for them.

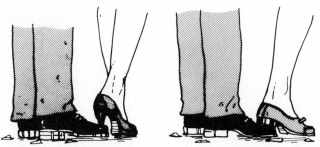

For example, if a $16\overline{0}$-lb woman rests her weight on a 4.00-in² heel, the pressure is

$$P = \frac{F}{A} = \frac{16\overline{0} \text{ lb}}{4.00 \text{ in}^2} = 40.0 \text{ lb/in}^2$$

But if she rests her weight on a pointed heel of $\frac{1}{4}$ in², which is a common area of a pointed heel, the pressure is

$$P = \frac{F}{A} = \frac{16\overline{0} \text{ lb}}{\frac{1}{4} \text{ in}^2} = 640 \text{ lb/in}^2$$

A similar comparison may be shown using metric units. If a 70.0-kg woman rests her weight on a 25.0-cm² heel, the pressure is

$$P = \frac{F}{A}$$

$$P = \frac{(70.0 \text{ kg})(9.80 \text{ m/s}^2)}{25.0 \text{ cm}^2 \times \dfrac{1 \text{ m}^2}{10^4 \text{ cm}^2}} = 2.74 \times 10^5 \text{ N/m}^2$$

$$= 2.74 \times 10^5 \text{ Pa} = 274 \text{ kPa}$$

But if she rests her weight on a pointed heel of 1.00 cm², the pressure is

$$P = \frac{F}{A}$$

$$P = \frac{(70.0 \text{ kg})(9.80 \text{ m/s}^2)}{1.00 \text{ cm}^2 \times \dfrac{1 \text{ m}^2}{10^4 \times \text{ cm}^2}} = 6.86 \times 10^6 \text{ N/m}^2$$

$$= 6.86 \times 10^6 \text{ Pa}$$

$$= 6860 \text{ kPa}$$

Since the pascal is a relatively small unit, the kilopascal (kPa) is a commonly used unit of pressure.

Liquids present a slightly different situation. As you probably know already, the pressure increases as you go deeper in water. Liquids are different in this respect from solids in that, where solids exert only a downward force due to gravity, the force exerted by liquids is in all directions. The pressure a liquid exerts on a submerged object is called *hydrostatic pressure*.

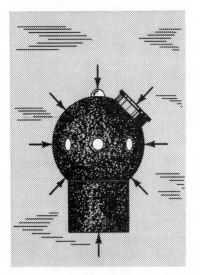

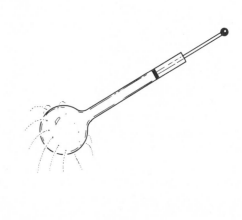

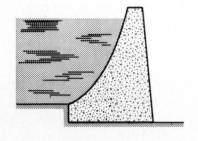

The pressure in a liquid depends only on the depth and weight density of the liquid and not on the length or width of the catchment area. Because the pressure exerted by water increases with depth, dams are built much thicker at the base than at the top.

To find the pressure at a given depth in a liquid, use the formula

$$\boxed{P = hD_w}$$

where: $P$ = pressure
$h$ = height (or depth)
$D_w$ = weight density of the liquid

EXAMPLE 1

Find the pressure at the bottom of a water-filled drum 4.00 ft high.

SKETCH:

4.00 ft

DATA:

$$h = 4.00 \text{ ft}$$
$$D_w = 62.4 \text{ lb/ft}^3$$
$$P = ?$$

BASIC EQUATION:

$$P = hD_w$$

WORKING EQUATION: Same

SUBSTITUTION:

$$P = (4.00 \text{ ft}) \left( 62.4 \frac{\text{lb}}{\text{ft}^3} \right)$$

$$= 25\overline{0} \frac{\text{lb}}{\text{ft}^2} \cdot \frac{1 \text{ ft}^2}{144 \text{ in}^2}$$

$$= 1.74 \text{ lb/in}^2$$

EXAMPLE 2

Find the depth in a lake at which the pressure is $10\overline{0}$ lb/in²

SKETCH: None needed

DATA:

$$P = 10\overline{0} \text{ lb/in}^2$$
$$D_w = 62.4 \text{ lb/ft}^3$$
$$h = ?$$

BASIC EQUATION:

$$P = hD_w$$

WORKING EQUATION:

$$h = \frac{P}{D_w}$$

SUBSTITUTION:

$$h = \frac{10\overline{0} \text{ lb/in}^2}{62.4 \text{ lb/ft}^3}$$

$$= 1.60 \frac{\text{lb}}{\text{in}^2} \cdot \frac{\text{ft}^3}{\text{lb}}$$

$$= 1.60 \frac{\text{ft}^3}{\text{in}^2} \cdot \frac{144 \text{ in}^2}{1 \text{ ft}^2}$$

$$= 23\overline{0} \text{ ft}$$

EXAMPLE 3

Find the height of a water column where the pressure at the bottom of the column is $40\overline{0}$ kPa and the weight density of water is $98\overline{0}0$ N/m³.

SKETCH: None needed

DATA:

$$P = 40\bar{0} \text{ kPa}$$
$$D_w = 98\bar{0}0 \text{ N/m}^3$$
$$h = ?$$

BASIC EQUATION:

$$P = hD_w$$

WORKING EQUATION:

$$h = \frac{P}{D_w}$$

SUBSTITUTION:

$$h = \frac{40\bar{0} \text{ kPa}}{98\bar{0}0 \text{ N/m}^3}$$

$$= \frac{40\bar{0} \text{ kPa}}{98\bar{0}0 \text{ N/m}^3} \times \frac{10^3 \text{ N/m}^2}{1 \text{ kPa}} \quad (Recall: 1 \text{ kPa} = 10^3 \text{ N/m}^2)$$

$$= 40.8 \text{ m}$$

**13-2**
**TOTAL FORCE**
**EXERTED BY LIQUIDS**

The *total force* exerted by a liquid on a horizontal surface (such as the bottom of a barrel) depends on the area of the surface, the depth of the liquid, and the weight density of the liquid. By formula:

$$\boxed{F_t = AhD_w}$$

where:  $F_t$ = total force
   $A$ = area
   $h$ = height or depth of the liquid
   $D_w$ = weight density

The total force on a vertical surface $F_s$ (such as the *side* of a barrel) is found by using half the vertical height (average height):

$$\boxed{F_s = \tfrac{1}{2} AhD_w}$$

EXAMPLE
___

What is the total force on the bottom of a rectangular tank 10.0 ft by 5.00 ft by 4.00 ft deep?

SKETCH:

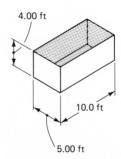

4.00 ft

10.0 ft

5.00 ft

DATA:

$$A = lw = (10.0 \text{ ft})(5.00 \text{ ft}) = 50.0 \text{ ft}^2$$
$$h = 4.00 \text{ ft}$$
$$D_w = 62.4 \text{ lb/ft}^3$$
$$F_t = ?$$

BASIC EQUATION:

$$F_t = AhD_w$$

WORKING EQUATION:   Same

SUBSTITUTION:

$$F_t = (50.0 \text{ ft}^2)(4.00 \text{ ft}) \left( 62.4 \frac{\text{lb}}{\text{ft}^3} \right)$$

$$= 12,500 \text{ lb}$$

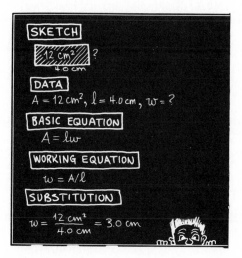

## PROBLEMS

1. A packing crate is 2.50 m × 0.80 m × 0.45 m and weighs 14.1 × 10⁴ N. Find the pressure (in kPa) exerted by the crate on the floor in each of its three possible positions.
2. A packing crate is 2.50 m × 2$\bar{0}$ cm × 3$\bar{0}$ cm and has a mass of 950 kg. Find the pressure (in kPa) exerted by the crate on the floor in each of its three possible positions.
3. What is the water pressure at the bottom of a water tower 50.0 ft high?
4. If the water pressure is 20.0 lb/in², how high can a column of water be raised?
5. What is the density of a liquid which exerts a pressure of 0.400 lb/in² at a depth of 42.0 in.?
6. What is the total force on the bottom of a water-filled circular cattle tank 2.50 ft high with a radius of 4.00 ft?
7. What is the total force on the side of the tank in Problem 6?
8. What must the water pressure be to supply water to the third floor of a building (35.0 ft up) with a pressure of 40.0 lb/in² at that level?
9. A small tank 3.00 in. by 5.00 in. is filled with mercury. If the total force on the bottom of the tank is 14.5 lb, how deep is the mercury? (Weight density of mercury = 0.490 lb/in³.)
10. What is the total force on the largest side of the tank in Problem 9?
11. What is the water pressure at the 25.0-m level of a 50.0-m water tower?
12. If the water pressure is 40.0 N/m², how high can a column of water be raised?
13. What is the density of a liquid that exerts a pressure of 60.0 N/cm² at a depth of 64.0 m?
14. What is the total force on the bottom of a gasoline storage tank 15.0 m high with a radius of 23.0 m?
15. What is the total force on the side of the tank in Problem 14?
16. What must the water pressure be to supply the second floor (18.0 ft up) with a pressure of 40.0 lb/in² at that level?

**13-3**
**HYDRAULIC PRINCIPLE**

The hydraulic jack or press illustrates an important principle that a liquid can be used as a simple machine to multiply force.

> *HYDRAULIC PRINCIPLE (PASCAL'S PRINCIPLE)*
>
> Pressure applied to a confined liquid is transmitted without loss throughout the entire liquid to the walls of the container.

If we apply a force to the small piston of the hydraulic jack on page 192, the pressure is transmitted without measurable loss in all directions. The reason for this is the virtual noncompressibility of liquids.

The *pressure* on the large piston is the same as on the small piston; however, the *total force* on the large piston is greater because of its larger surface area.

In the diagram of the hydraulic jack:
(a) What is the pressure on the small piston?
(b) What is the pressure on the large piston?
(c) What is the total force on the large piston?
(d) What is the mechanical advantage of the jack?

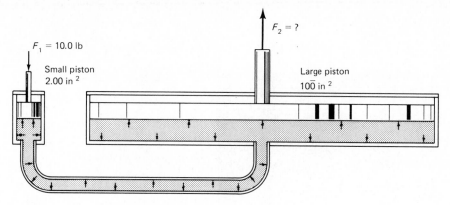

DATA:

$$F_1 = 10.0 \text{ lb}$$
$$A_{\text{small piston}} = 2.00 \text{ in}^2$$
$$A_{\text{large piston}} = 10\bar{0} \text{ in}^2$$
$$P_1 = ?$$
$$P_2 = ?$$
$$F_2 = ?$$
$$\text{MA} = ?$$

(a) BASIC EQUATION:

$$P_1 = \frac{F_1}{A}$$

WORKING EQUATION:   Same

SUBSTITUTION:

$$P_1 = \frac{10.0 \text{ lb}}{2.00 \text{ in}^2}$$
$$= 5.00 \text{ lb/in}^2$$

(b) Applying Pascal's Principle:

$$P_2 = 5.00 \text{ lb/in}^2$$

(c) BASIC EQUATION:

$$P_2 = \frac{F_2}{A}$$

WORKING EQUATION:

$$F_2 = P_2 A$$

SUBSTITUTION:

$$F_2 = \left( 5.00 \frac{\text{lb}}{\text{in}^2} \right) (10\bar{0} \text{ in}^2)$$
$$= 50\bar{0} \text{ lb}$$

(d) BASIC EQUATION:

$$\text{MA} = \frac{F_R}{F_E}$$

WORKING EQUATION: Same

SUBSTITUTION:

$$MA = \frac{50\overline{0} \cdot \cancel{lb}}{10.0 \ \cancel{lb}}$$

$$= 50.0$$

## EXAMPLE 2

The small piston of a hydraulic press has an area $(A_1)$ of 10.0 cm². If the applied force $(F_1)$ is 50.0 N, what must the area of the large piston $(A_2)$ be to exert a pressing force of 4800 N $(F_2)$?

SKETCH: None needed

DATA:

$$A_1 = 10.0 \text{ cm}^2$$
$$F_1 = 50.0 \text{ N}$$
$$F_2 = 4800 \text{ N}$$
$$A_2 = ?$$

BASIC EQUATIONS:

$$P_1 = P_2, \quad P = \frac{F}{A}, \quad \text{and} \quad \frac{F_1}{A_1} = \frac{F_2}{A_2}$$

WORKING EQUATION:

$$A_2 = \frac{A_1 F_2}{F_1}$$

or

$$A_2 = \frac{(10.0 \text{ cm}^2)(4800 \ \cancel{N})}{50.0 \ \cancel{N}}$$

$$= 960 \text{ cm}^2$$

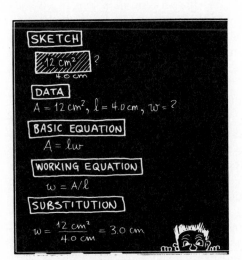

## PROBLEMS

1. The area of the small piston in a hydraulic jack is 0.750 in². The area of the large piston is 3.00 in². If a force of 15.0 lb is applied to the small piston, what weight can be lifted by the large one?

2. If the mechanical advantage of a hydraulic press is 25, what applied force is necessary to produce a pressing force of 1200 N?

3. What is the mechanical advantage of a hydraulic press that produces a pressing force of 7500 N when the applied force is $6\overline{0}0$ N?

4. If the mechanical advantage of a hydraulic press is $2\overline{0}$, what applied force is necessary to produce a pressing force of 325 lb?

5. What is the mechanical advantage of a hydraulic press that produces a pressing force of $10\overline{0}0$ lb when the applied force is $15\overline{0}$ lb?

6. The small piston of a hydraulic press has an area of 8.00 cm². If the applied force is 25.0 N, what must the area of the large piston be to exert a pressing force of $36\overline{0}0$ N?

7. The MA of a hydraulic jack is 360. What force must be applied to lift an automobile weighing 12,000 N?

8. The small piston of a hydraulic press has an area of 4.00 in². If the applied force is 10.0 lb, what must the area of the large piston be to exert a pressing force of $10\overline{0}0$ lb?

9. The MA of a hydraulic jack is 450. What is the weight of the heaviest automobile that can be lifted by an applied force of 8.00 lb?

10. The pistons of a hydraulic press have radii of 2.00 cm and 12.0 cm.
    (a) What force must be applied to the smaller piston to exert a force of 50$\overline{0}$0 N on the larger?
    (b) What is the pressure on each piston?
    (c) What is its mechanical advantage?

11. The mechanical advantage of a hydraulic jack is 45$\overline{0}$. What is the weight of the heaviest automobile that can be lifted by an applied force of 40.0 N?

12. The small circular piston of a hydraulic press has an area of 8.00 cm². If the applied force is 25.0 N, what must the area of the large piston be to exert a pressing force of 36$\overline{0}$0 N?

**13-4**
**AIR PRESSURE**

We saw that air has weight and, as any fluid, it must exert pressure. For example, when a straw is used to drink, the air pressure inside the straw is reduced. As a result, the outside air pressure is higher than in the straw, which forces the fluid up the straw.

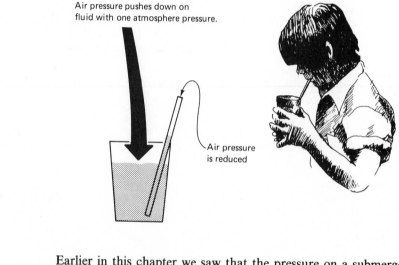

Air pressure pushes down on fluid with one atmosphere pressure.

Air pressure is reduced

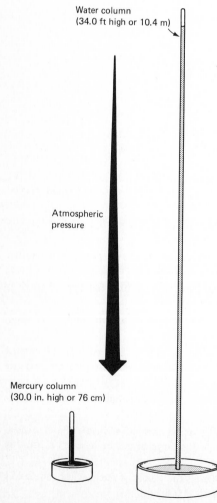

Water column (34.0 ft high or 10.4 m)

Atmospheric pressure

Mercury column (30.0 in. high or 76 cm)

Earlier in this chapter we saw that the pressure on a submerged body increases as the body goes deeper into the liquid. Some creatures live near the bottom of the ocean where the pressure of the water above is so great that it would collapse any human body and most submarines. But through the process of evolution, such creatures have adapted to this tremendous pressure.

Similarly, we on earth live at the bottom of a fluid, air, which is several miles deep. The pressure from this fluid is normally 14.7 lb/in² or 101.32 kPa at sea level. We do not feel this pressure because it normally is almost the same from all directions and also because our bodies have become accustomed to it. When the air pressure becomes unequal, its force becomes quite evident in the form of wind. This wind may be a cool summer breeze or the tremendous concentrated force of a tornado.

What is the pressure of our atmosphere equivalent to? Experiments have shown that the atmosphere supports a column of water 34.0 ft high in a tube in which the air has been removed. The atmosphere supports 30.0 in. of mercury in a similar tube. This is not surprising, as mercury is 13.6 times as dense as water and

$$\frac{1}{13.6} \times 34.0 \text{ ft} \times \frac{12.0 \text{ in.}}{1.00 \text{ ft}} = 30.0 \text{ in.}$$

The height of the mercury column in a barometer is independent of the width (or diameter or cross-sectional area) of the barometer tube.

This measurement was standard for many years on TV weather programs but is increasingly being replaced by the metric standard measurement in kilopascals (kPa).

The pressure of the atmosphere can be expressed in terms of the pressure of an equivalent column of mercury. Air pressure at sea level is normally 29.9 in. or 76.0 cm or $76\overline{0}$ mm of mercury.

How do we arrive at the 14.7 lb/in² measurement? In terms of mercury, its height is 29.9 in. or 2.49 ft. Its density is $13.6 \times 62.4$ lb/ft³. Therefore,

$$P = hD_w$$

$$P = 2.49 \text{ ft} \times 13.6 \times 62.4 \frac{\text{lb}}{\text{ft}^3} \times \frac{1.00 \text{ ft}^2}{144 \text{ in}^2}$$

$$= 14.7 \text{ lb/in}^2$$

The pressure of 2 atm would be 29.4 lb/in² or 202.64 kPa. If the pressure is $\frac{1}{2}$ atm at one point in the sky, it would be 7.35 lb/in² or 50.66 kPa.

When we purchase bottled gas, the amount of gas and its density vary with the pressure. If the pressure is low, the amount of gas in the bottle is low. If the pressure is high, the bottle is "nearly full."

The gauge that is usually used for checking the pressure in bottles and tires shows a reading of zero at normal atmospheric pressure. The pressure of the atmosphere is not included in this reading. The actual pressure, called *absolute pressure*, is the gauge pressure reading plus the normal atmospheric pressure 101.32 kPa or 14.7 lb/in². That is,

$$\text{absolute pressure} = \text{gauge pressure} + \text{atmospheric pressure}$$

or

$$\boxed{P_{\text{abs}} = P_{\text{ga}} + P_{\text{atm}}}$$

where $P_{\text{atm}} = 101.32$ kPa or 14.7 lb/in².

## EXAMPLE

What is the absolute pressure in a tire inflated to 32.0 lb/in²

(a) in lb/in²?

(b) in kPa?

(a) SKETCH:   None needed

DATA:

$$P_{\text{ga}} = 32.0 \text{ lb/in}^2$$
$$P_{\text{atm}} = 14.7 \text{ lb/in}^2$$
$$P_{\text{abs}} = ?$$

BASIC EQUATION:

$$P_{\text{abs}} = P_{\text{ga}} + P_{\text{atm}}$$

WORKING EQUATION:   Same

SUBSTITUTION:

$$P_{\text{abs}} = 32.0 \text{ lb/in}^2 + 14.7 \text{ lb/in}^2$$
$$= 46.7 \text{ lb/in}^2$$

(b) We use the conversion factor:

$$101.32 \text{ kPa} = 14.7 \text{ lb/in}^2$$

Therefore,

$$P_{abs} = 46.7 \text{ lb/in}^2 \times \frac{101.32 \text{ kPa}}{14.7 \text{ lb/in}^2}$$

$$= 322 \text{ kPa}$$

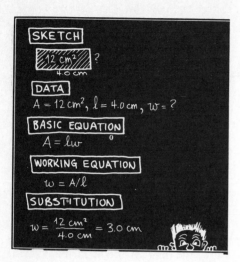

SKETCH

12 cm² ?
4.0 cm

DATA

$A = 12 \text{ cm}^2, \ell = 4.0 \text{ cm}, w = ?$

BASIC EQUATION

$A = \ell w$

WORKING EQUATION

$w = A/\ell$

SUBSTITUTION

$w = \dfrac{12 \text{ cm}^2}{4.0 \text{ cm}} = 3.0 \text{ cm}$

## PROBLEMS

1. Convert $80\overline{0}$ kPa to lb/in².
2. Convert 64.3 lb/in² to kPa.
3. What is the absolute pressure in a bicycle tire with a gauge pressure of 70.0 lb/in²?
4. What is the absolute pressure of a motorcycle tire with a gauge pressure of $25\overline{0}$ kPa?
5. What is the gauge pressure of a tire with an absolute pressure of 45.0 lb/in²?
6. Find the gauge pressure of a tire with an absolute pressure of $45\overline{0}$ kPa.

**13-5**
BUOYANCY

A floating boat displaces its own weight of the water in which it floats. If two or more people get in the boat, the boat rides lower due to the increased weight in the boat.

Archimedes, a Greek philosopher, was one of the first to study fluids and formulated what is now called

> *ARCHIMEDES' PRINCIPLE*
>
> Any object placed in a fluid apparently loses weight equal to the weight of the displaced liquid.

Weight of object = Weight of displaced water

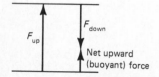

The object does not actually lose weight, but because of the depth difference, the force pushing up on the bottom of the object is greater than the force pushing down on the top of the object. This net upward force is called the *buoyant force*.

Archimedes' principle applies to gases as well as liquids. Lighter-than-air craft (e.g., Goodyear's blimps) operate on this principle. Since they are filled with a gas lighter than air (helium), the buoyant force on them causes them to be supported by the air. Being "submerged" in the air, a blimp is buoyed up by the weight of the air it displaces, which equals the buoyant force of the air on the balloon.

## EXAMPLE

A solid concrete block 15.0 cm × 20.0 cm × 10.0 cm weighs 67.6 N. When it is lowered into the water, its weight registers 38.2 N. The buoyant force is 67.6 N − 38.2 N = 29.4 N. The volume of the displaced water is

$$
\begin{aligned}
V &= lwh \\
&= 15.0 \text{ cm} \times 20.0 \text{ cm} \times 10.0 \text{ cm} \\
&= 30\overline{0}0 \text{ cm}^3 = 3.00 \times 10^{-3} \text{ m}^3
\end{aligned}
$$

The mass of the displaced water is

$$
\begin{aligned}
M &= D_m V \\
&= (100\overline{0} \text{ kg/m}^3)(3.00 \times 10^{-3} \text{ m}^3) \\
&= 3.00 \text{ kg}
\end{aligned}
$$

The weight of the displaced water is then

$$
\begin{aligned}
F_w &= mg \\
&= (3.00 \text{ kg})(9.80 \text{ m/s}^2) \\
&= 29.4 \text{ N}
\end{aligned}
$$

which equals the buoyant force.

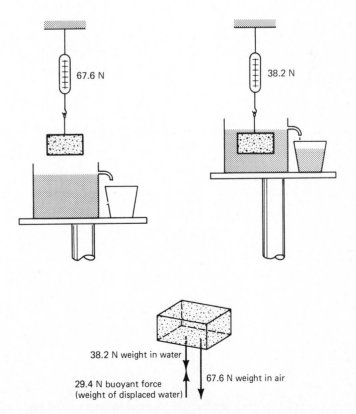

Weight in water = weight in air − buoyant force (weight of displaced water)

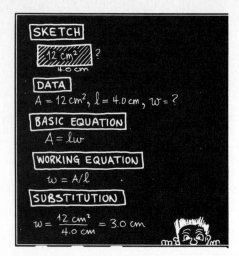

SKETCH

12 cm² ?

4.0 cm

DATA

$A = 12 \text{ cm}^2, \ell = 4.0 \text{ cm}, w = ?$

BASIC EQUATION

$A = \ell w$

WORKING EQUATION

$w = A/\ell$

SUBSTITUTION

$w = \dfrac{12 \text{ cm}^2}{4.0 \text{ cm}} = 3.0 \text{ cm}$

## PROBLEMS

*Use Archimedes' principle to do the following.*

1. If a metal alloy weighs 81.0 lb in air and 68.0 lb when under water, what is the buoyant force of the water?

2. If a piece of metal weighs 67.0 N in air and 62.0 N in water, what is the buoyant force of the water?

3. If a rock weighs 25.7 N in air and 21.8 N in water, what is the buoyant force of the water?

4. If a rock displaces 1.21 ft³ of water, what is the buoyant force of the water?

5. If a metal casting displaces 327 cm³ of water, what is the buoyant force of the water?

6. If a piece of metal displaces 657 cm³ of water, what is the buoyant force of the water?

7. A flat-bottom river barge is 30.0 ft wide, 85.0 ft long, and 15.0 ft deep. How many ft³ of water will it displace while the top stays 3.00 ft above the water? And what load in tons will the barge contain under these conditions if the barge weighs 160 tons in dry dock?

8. A flat-bottom river barge is 12.0 m wide, 30.0 m long, and 6.00 m deep. How many m³ of water will it displace while the top stays 1.00 m above the water? What load in newtons will the barge contain under these conditions if the barge weighs $3.55 \times 10^6$ N in dry dock? The metric weight density of water is $980\overline{0}$ N/m³.

**13-6**
SPECIFIC GRAVITY

When we check the antifreeze in a radiator in winter, we are really determining the specific gravity of the liquid. *Specific gravity is a comparison of the density of a substance to that of water.* Because the density of antifreeze is different from the density of water, we can find the concentration of antifreeze (and thus the amount of protection from freezing) by measuring the specific gravity of the solution in the radiator.

A hydrometer is a sealed glass tube weighted at one end so that it floats vertically in a liquid. It sinks in the liquid until it displaces an amount of liquid equal to its own weight. The densities of the displaced liquids are inversely proportional to the depths to which the tube sinks. That is, the greater the density of the liquid, the less the tube sinks, and the lesser the density of the liquid, the more the tube sinks.

A hydrometer usually has a scale inside the tube and is calibrated so that it floats in water at the 1.000 mark.

Hydrometers are commonly used to measure the specific gravities of battery acid and antifreeze in radiators. In a lead storage battery, the electrolyte is a solution of sulfuric acid and water and the specific gravity of the solution varies with the amount of charge of the battery. The following chart gives common specific gravities of conditions of a lead storage battery.

| Condition | Specific gravity |
|---|---|
| New (fully charged) | 1.30 |
| Old (discharged) | 1.15 |

The next chart gives various specific gravities and the corresponding temperatures below which the antifreeze and water solution will freeze.

| Temperature (°C) | Specific gravity |
|---|---|
| −1.24 | 1.00 |
| −2.99 | 1.01 |
| −6.89 | 1.02 |
| −19.82 | 1.05 |
| −44.83 | 1.07 |
| −51.23 | 1.08 |

Common battery tester and antifreeze tester hydrometers are shown below.

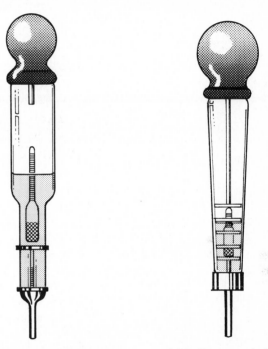

(a) Common battery tester hydrometer        (b) Antifreeze tester hydrometer

One other factor must be considered in the use of the hydrometer—that of temperature. Significant differences in readings will occur over a range of temperatures.

Specific gravities of some common liquids at room temperature are given in the following table.

| Liquid | Specific gravity |
|---|---|
| Benzene | 0.9 |
| Ethyl alcohol | 0.79 |
| Gasoline | 0.68 |
| Kerosene | 0.82 |
| Mercury | 13.6 |
| Seawater | 1.025 |
| Sulfuric acid | 1.84 |
| Turpentine | 0.87 |
| Water | 1.000 |

Think for a minute about the motion of water flowing down a fast-moving mountain stream that contains boulders and rapids and about the motion of the air during a thunderstorm or during a tornado. These types of motion are complex, indeed. We will limit our discussion to the simpler examples of fluid flow.

*Streamline flow* is the smooth flow of a fluid through a tube or pipe. By smooth flow we mean that each particle of the fluid follows the same uniform path as all the others. If the speed of the flow becomes too great or if the tube changes direction or diameter too abruptly, the motion of the flow is described as *turbulent.*

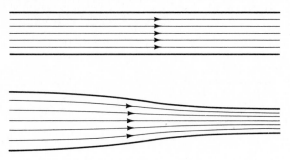

Streamline flow of a fluid through a tube or pipe

The *flow rate* of a fluid is the volume of fluid flowing past a given point in a pipe per unit time. Assume that we have a streamline flow through a straight section of pipe at speed *v*. During a time interval of *t,* each particle of fluid travels a distance *vt*. If *A* is the cross-sectional area of the pipe, the volume of fluid passing a given point during time interval *t* is *vtA*. Thus, the flow rate, *Q,* is given by

$$Q = \frac{vtA}{t}$$

or

$$\boxed{Q = vA}$$

where: $Q$ = flow rate
$v$ = velocity of the fluid
$A$ = cross-sectional area of the tube or pipe

EXAMPLE

---

Water flows through a fire hose of diameter 6.40 cm at a velocity of 5.90 m/s. Find the flow rate of the fire hose in L/min.

DATA:

$$v = 5.90 \text{ m/s}$$
$$r = 3.20 \text{ cm} = 0.0320 \text{ m}$$
$$A = \pi r^2 = \pi(0.0320 \text{ m})^2$$
$$Q = ?$$

BASIC EQUATION:

$$Q = vA$$

WORKING EQUATION:   Same

SUBSTITUTION:

$$Q = (5.90 \text{ m/s})\pi(0.0320 \text{ m})^2 \times \frac{10^3 \text{ L}}{1 \text{ m}^3} \times \frac{60 \text{ s}}{1 \text{ min}}$$

$$= 1140 \text{ L/min}$$

*For an incompressible fluid, the flow rate is constant throughout the pipe.* If the cross-sectional area of the pipe changes and streamline flow is maintained, the flow rate is the same all along the pipe.

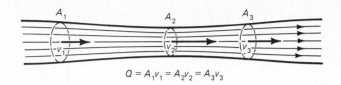

$$Q = A_1 v_1 = A_2 v_2 = A_3 v_3$$

That is, as the cross-sectional area increases, the velocity decreases, and vice versa.

**13-8**
**BERNOULLI'S PRINCIPLE**

What happens to the pressure as the cross-sectional area of the pipe changes? This concept can be illustrated by use of a Venturi meter.

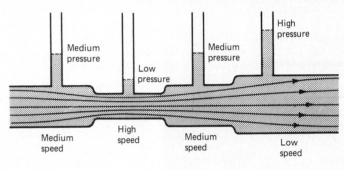

Venturi meter

Here the vertical tubes act like pressure gauges; the higher the column, the higher the pressure. As you can see, the higher the speed, the lower the pressure, and vice versa.

The explanation of this change in pressure of a fluid in streamline flow was found by Daniel Bernoulli (1700–1782).

---

*BERNOULLI'S PRINCIPLE*

For the horizontal flow of a fluid through a tube, the sum of the pressure and energy of motion (kinetic energy) per unit volume of the fluid is constant.

---

Another application of Bernoulli's principle involves airplane travel. The figure shows the flow of air rushing past the wing of an airplane. The velocity, $v_1$, of the air above the wing is greater than the velocity of the air below, $v_2$, because it has farther to travel in a given time. Thus, the pressure at point 2 is greater, which causes a lift on the wing.

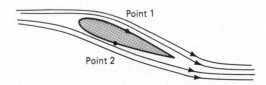

Two examples that illustrate Bernoulli's principle are a curving baseball and a paint spray gun.

The spin of the ball causes the air on either side of the ball to be moving past the ball at different speeds. The air pressure on the faster side is reduced, so the ball curves that way.

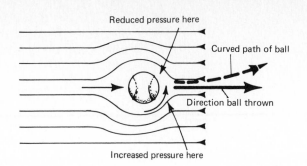

Reduced pressure here

Curved path of ball

Direction ball thrown

Increased pressure here

When the air in a spray gun is accelerated through a narrowing in the line, the pressure is reduced and paint is drawn into the airstream and is then forced from the gun.

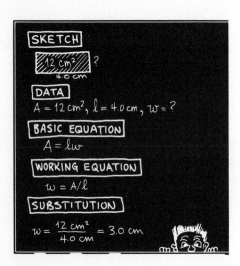

## PROBLEMS

1. Water flows through a hose of diameter 3.80 cm at a velocity of 5.00 m/s. Find the flow rate of the hose in L/min.

2. Water flows through a 15.0-cm fire hose at a rate of 5.00 m/s.
   (a) Find the rate of flow through the hose in L/min.
   (b) How many litres pass through the hose in 1.00 min?

3. Water flows from a pipe at 650 L/min.
   (a) What is the diameter of the pipe if the velocity of the water is 1.5 m/s?
   (b) What is the velocity of the water if the diameter of the pipe is 20.0 cm?

4. Water flows through a pipe of diameter 3.00 in. at 45.0 ft/min. Find the flow rate (a) in ft³/min and (b) in gal/min.

5. A pump is rated to deliver 50.0 gal/min. What is the velocity of water in (a) a 6.00 in. diameter pipe and (b) a 3.00 in. diameter pipe?

6. What size pipe needs to be attached to a pump rated at 36.0 gal/min if the desired velocity is 10.0 ft/min? Give the inner diameter in inches.

# CHAPTER
# *14*

# *Temperature and Heat*

**14-1**
INTRODUCTION

An understanding of temperature and heat is very important to all technicians. An automotive technician is concerned with the heat energy released by a fuel mixture in a combustion chamber. The excess heat produced by an engine must be transferred to the atmosphere. The highway technician is concerned with the expansion of bridges and roads when the temperature changes.

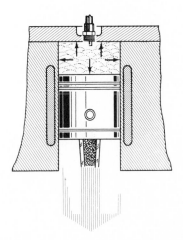

**14-2**
TEMPERATURE

Basically, temperature is a measure of the hotness or coldness of an object. Temperature could be measured in a simple way by using your hand to sense the hotness or coldness of an object. However, the range of temperatures which your hand can withstand is too small, and your hand is not precise enough to measure temperature. Therefore, other methods are used for measuring temperature.

We will use the fact that certain properties of matter vary with their temperature. For example, we could use the fact that when objects are heated, they give off light of different colors. When an object is heated, in the absence of chemical reactions, it first gives off red light. As it is heated more, it appears white (see the table on page 204).

**METALLURGY AND HEAT TREATMENT**
**Temperatures, Steel Colors, and Related Processes[a]**

| | Colors | °F | °C | Processes |
|---|---|---|---|---|
| Heat colors | White | 2500 | 1371 | Welding |
| | | 2400 | 1315 | High-speed steel hardening (2150–2450°F) |
| | Yellow white | 2300 | 1259 | |
| | | 2200 | 1204 | |
| | | 2100 | 1149 | |
| | Yellow | 2000 | 1093 | |
| | | 1900 | 1036 | |
| | Orange red | 1800 | 981 | |
| | | 1700 | 926 | Alloy tool steel hardening (1500–1950°F) |
| | | 1600 | 871 | |
| | Light cherry red | 1500 | 815 | |
| | Cherry red | 1400 | 760 | Carbon tool steel hardening (1350–1550°F) |
| | | 1300 | 704 | |
| | Dark red | 1200 | 648 | |
| | | 1100 | 593 | |
| | Very dark red | 1000 | 538 | High-speed steel tempering (1000–1100°F) |
| | | 900 | 482 | |
| | Black red in dull light or darkness | 800 | 426 | |
| | | 700 | 371 | |
| Temper colors | Pale blue (590°F) | 600 | 315 | Carbon tool steel tempering (300–1050°F) |
| | Violet (545°F) | | | |
| | Purple (525°F) | 500 | 260 | |
| | Yellowish brown (490°F) | | | |
| | Dark straw (465°F) | 400 | 204 | |
| | Light straw (425°F) | | | |
| | | 300 | 149 | |
| | | 200 | 93 | |
| | | 100 | 38 | |
| | | 0 | 18 | |

[a] Allegheny Ludlum Steel Corp. Reprinted by permission.

Chemical reactions sometimes cause different colors to be given off. When carbon steel is heated and exposed to air, several colors are given off before it appears red. This is due to a chemical reaction involving the carbon. If we could measure the color of the light given off, we could then determine the temperature. Although this works only for high temperatures, it is used in the production of metal alloys. The temperature of hot molten metals is determined this way.

Another property of matter that can be used is that the volume of a liquid or a solid changes as its temperature changes. The liquid in glass thermometers is an example. This type of thermometer consists of a hollow glass bulb and a hollow glass tube joined together as shown. A small amount of liquid mercury or alcohol is placed in the bulb. The air is removed from the tube. When the liquid is heated, it expands and rises up the glass tube. The height to which the liquid rises indicates the temperature.

The thermometer is standardized by marking two points on the glass which indicate the liquid level at two known temperatures. The temperatures used are the melting point of ice (called the *ice point*) and the boiling point of water at sea level. The distance between these marks is then divided up into equal segments called *degrees*.

There are four temperature scales which we will study. On the Fahrenheit scale, the ice point is 32°F and the boiling point is 212°F. When we write a certain temperature, we give a number, then the degree symbol (°) and a capital letter which indicates the scale. The letter F following the degree symbol is not to be considered as a unit when working problems. Temperatures below zero on a scale are written as negative numbers. Thus, 30° below zero on the Fahrenheit scale is written as −30°F.

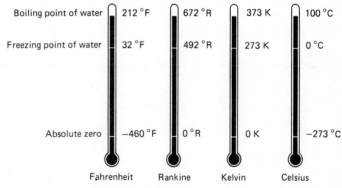

Four Basic Temperature Scales

The metric temperature scale is the Celsius scale, which has replaced the centigrade scale. The ice point is 0°C and the boiling point is 100°C. The relationship between Fahrenheit temperatures ($T_F$) and Celsius temperatures ($T_C$) is as follows:

$$T_F = \frac{9 T_C}{5} + 32°$$

$$T_C = \frac{5(T_F - 32°)}{9}$$

where:  $T_F$ = Fahrenheit temperature
  $T_C$ = Celsius temperature

## EXAMPLE 1

The average human body temperature is 98.6°F. What is it in Celsius?

DATA:

$$T_F = 98.6°F$$
$$T_C = ?$$

BASIC EQUATION:

$$T_C = \frac{5(T_F - 32°)}{9}$$

WORKING EQUATION:   Same

SUBSTITUTION:

$$T_C = \frac{5(98.6° - 32°)}{9}$$

$$= \frac{5(66.6°)}{9}$$

$$= 37.0°C$$

In some technical work it is necessary to use the absolute temperature scales, which are the Rankine scale and the Kelvin scale. These are called absolute scales because 0 on either scale refers to the lowest limit of temperature, called absolute zero. Absolute zero and lower temperatures can never be attained. The reasons are discussed in the next section. There is no such limit on high temperatures.

The Rankine scale is closely related to the Fahrenheit scale. The relationship is

$$\boxed{T_R = T_F + 46\overline{0}°}$$

The Kelvin scale is closely related to the Celsius scale. The relationship is

$$\boxed{T_K = T_C + 273*}$$

## EXAMPLE 2

The lower limit on temperatures is 0°R. Find this limit on the Fahrenheit scale.

DATA:

$$T_R = 0°R$$
$$T_F = ?$$

BASIC EQUATION:

$$T_R = T_F + 46\overline{0}°$$

WORKING EQUATION:

$$T_F = T_R - 46\overline{0}°$$

SUBSTITUTION:

$$T_F = 0° - 46\overline{0}°$$
$$= -46\overline{0}°F$$

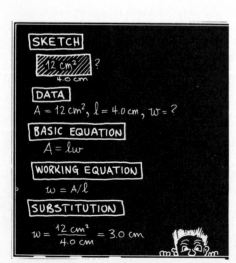

## PROBLEMS

*Convert the following temperatures as indicated.*

1. $T_F = 7\overline{0}°F$, $T_C =$ _____
2. $T_F = 12\overline{0}°F$, $T_C =$ _____
3. $T_F = 25\overline{0}°F$, $T_C =$ _____
4. $T_C = 17°C$, $T_F =$ _____
5. $T_C = 125°C$, $T_F =$ _____
6. $T_C = 5°C$, $T_F =$ _____
7. $T_F = 1\overline{0}°F$, $T_C =$ _____
8. $T_F = 2\overline{0}°F$, $T_C =$ _____
9. $T_C = 5°C$, $T_F =$ _____
10. $T_F = -5\overline{0}°F$, $T_C =$ _____
11. $T_F = 15\overline{0}°F$, $T_R =$ _____
12. $T_F = 55°F$, $T_R =$ _____
13. $T_R = 60\overline{0}°R$, $T_F =$ _____
14. $T_C = 77°C$, $T_K =$ _____

* The degree symbol (°) is *not* used when writing a temperature on the Kelvin scale.

**15.** $T_C = -5\bar{0}°C$, $T_K =$ _____          **16.** $T_K = 20\bar{0}$ K, $T_C =$ _____

**17.** $T_K = 600\bar{0}$ K, $T_C =$ _____

**18.** The melting point of pure iron is 1505°C. What Fahrenheit temperature is this?

**19.** Steel heated to 1650°F is cherry red. What Celsius temperature is this?

**20.** A welding white heat is approximately 14$\bar{0}$0°C. What is this temperature expressed in Fahrenheit degrees?

**14-3**

**HEAT**

When a machinist drills a hole in a metal block, it becomes very hot. As the drill does mechanical work on the metal, the temperature of the metal increases. How can we explain this? We need to look at the difference between the metal at low temperatures and at high temperatures. At high temperatures the atoms in the metal vibrate more rapidly than at low temperatures. Their velocity is higher at high temperatures, and thus their kinetic energy (KE $= \frac{1}{2}\ mv^2$) is greater.

To raise the temperature of a material, we must speed up the atoms; that is, we must add energy to them. *Heat is the name given to this energy which is being added to or taken from a material.*

Drilling a hole in a metal block causes a temperature increase. As the drill turns, it collides with atoms of the metal, causing them to speed up. This mechanical work done on the metal has caused an increase in the energy (speed) of the atoms. For this reason, any friction between two surfaces results in a temperature rise of the materials.

Since heat is a form of energy, we could measure it in ft lb or joules, which are energy units. However, it was not always known that heat was a form of energy, and special units for heat were developed and are still in use.

These units are the Btu (British thermal unit) in the English system and the calorie and the kilocalorie in the metric system. The Btu is the amount of heat (energy) necessary to raise the temperature of 1 lb of water 1°F. The calorie (cal) is the amount of heat (energy) necessary to raise the temperature of 1 gram of water 1°C. The kilocalorie (kcal) is the amount of heat necessary to raise the temperature of 1 kilogram of water 1°C. *Note:* One food calorie is the same as 1 kilocalorie.

We said in Section 14-2 that temperatures below absolute zero cannot be reached. Now we can see the reason for this. To lower the temperature of a substance, we need to remove some of the energy of motion of the molecules (heat). When we have removed all the heat possible (when the molecules are moving as slowly as possible), we have reached the lowest possible temperature called absolute zero. Lower temperatures cannot be reached because all the heat has been removed. However, there is no upper limit on temperature because we can always add more heat (energy) to a substance to increase its temperature.

As we saw above, heat and work are somehow related. James Prescott Joule (1818–1889), an English scientist, determined by experiments the relationship between heat and work. He found that

1. 1 cal of heat is produced by 4.19 J of work.
2. 1 kcal of heat is produced by 4190 J of work.
3. 1 Btu of heat is produced by 778 ft lb of work.

These relationships are known as the *mechanical equivalent of heat.*

Of the many examples by which heat is converted into useful work, we present the following:

1. *In our bodies:* When food is oxidized, heat energy is produced which can be converted into muscular energy, which in turn can be turned

into work. Experiments have shown that only about 25% of the heat energy from our food is converted into muscular energy. That is, our bodies are less than 25% efficient.

2. *By burning gases.* When a gas is burned, the gas expands and builds up a tremendous pressure which may convert heat to work by exerting a force to move a piston or turn the blades of a turbine. Since the burning of the fuel occurs within the cylinder or turbine, such engines are called *internal combustion engines.*

3. *By steam.* Heat from burning oil, coal, or wood may be used to generate steam. When water changes to steam under normal atmospheric pressure, it expands about 1700 times. When confined to a boiler, the pressure exerts a force against the piston in a steam engine or against the blades of a steam turbine. Since the fuel burns outside the engine, most steam engines or steam turbines are *external combustion engines.*

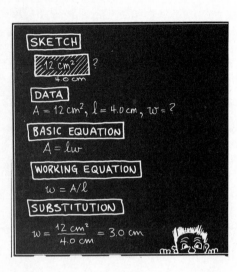

## PROBLEMS

1. Find the amount of heat in cal generated by 95 J of work.
2. Find the amount of heat in kcal generated by 7510 J of work.
3. Find the amount of work that is equivalent to $15\overline{0}0$ Btu.
4. Find the amount of work that is equivalent to $38\overline{0}0$ kcal.
5. Find the amount of heat energy that must be produced by the body to be converted into muscular energy and then into $10\overline{0}0$ ft lb of work. Assume that the body is only 25% efficient.

**14-4**
**HEAT TRANSFER**

The movement of heat from a hot engine to the air is necessary to keep the engine from overheating. The heat produced by a furnace is transferred to the various rooms in a house. The movement of heat is a major technical problem.

The transfer of heat from one object to another is always from the warmer object to the colder one or from the warmer part of an object to a colder part.

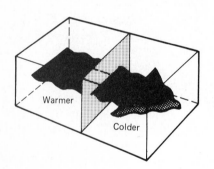

There are three methods of heat transfer: conduction, convection, and radiation. The usual method of heat transfer in solids is *conduction.* When one end of a metal rod is heated, the molecules in that end move faster than before. These molecules collide with other molecules and cause them to move faster also. In this way the heat is transferred from one end of the metal to the other. Another example of conduction is the transferring of the excess heat produced in the combustion chamber of an engine through the engine block into the water coolant.

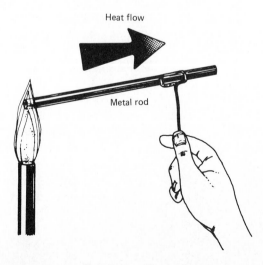

Heat flow

Metal rod

The conduction of heat through some materials is better than through others. A poor conductor of heat is called an insulator. A list of several good conductors and poor conductors is given here.

| Good heat conductors | Poor heat conductors |
| --- | --- |
| Copper | Asbestos |
| Aluminum | Glass |
| Steel | Wood |
| | Air |

Another method of heat transfer is called *convection*. This is the movement of warm gases or liquids from one place to another. The wind carries heat along with it. The water coolant in an engine carries hot water from the engine block to the radiator by a convection process. The wind is a natural convection process. The engine coolant is a forced convection process because it depends on a pump.

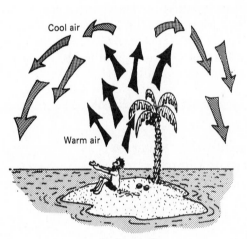

Cool air

Warm air

Heat convection

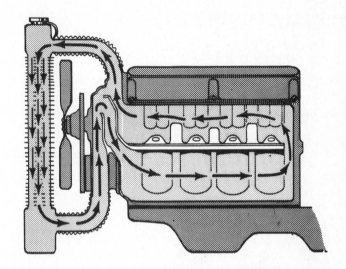

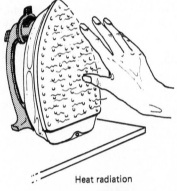

Heat radiation

Convection currents are caused by the expansion of liquids or gases as they are heated or cooled. This expansion makes the hot gas or liquid less dense than the surrounding fluid. The lighter fluid is then forced upward by the heavier surrounding fluid, which then flows in to replace it. This type of behavior occurs in a fireplace as hot air goes up the chimney and is replaced by cool air from the adjacent room. The cool air draft, as this is called, is eventually supplied from outside air. This is why a fireplace is not very effective in heating a house. An airtight woodburning stove, however, draws little air from the inside of a house and is therefore much more efficient at heating the house.

The third method of heat transfer is *radiation*. Put your hand several inches from a hot iron. The heat you feel is not transferred by conduction because air is a poor conductor. It is not transferred by convection because the hot air rises. This kind of heat transfer is called radiation. This radiant heat is similar to light and passes through air, glass, and the vacuum of space. The energy that comes to us from the sun is in the form of radiant energy.

Dark objects absorb radiant heat and light objects reflect radiant heat. This is why we feel cooler on a hot day in light-colored clothing than in dark clothing.

Calculations of heat flow are of great importance because of concern about energy conservation. All three of the heat-transfer mechanisms discussed here must be accounted for in any estimation of heat loss from a building. In addition, one must account for infiltration losses that arise from leakage through cracks and openings near doors, windows, and other such areas. Infiltration losses are, of course, a form of convective transfer but are often treated separately. We will discuss methods for calculating heat loss by conduction only.

The equations describing the flow of heat through an object are very similar to those for the flow of electricity, which will be developed in later chapters. The driving potential for heat flow is the temperature difference between the hot and cold sides of the object. Heat flow is similar to the flow of electrical charge.

The ability of a material to transfer heat by conduction depends on its thermal conductivity. Metals are good conductors of heat. Glass and air are poor conductors. The rate at which heat is transferred through an object depends on the following factors:

1. The thermal conductivity $K$.
2. The cross-sectional area $A$ through which the heat flows.
3. The thickness of the material $L$.
4. The temperature difference between the two sides of the material.

The total amount of heat transferred is given by the equation

$$Q = \frac{KAt\,(T_2 - T_1)}{L}$$

where:  $Q$ = heat transferred in Btu or cal
$K$ = thermal conductivity
$A$ = cross-sectional area
$t$ = total time
$T_2$ = temperature of the hot side
$T_1$ = temperature of the cool side
$L$ = thickness of the material

The following table gives the thermal conductivities of some common materials.

**THERMAL CONDUCTIVITIES**

| Substance | Btu/ft °F h | cal/cm °C h |
|---|---|---|
| Air | 0.011 | 0.16 |
| Aluminum | 140 | 2$\bar{0}$00 |
| Brass | 62 | 930 |
| Brick | 0.42 | 6.2 |
| Cellulose fiber (loose fill) | 0.023 | 0.34 |
| Concrete blocks | 0.35 | 5.1 |
| Copper | 220 | 3300 |
| Corkboard | 0.022 | 0.32 |
| Glass | 0.50 | 7.4 |
| Gypsum board (sheetrock) | 0.092 | 1.4 |
| Mineral wool | 0.026 | 0.39 |
| Plaster | 0.083 | 1.2 |
| Polystyrene (expanded) | 0.020 | 0.30 |
| Polyurethane (expanded) | 0.014 | 0.21 |
| Steel | 26 | 390 |

EXAMPLE 1

Determine the heat flow in an 8.0 h period through a 36 in. × 36 in. pane of glass (0.125 in. thick) if the temperature of the inner surface of the glass is $5\bar{0}°F$ and the temperature of the outer surface is 25°F.

DATA:

$$K = 0.50 \text{ Btu/ft °F h}$$
$$A = 3.0 \text{ ft} \times 3.0 \text{ ft} = 9.0 \text{ ft}^2$$
$$t = 8.0 \text{ h}$$
$$T_2 = 5\bar{0}°F$$
$$T_1 = 25°F$$
$$L = 0.125 \text{ in.} \times \left(\frac{1 \text{ ft}}{12 \text{ in.}}\right) = 0.0104 \text{ ft}$$
$$Q = ?$$

BASIC EQUATION:

$$Q = \frac{KAt(T_2 - T_1)}{L}$$

WORKING EQUATION: Same

SUBSTITUTION:

$$Q = \frac{(0.50 \text{ Btu/ft °F h})(9.0 \text{ ft}^2)(8.0 \text{ h})(5\bar{0}°F - 25°F)}{0.0104 \text{ ft}}$$
$$= 8.7 \times 10^4 \text{ Btu}$$

It is common to find that the insulation value of construction material is expressed in terms of the "$R$ value," which is indicative of the ability of the material to resist the flow of heat. The $R$ value is inversely proportional to the thermal conductivity and directly proportional to the thickness. Low thermal conductivity is characteristic of good insulators. This is described by the equation

$$\boxed{R = \frac{L}{K}}$$

where: $R = R$ value
$K =$ thermal conductivity
$L =$ thickness of the material

EXAMPLE 2

Calculate the $R$ value of 6.0 in. of mineral wool insulation.

DATA:

$$L = 6.0 \text{ in.} = 0.50 \text{ ft}$$
$$K = 0.026 \text{ Btu/ft °F h}$$
$$R = ?$$

BASIC EQUATION:

$$R = \frac{L}{K}$$

WORKING EQUATION: Same

$$R = \frac{0.50 \text{ ft}}{0.026 \text{ Btu/ft } ^\circ\text{F h}}$$

$$= 19 \text{ ft}^2 \ ^\circ\text{F h/Btu}$$

$$\frac{\dfrac{\text{ft}}{\text{Btu}}}{\dfrac{}{\text{ft } ^\circ\text{F h}}} = \text{ft} \div \frac{\text{Btu}}{\text{ft } ^\circ\text{F h}} = \text{ft} \cdot \frac{\text{ft } ^\circ\text{F h}}{\text{Btu}} = \frac{\text{ft}^2 \ ^\circ\text{F h}}{\text{Btu}}$$

## PROBLEMS

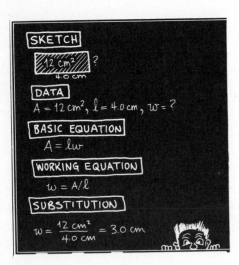

1. Find the $R$ value of a pane of 0.125-in.-thick glass.
2. Find the $R$ value of a brick wall 4.0 in. thick.
3. Find the $R$ value of 0.50-in.-thick sheetrock.
4. Find the thermal conductivity of a piece of building material 0.25 in. thick which has an $R$ value of 1.6 ft² °F h/Btu.
5. Find the $R$ value of 0.50-in. corkboard.
6. Determine the amount of heat transferred by conductivity through the walls of a building in 24 h if the $R$ value of the walls is 11 ft² °F h/Btu. Assume the dimensions of each of the building's four walls are $2\overline{0}$ ft × $10\overline{0}$ ft. Also assume that the average outer wall temperature is $2\overline{0}$°F and that the average inner wall temperature is 55°F.
7. Determine the heat flow during $3\overline{0}$ days through a glass window of thickness 0.20 in. with an area of 15 ft² if the average outer surface temperature is 25°F and the average inner glass surface temperature is $5\overline{0}$°F.
8. Determine the heat flow in $3\overline{0}$ days through a 0.25-cm-thick steel plate with a cross section of 45 cm × 75 cm. Assume a temperature differential of 95°C.
9. Determine the heat flow through a steel rod in 75 s if the length is 85 cm and the diameter is 0.50 cm. Assume that the temperature of the hot end of the rod is $11\overline{0}$°C and the temperature of the cool end is $-25$°C.
10. Determine the heat flow in 15 min through a 0.10-cm-thick copper plate with a cross-sectional area of 150 cm² if the temperature of the hot side is $99\overline{0}$°C and the temperature of the cool side is 5°C.

**14-5**

**SPECIFIC HEAT**

If we place a piece of steel and a pan of water in the direct sunlight, we would notice that the water becomes only slightly warmer while the iron gets quite hot. Why should one get so much hotter than the other? If equal masses of steel and water were placed over the same flame for 1 minute, the temperature of the steel would increase almost 10 times more than that of the water. The water has a greater capacity to absorb heat.

The specific heat of a substance is a measure of its capacity to absorb or give off heat per degree change in temperature. This property of water to absorb or give up large amounts of heat makes it an effective substance to transfer heat in industrial processes.

The specific heat of a substance is the amount of heat necessary to change the temperature of 1 lb 1°F (1 kg 1°C in the metric system). By formula,

$$c = \frac{Q}{w\Delta T} \qquad \text{(English system)}$$

or

$$c = \frac{Q}{m\Delta T} \qquad \text{(metric system)}$$

where:  $c$ = specific heat
$Q$ = heat
$w$ = weight
$m$ = mass
$\Delta T$ = change in temperature

These equations can be rearranged to give the amount of heat added or taken away from a material to produce a certain temperature change.

$$Q = cw\Delta T \qquad \text{(English)}$$
$$Q = cm\Delta T \qquad \text{(metric)}$$

A short table of specific heats follows. See also Table 14 in Appendix C.

| Substance | Specific heat[a] |
|-----------|------------------|
| Air | 0.24 |
| Aluminum | 0.22 |
| Brass | 0.091 |
| Copper | 0.093 |
| Ice | 0.51 |
| Steam | 0.48 |
| Steel | 0.115 |
| Water | 1.00 |

[a] Btu/lb °F (English system) or kcal/kg °C or cal/g °C (metric system).

### EXAMPLE

How much heat must be added to 10.0 kg of steel to raise its temperature $15\bar{0}$°C?

DATA:

$$m = 10.0 \text{ kg}$$
$$\Delta T = 15\bar{0}°C$$
$$c = 0.115 \text{ kcal/kg °C} \qquad \text{(from table)}$$
$$Q = ?$$

BASIC EQUATION:

$$Q = cm\Delta T$$

WORKING EQUATION:  Same

SUBSTITUTION:

$$Q = 0.115 \frac{\text{kcal}}{\text{kg °C}} \times 10.0 \text{ kg} \times 15\bar{0}°C$$
$$= 173 \text{ kcal}$$

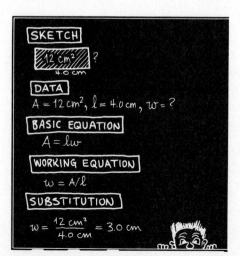

SKETCH

$12 \text{ cm}^2$  ?
$4.0 \text{ cm}$

DATA
$A = 12 \text{ cm}^2, \ell = 4.0 \text{ cm}, w = ?$

BASIC EQUATION
$A = \ell w$

WORKING EQUATION
$w = A/\ell$

SUBSTITUTION
$w = \frac{12 \text{ cm}^2}{4.0 \text{ cm}} = 3.0 \text{ cm}$

## PROBLEMS

*Find* Q *for the following materials. (1–11).*

1. Steel, $w = 3.00$ lb, $\Delta T = 50\bar{0}$°F, $Q =$ ____ Btu
2. Copper, $m = 155$ kg, $\Delta T = 17\bar{0}$°C, $Q =$ ____ kcal
3. Water, $w = 19.0$ lb, $\Delta T = 20\bar{0}$°F, $Q$ ____ Btu
4. Water, $m = 25\bar{0}$ g, $\Delta T = 17$°C, $Q =$ ____ cal
5. Ice, $m = 5.00$ kg, $\Delta T = 2\bar{0}$°C, $Q =$ ____ kcal

6. Steam, $w = 5.00$ lb, $\Delta T = 4\overline{0}°F$, $Q =$ _____ Btu
7. Aluminum, $m = 79.0$ g, $\Delta T = 16°C$, $Q =$ _____ cal
8. Brass, $m = 75\overline{0}$ kg, $\Delta T = 125°C$, $Q =$ _____ kcal
9. Steel, $m = 1250$ g, $\Delta T = 5\overline{0}°C$, $Q =$ _____ kcal
10. Aluminum, $m = 85\overline{0}$ g, $\Delta T = 115°C$, $Q =$ _____ kcal
11. Water, $m = 80\overline{0}$ g, $\Delta T = 8\overline{0}°C$, $Q =$ _____ kcal
12. How much heat must be added to $75\overline{0}$ kg of steel to raise its temperature from $75°C$ to $30\overline{0}°C$?
13. How much heat must be added to $12\overline{0}0$ lb of copper to raise its temperature from $10\overline{0}°F$ to $45\overline{0}°F$?
14. How much heat is given off by $50\overline{0}$ lb of aluminum when it cools from $65\overline{0}°F$ to $75°F$?
15. How many kcal of heat must be added to $12\overline{0}0$ kg of copper to raise its temperature from $25°C$ to $275°C$?
16. How many joules of heat are absorbed by an electric freezer in lowering the temperature of $185\overline{0}$ g of water from $8\overline{0}°C$ to $1\overline{0}°C$?
17. How many joules of heat are required to raise the temperature of $75\overline{0}$ kg of water from $15°C$ to $75°C$?
18. A 520-kg steam boiler is made of steel and contains $30\overline{0}$ kg of water at $4\overline{0}°C$. Assuming that 75% of the heat is delivered to the boiler and water, how many calories are required to raise the temperature of both the boiler and water to $10\overline{0}°C$?

**14-6
METHOD
OF MIXTURES**

When two substances at different temperatures are "mixed" together, heat flows from the warmer body to the cooler body until they reach the same temperature. Part of the heat lost by the hotter body is transferred to the colder body and part is lost to the surrounding objects or the air. In many cases most all the heat is transferred to the colder body. We will assume that all the heat lost by the warmer body is transferred to the cooler body. That is, the heat lost by the warmer body equals the heat gained by the cooler body. The amount of heat lost or gained by a body is

$$Q = cw\Delta T \quad \text{or} \quad Q = cm\Delta T$$

By formula,

$$Q_{\text{lost}} = Q_{\text{gained}}$$

$$c_l w_l (T_l - T_f) = c_g w_g (T_f - T_g)$$

where the subscript $l$ refers to the warmer body which *loses* heat, the subscript $g$ refers to the cooler body which *gains* heat, and $T_f$ is the final temperature of the mixture.

### EXAMPLE 1

A 10.0-lb piece of hot copper is dropped into 30.0 lb of water at $5\overline{0}°F$. If the final temperature of the mixture is $65°F$, what was the initial temperature of the copper?

DATA:

| | |
|---|---|
| $w_l = 10.0$ lb | $w_g = 30.0$ lb |
| $c_l = 0.093$ Btu/lb °F | $c_g = 1$ Btu/lb °F |
| $T_l = ?$ | $T_g = 5\overline{0}°F$ |
| $T_f = 65°F$ | |

BASIC EQUATION:

$$c_l w_l (T_l - T_f) = c_g w_g (T_f - T_g)$$

WORKING EQUATION:  In this type of problem, it is advisable to substitute directly into the basic equation.

SUBSTITUTION:

$$0.093 \frac{Btu}{lb \, °F} \times 10.0 \, lb(T_l - 65°F) = 1 \frac{Btu}{lb \, °F} \times 30.0 \, lb(65°F - 5\overline{0}°F)$$

$$0.93 \, T_l \, Btu/°F - 6\overline{0} \, Btu = 450 \, Btu$$

$$0.93 \, T_l \, Btu/°F = 510 \, Btu$$

$$T_l = \frac{510 \, Btu}{0.93 \, Btu/°F}$$

$$T_l = 550°F$$

EXAMPLE 2

If $20\overline{0}$ g of steel at $22\overline{0}°C$ is added to $50\overline{0}$ g of water at $10.0°C$, what is the final temperature of this mixture?

DATA:

$$c_l = 0.115 \, cal/g \, °C \qquad c_g = 1.00 \, cal/g \, °C$$
$$m_l = 20\overline{0} \, g \qquad m_g = 50\overline{0} \, g$$
$$T_l = 22\overline{0}°C \qquad T_g = 10.0°C$$
$$T_f = ?$$

BASIC EQUATION:

$$c_l m_l(T_l - T_f) = c_g m_g(T_f - T_g)$$

WORKING EQUATION:  In this type of problem, it is advisable to substitute directly into the basic equation.

SUBSTITUTION:

$$0.115 \frac{cal}{g \, °C} \times 20\overline{0} \, g(22\overline{0}°C - T_f) = 1.00 \frac{cal}{g \, °C} \times 50\overline{0} \, g(T_f - 10.0°C)$$

$$5060 \frac{cal \, °C}{°C} - 23.0 \frac{cal}{°C} T_f = 50\overline{0} \frac{cal}{°C} T_f - 50\overline{0}0 \frac{cal \, °C}{°C}$$

$$10,060 \, cal = 523 \frac{cal}{°C} T_f$$

$$\frac{10,060 \, cal}{523 \, cal/°C} = T_f$$

$$19.2°C = T_f$$

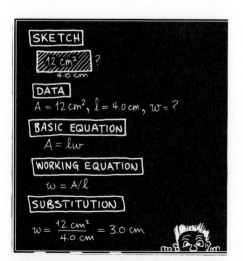

## PROBLEMS

1. A 2.50-lb piece of steel is dropped into 11.0 lb of water at 75.0°F. The final temperature is 84.0°F. What was the initial temperature of the steel?

2. Mary mixes 5.00 lb of water at $20\overline{0}°F$ with 7.00 lb of water at 65.0°F. What is the final temperature of the mixture?

3. A $25\overline{0}$-g piece of tin at 99°C is dropped in $10\overline{0}$ g of water at $1\overline{0}°C$. If the final temperature of the mixture is $2\overline{0}°C$, what is the specific heat of the tin?

4. How many grams of water at $2\overline{0}°C$ are necessary to change $80\overline{0}$ g of water at $9\overline{0}°C$ to $5\overline{0}°C$?

5. A 159-lb piece of aluminum at $50\overline{0}°F$ is dropped into $40\overline{0}$ lb of water at $6\overline{0}°F$. What is the final temperature?

6. A 42.0-lb piece of steel at $67\overline{0}°F$ is dropped into $10\overline{0}$ lb of water at 75.0°F. What is the final temperature of the mixture?

7. If 1250 g of copper at 20.0°C is mixed with $50\overline{0}$ g of water at 95.0°C, what is the final temperature of the mixture?

8. If $50\overline{0}$ g of brass at $20\overline{0}°C$ and $30\overline{0}$ g of steel at $15\overline{0}°C$ are added to $90\overline{0}$ g of water in a $15\overline{0}$-g aluminum pan and both are at $20.0°C$, what is the final temperature of this mixture assuming no loss of heat to the surroundings?

9. The following data were collected in the laboratory to determine the specific heat of an unknown metal.

| | |
|---|---|
| Mass of copper calorimeter | 153 g |
| Specific heat of calorimeter | 0.092 cal/g °C |
| Mass of water | 275 g |
| Specific heat of water | 1.00 cal/g °C |
| Mass of metal | 236 g |
| Initial temperature of water and calorimeter | 16.2°C |
| Initial temperature of metal | 99.6°C |
| Final temperature of calorimeter, water and metal | 22.7°C |

Find the specific heat of the unknown metal.

10. The following data were collected in the laboratory to determine the specific heat of an unknown metal.

| | |
|---|---|
| Mass of aluminum calorimeter | 132 g |
| Specific heat of calorimeter | $92\overline{0}$ J/kg °C |
| Mass of water | 285 g |
| Specific heat of water | 4190 J/kg °C |
| Mass of metal | 215 g |
| Initial temperature of water and calorimeter | 12.6°C |
| Initial temperature of metal | 99.1°C |
| Final temperature of calorimeter, water and metal | 18.6°C |

Find the specific heat of the unknown metal.

# CHAPTER
# *15*

## *Thermal Expansion of Solids and Liquids*

**15-1**
EXPANSION
OF SOLIDS

Most solids expand when heated and contract when cooled. They expand or contract in all three dimensions—length, width, and thickness.

When a solid is heated, the expansion is due to the increased length of the vibrations of the atoms and molecules. This results in the solid expanding in all directions. This increase in volume results in a decrease in weight density, which is discussed in Chapter 12. Engineers, technicians, and designers must know the effects of thermal expansion.

You have no doubt heard of highway pavements buckling on a hot summer day.

Bridges are built with special joints that allow for expansion and contraction of the bridge deck.

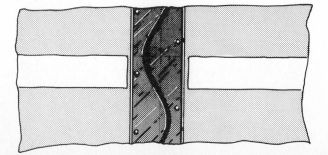

The clicking noise of a train's wheels passing over the rails can be heard more in winter than in summer. The space between rails is larger in winter than in summer. If the rails were placed snugly end to end in the winter, they would buckle in the summer.

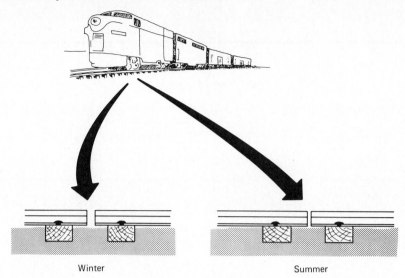

Winter              Summer

Similarly, TV towers, pipelines, and buildings must be designed and built to allow for this expansion and contraction.

There are some advantages to solids expanding. A bimetallic strip is made by fusing two different metals together side by side as illustrated. When heated, the brass expands more than the steel, which makes the strip curve as below. If the bimetallic strip is cooled below room temperature, the brass will contract more than the steel, forcing the strip to curve in the opposite direction. The thermostat operates on this principle.

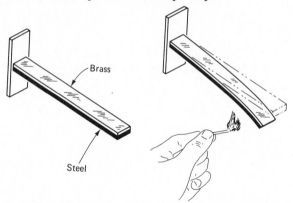

Brass

Steel

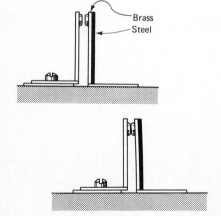

Brass
Steel

As shown at left, the basic parts of a thermostat are a bimetallic strip on the right and a regular metal strip on the left. The bimetallic strip of brass and steel bends with the temperature. The regular metal strip is moved by hand to set the temperature desired.

This particular bimetallic strip is made and placed so that it bends to the left when cooled. As a result, when it comes in contact with the strip on the left, it completes a circuit which turns on the furnace. When the room warms to the desired temperature, the bimetallic strip moves back to the right, which opens the contacts and shuts off the heat. Bimetallic strips are in spiral form in some thermostats.

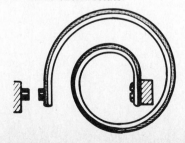

The amount that a solid expands depends on the following:

1. The material—different materials expand at different rates. Steel expands at a rate less than that of brass.
2. The length of the solid—the longer the solid, the larger the expansion. A 20.0-in. steel rod will expand twice as much as a 10.0-in. steel rod.
3. The amount of change in temperature—the greater the change in temperature, the greater the expansion.

This can be written by formula:

$$\Delta l = \alpha l \, \Delta T$$

where:   $\Delta l$ = change in length
$\alpha$ = a constant called the *coefficient of linear expansion**
$l$ = original length or any other linear measurement
$\Delta T$ = change in temperature

The following table lists the coefficients of linear expansion for some common solids.

| Material | $\alpha$ (English) | $\alpha$ (metric) |
|---|---|---|
| Aluminum | $1.3 \times 10^{-5}/F°$ | $2.3 \times 10^{-5}/C°$ |
| Brass | $1.0 \times 10^{-5}/F°$ | $1.9 \times 10^{-5}/C°$ |
| Concrete | $6.0 \times 10^{-6}/F°$ | $1.1 \times 10^{-5}/C°$ |
| Copper | $9.5 \times 10^{-6}/F°$ | $1.7 \times 10^{-5}/C°$ |
| Glass | $5.1 \times 10^{-6}/F°$ | $9.0 \times 10^{-6}/C°$ |
| Pyrex | $1.7 \times 10^{-6}/F°$ | $4.0 \times 10^{-6}/C°$ |
| Steel | $6.5 \times 10^{-6}/F°$ | $1.3 \times 10^{-5}/C°$ |
| Zinc | $1.5 \times 10^{-5}/F°$ | $2.6 \times 10^{-5}/C°$ |

Comparing the coefficients of linear expansion of common glass and Pyrex, we can see that Pyrex expands and contracts approximately one-third as much as glass. This is why it is used in cooking and chemical laboratories.

EXAMPLE 1

A steel railroad rail is 40.0 ft long at 0°F. How much will it expand when heated to $10\bar{0}°F$?

DATA:

$$l = 40.0 \text{ ft}$$
$$\Delta T = 10\bar{0}°F - 0°F = 10\bar{0}°F$$
$$\alpha = 6.5 \times 10^{-6}/F°$$
$$\Delta l = ?$$

BASIC EQUATION:

$$\Delta l = \alpha l \, \Delta T$$

WORKING EQUATION:   Same

SUBSTITUTION:

$$\Delta l = (6.5 \times 10^{-6}/\cancel{F}°)(40.0 \text{ ft})(10\bar{0}°\cancel{F})$$
$$= 0.026 \text{ ft or } 0.31 \text{ in.}$$

Pipes that undergo large temperature changes are usually installed as shown to allow for expansion and contraction.

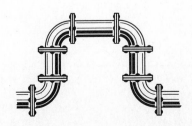

* Defined as the change in the unit length of a solid when its temperature is changed 1 degree.

EXAMPLE 2

What allowance for expansion must be made for a steel pipe $12\bar{0}$ m long that handles coolants and must undergo temperature changes of $20\bar{0}°C$?

DATA:

$$\alpha = 1.3 \times 10^{-5}/C°$$
$$l = 12\bar{0} \text{ m}$$
$$\Delta T = 20\bar{0}°C$$
$$\Delta l = ?$$

BASIC EQUATION:

$$\Delta l = \alpha l \Delta T$$

WORKING EQUATION: Same

SUBSTITUTION:

$$\Delta l = (1.3 \times 10^{-5}/\cancel{C°})(12\bar{0} \text{ m})(20\bar{0}°\cancel{C})$$
$$= 0.31 \text{ m} \quad \text{or} \quad 31 \text{ cm}$$

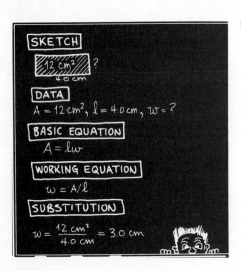

## PROBLEMS

1. Find the increase in length of copper tubing $20\bar{0}$ ft long at $4\bar{0}°F$ when it is heated to $20\bar{0}°F$.

2. Find the increase in length of a zinc rod 50.0 m long at 15°C when it is heated to $13\bar{0}°C$.

3. Compute the increase in length of 300.00 m of copper wire when its temperature changes from 14°C to 34°C.

4. A steel pipe 25.0 ft long is installed at $8\bar{0}°F$. Find the decrease in length when coolants at $-9\bar{0}°F$ pass through the pipe.

5. A steel tape measures 200.00 m at $1\bar{0}°C$. What is its length at $5\bar{0}°C$?

6. A brass rod 40.000 in. long expands 0.016 in. when it is heated. Find the temperature change.

7. The road bed on a bridge $50\bar{0}$ ft long is made of concrete. What allowance is needed for temperatures of $-4\bar{0}°F$ in winter and $14\bar{0}°F$ in summer?

8. An aluminum plug has a diameter of 10.003 cm at 40.0°C. At what temperature will it fit precisely into a hole of constant diameter 10.000 cm?

9. The diameter of a steel drill at $5\bar{0}°F$ is 0.750 in. Find its diameter at $35\bar{0}°F$.

10. A brass ball has a diameter 12.000 cm and is 0.011 cm too large to pass through a hole in a copper plate when the ball and plate are at a temperature of $2\bar{0}°C$. What is the temperature of the ball when it will just pass through the plate, assuming that the temperature of the plate does not change? What is the temperature of the plate when the ball will just pass through, assuming that the temperature of the ball does not change?

**15-3
AREA
AND VOLUME EXPANSION
OF SOLIDS**

Solids expand in width and thickness as well as in length when heated. The area of a hole cutout of a metal sheet will expand in the same way as the surrounding material. To allow for this expansion the following formulas are used:

$$\text{area expansion: } \Delta A = 2\alpha A \Delta T$$
$$\text{volume expansion: } \Delta V = 3\alpha V \Delta T$$

EXAMPLE

The top of a circular copper disc has an area of 64.2 in² at $2\bar{0}°F$. What is the change in area when the temperature is increased to $15\bar{0}°F$?

DATA:

$$\alpha = 9.5 \times 10^{-6}/\text{F}°$$
$$A = 64.2 \text{ in}^2$$
$$\Delta T = 15\bar{0}°\text{F} - 2\bar{0}°\text{F} = 13\bar{0}°\text{F}$$
$$\Delta A = ?$$

BASIC EQUATION:

$$\Delta A = 2\alpha A \, \Delta T$$

WORKING EQUATION:  Same

SUBSTITUTION:

$$\Delta A = 2(9.5 \times 10^{-6}/\cancel{\text{F}}°)(64.2 \text{ in}^2)(13\bar{0}°\cancel{\text{F}})$$
$$= 0.16 \text{ in}^2$$

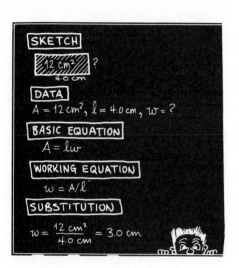

## PROBLEMS

1. A brass cylinder has a cross-sectional area of 74.8 in² at 1$\bar{0}$°F. Find its change in area when heated to 23$\bar{0}$°F.

2. The volume of the cylinder in Problem 1 is 237 in³ at 24$\bar{0}$°F. Find its change in volume when cooled to −75°F.

3. An aluminum pipe has a cross-sectional area of 88.4 cm² at 1$\bar{0}$°C. What is its cross-sectional area when heated to 15$\bar{0}$°C?

4. A steel pipe has a cross-sectional area of 27.2 in² at 3$\bar{0}$°F. What is its cross-sectional area when heated to 18$\bar{0}$°F?

5. A glass plug has a volume of 60.00 cm³ at 12°C. What is its volume at 76°C?

6. A section of concrete dam is a rectangular solid 20.0 ft by 50.0 ft by 80.0 ft at 11$\bar{0}$°F. What allowance for change in volume is necessary for a temperature of −2$\bar{0}$°F?

**15-4
EXPANSION
OF LIQUIDS**

Liquids also generally expand when heated and contract when cooled. The thermometer is made using this principle.

When a thermometer is placed under your tongue, the heat from your mouth causes the mercury in the bottom of the thermometer to expand. Mercury is then forced to rise up the thin calibrated tube. The formula for *volume expansion of liquids* is

$$\boxed{\Delta V = \beta V \Delta T}$$

where $\beta$ is the coefficient of volume expansion for liquids.

The following table lists the coefficients of volume expansion for some common liquids.

| Liquid | $\beta$ (English) | $\beta$ (metric) |
|---|---|---|
| Acetone | $8.28 \times 10^{-4}/\text{F}°$ | $1.49 \times 10^{-3}/\text{C}°$ |
| Alcohol, ethyl | $6.62 \times 10^{-4}/\text{F}°$ | $1.12 \times 10^{-3}/\text{C}°$ |
| Carbon tetrachloride | $6.89 \times 10^{-4}/\text{F}°$ | $1.24 \times 10^{-3}/\text{C}°$ |
| Mercury | $1.0 \times 10^{-4}/\text{F}°$ | $1.8 \times 10^{-4}/\text{C}°$ |
| Petroleum | $5.33 \times 10^{-4}/\text{F}°$ | $9.6 \times 10^{-4}/\text{C}°$ |
| Turpentine | $5.39 \times 10^{-4}/\text{F}°$ | $9.7 \times 10^{-4}/\text{C}°$ |
| Water | $1.17 \times 10^{-4}/\text{F}°$ | $2.1 \times 10^{-4}/\text{C}°$ |

## EXAMPLE 1

If petroleum at 0°C occupies $25\overline{0}$ L, what is its volume at $5\overline{0}$°C?

DATA:

$$\beta = 9.6 \times 10^{-4}/C°$$
$$V = 25\overline{0} \text{ L}$$
$$\Delta T = 5\overline{0}°C$$
$$\Delta V = ?$$

BASIC EQUATION:

$$\Delta V = \beta V \Delta T$$

WORKING EQUATION:   Same

SUBSTITUTION:

$$\Delta V = (9.6 \times 10^{-4}/\cancel{C}°)(25\overline{0} \text{ L})(5\overline{0}°\cancel{C})$$
$$= 12 \text{ L}$$
$$\text{volume at } 5\overline{0}°C = 25\overline{0} \text{ L} + 12 \text{ L} = 262 \text{ L}$$

## EXAMPLE 2

What is the increase in volume of 18.2 in³ of water when heated from $4\overline{0}$°F to $18\overline{0}$°F?

DATA:

$$\beta = 1.17 \times 10^{-4}/F°$$
$$V = 18.2 \text{ in}^3$$
$$\Delta T = 18\overline{0}°F - 4\overline{0}°F = 14\overline{0}°F$$
$$\Delta V = ?$$

BASIC EQUATION:

$$\Delta V = \beta V \Delta T$$

WORKING EQUATION:   Same

SUBSTITUTION:

$$\Delta V = (1.17 \times 10^{-4}/\cancel{F}°)(18.2 \text{ in}^3)(14\overline{0}°\cancel{F})$$
$$= 0.298 \text{ in}^3$$

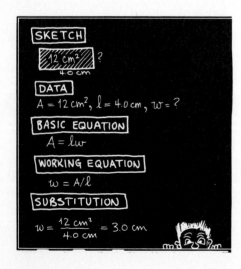

## PROBLEMS

1. A quantity of carbon tetrachloride occupies 625 L at 12°C. What is its volume at 48°C?

2. Some mercury occupies 157 in³ at $-3\overline{0}$°F. What is the change in volume when heated to $9\overline{0}$°F?

3. Some petroleum occupies 287 ft³ at 0°F. What volume does it occupy at $8\overline{0}$°F?

4. What is the increase in volume of 35 L of acetone when heated from 28°C to 38°C?

5. Some water at $18\overline{0}$°F occupies $378\overline{0}$ ft³. What is its volume at 122°F?

6. A $120\overline{0}$-L tank of petroleum is completely filled at $1\overline{0}$°C. How much spills over if the temperature rises to 45°C?

7. Calculate the increase in volume of 215 cm³ of mercury when its temperature increases from $1\overline{0}$°C to 25°C.

8. What is the drop in volume of alcohol in a railroad tank car that contains $200\overline{0}$ ft³ if the temperature drops from 75°F to 54°F?

9. A gasoline service station owner receives a truckload of $33,00\overline{0}$ L of gasoline which is at 32°C. It cools to 15°C in the underground tank. At 40 cents/L, how much money is lost as a result of the contraction of the gasoline?

Water is unusual in its expansion characteristics. Most of us have seen the mound in the middle of each ice cube in ice cube trays. This evidence shows the expansion of water during its change of state from liquid to solid form. Nearly all liquids are densest at their lowest temperature before a change of state to become solids. As the temperature drops, the molecular motion slows and the substance becomes denser.

Water does not follow the general rule stated above. Because of its unusual structural characteristics, water is densest at 4°C or 39.2°F instead of 0°C or 32°F. A graph of its change in density with an increase in temperature follows. As ice melts and the water temperature is slightly increased, there are still groups of molecules that have the open crystallographic structure of ice, which is less dense than water. As the water is heated to 4°C, these groupings disappear and the water becomes denser. Above 4°C, water then expands normally and the temperature is raised.

When ice melts at 0°C or 32°F, the water formed *contracts* as the temperature is raised to 4°C or 39.2°F. Then it begins to *expand,* as do most other liquids.

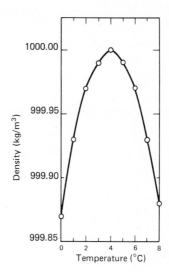

# Gas Laws

**16-1
CHARLES' LAW**

Suppose before making a long trip in the summer you go to your car and notice that the tires are low. So you stop at a gas station around the corner and have the attendant add air to 28.0-lb/in² gauge pressure. Then later in the afternoon you need to stop for gas. Since your tires were low that morning, you decide to have them checked again. Now you notice that they look a bit larger and the gauge pressure is 40.0 lb/in². What happened?

When a gas is heated, the increased kinetic energy causes the volume to increase, the pressure to increase, or both to increase.

---

*CHARLES' LAW*

If the pressure on a gas is constant, the volume is directly proportional* to its Kelvin or Rankine temperature.

---

By formula,

$$\frac{V}{T} = \frac{V'}{T'} \quad \text{or} \quad VT' = V'T$$

where:  $V =$ original volume
 $T =$ original temperature
 $V' =$ final volume
 $T' =$ final temperature

## EXAMPLE 1

A gas occupies $45\overline{0}$ cm³ at $3\overline{0}$°C. At what temperature will the gas occupy $48\overline{0}$ cm³?

---

* Directly proportional means that as temperature increases, volume increases, and as temperature decreases, volume decreases.

**225**

*Gas Laws*

DATA:

$$V = 45\overline{0} \text{ cm}^3$$

$$T = 3\overline{0}° + 273 = 303 \text{ K} \qquad (\textit{Note}: \text{We must use Kelvin temperature})$$

$$V' = 48\overline{0} \text{ cm}^3$$

$$T' = ?$$

BASIC EQUATION:

$$\frac{V}{T} = \frac{V'}{T'}$$

WORKING EQUATION:

$$T' = \frac{TV'}{V}$$

SUBSTITUTION:

$$T' = \frac{(303 \text{ K})(48\overline{0} \text{ cm}^3)}{45\overline{0} \text{ cm}^3}$$

$$= 323 \text{ K} \quad \text{or} \quad 5\overline{0}°\text{C}$$

## EXAMPLE 2

At $4\overline{0}°$F, a gas occupies 15.0 ft³. What will be the volume of this gas at $9\overline{0}°$F?

DATA:

$$V = 15.0 \text{ ft}^3$$

$$T = 4\overline{0}° + 46\overline{0}° = 50\overline{0}°\text{R} \qquad (\textit{Note}: \text{We must use Rankine temperature})$$

$$T' = 9\overline{0}° + 46\overline{0}° = 55\overline{0}°\text{R}$$

$$V' = ?$$

BASIC EQUATION:

$$\frac{V}{T} = \frac{V'}{T'}$$

WORKING EQUATION:

$$V' = \frac{VT'}{T}$$

SUBSTITUTION:

$$V' = \frac{(15.0 \text{ ft}^3)(55\overline{0}°\text{R})}{50\overline{0}°\text{R}}$$

$$= 16.5 \text{ ft}^3$$

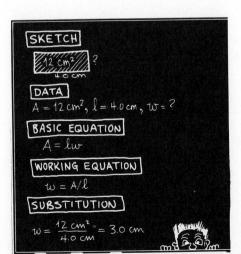

## PROBLEMS

1. Change 15°C to K.
2. Change −13°C to K.
3. Change 317 K to °C.
4. Change 235 K to °C.
5. Change 72°F to °R.
6. Change −5$\overline{0}$°F to °R.
7. Change 55$\overline{0}$°R to °F.
8. Change 375°R to °F.

*Use $V/T = V'/T'$ to find each of the following:*

9. $T = 30\overline{0}$ K, $V' = 20\overline{0}$ cm³, $T' = 27\overline{0}$ K, find $V$.
10. $T = 60\overline{0}$°R, $V = 60.3$ in³, $T' = 45\overline{0}$°R, find $V'$.

11. $V = 20\overline{0}$ ft³, $T' = 95°F$, $V' = 25\overline{0}$ ft³, find $T$.

12. $V = 19.7$ L, $T = 51°C$, $V' = 25.2$ L, find $T'$.

13. Some gas occupies a volume of 325 m³ at 41°C. What is its volume at 94°C?

14. Some oxygen occupies 275 in³ at $3\overline{0}°F$. What is its volume at 95°F?

15. Some methane occupies $150\overline{0}$ L at 45°C. What is its volume at $2\overline{0}°C$?

16. Some helium occupies $120\overline{0}$ ft³ at $7\overline{0}°F$. At what temperature will its volume be $60\overline{0}$ ft³?

**16-2**
**BOYLE'S LAW**

---

*BOYLE'S LAW*

If the temperature on a gas is constant, the volume is inversely proportional* to the pressure.

---

By formula,

$$\frac{V}{V'} = \frac{P'}{P} \quad \text{or} \quad VP = V'P'$$

where:  $V =$ original volume
$V' =$ final volume
$P =$ original pressure
$P' =$ final pressure

*Note*: The pressure must be expressed in terms of *absolute pressure*.

## EXAMPLE 1

Some oxygen occupies $50\overline{0}$ in³ at a pressure of 40.0 lb/in² (psi). What is its volume at a pressure of $10\overline{0}$ psi?

DATA:

$$V = 50\overline{0} \text{ in}^3$$
$$P = 40.0 \text{ psi}$$
$$P' = 10\overline{0} \text{ psi}$$
$$V' = ?$$

BASIC EQUATION:

$$\frac{V}{V'} = \frac{P'}{P}$$

WORKING EQUATION:

$$V' = \frac{VP}{P'}$$

SUBSTITUTION:

$$V' = \frac{(50\overline{0} \text{ in}^3)(40.0 \text{ psi})}{10\overline{0} \text{ psi}}$$
$$= 20\overline{0} \text{ in}^3$$

## EXAMPLE 2

Some nitrogen occupies 20.0 m³ at a pressure of $30\overline{0}$ kPa. What is its volume at $20\overline{0}$ kPa?

---

* Inversely proportional means that as volume increases, pressure decreases, and as volume decreases, pressure increases.

DATA:

$$V = 20.0 \text{ m}^3$$
$$P = 30\overline{0} \text{ kPa}$$
$$P' = 20\overline{0} \text{ kPa}$$
$$V' = ?$$

BASIC EQUATION:

$$\frac{V}{V'} = \frac{P'}{P}$$

WORKING EQUATION:

$$V' = \frac{VP}{P'}$$

SUBSTITUTION:

$$V' = \frac{(20.0 \text{ m}^3)(30\overline{0} \text{ kPa})}{(20\overline{0} \text{ kPa})}$$
$$= 30.0 \text{ m}^3$$

**16-3**
**DENSITY**
**AND PRESSURE**

If the pressure of a given amount of gas is increased, its density increases as the gas molecules are forced closer together. Also, if the pressure is decreased, the density decreases. That is, the density of a gas is directly proportional to its pressure. In equation form,

$$\boxed{\frac{D}{D'} = \frac{P}{P'} \quad \text{or} \quad DP' = D'P}$$

where:  $D$ = original density
$D'$ = final density
$P$ = original pressure (absolute)
$P'$ = final pressure (absolute)

EXAMPLE 1

A given gas has a density of 1.60 kg/m³ at a pressure of 95.0 kPa. What is the density when the pressure is decreased to 80.0 kPa?

DATA:

$$D = 1.60 \text{ kg/m}^3$$
$$P = 95.0 \text{ kPa}$$
$$P' = 80.0 \text{ kPa}$$
$$D' = ?$$

BASIC EQUATION:

$$\frac{D}{D'} = \frac{P}{P'}$$

WORKING EQUATION:

$$D' = \frac{DP'}{P}$$

SUBSTITUTION:

$$D' = \frac{(1.60 \text{ kg/m}^3)(80.0 \text{ kPa})}{95.0 \text{ kPa}}$$
$$= 1.35 \text{ kg/m}^3$$

EXAMPLE 2

A gas has a density of 2.00 kg/m³ at a gauge pressure of 16$\overline{0}$ kPa. What is the density at a gauge pressure of 30$\overline{0}$ kPa?

DATA:

$$D = 2.00 \text{ kg/m}^3$$
$$P = 16\overline{0} \text{ kPa} + 101 \text{ kPa} = 261 \text{ kPa}$$
$$P' = 30\overline{0} \text{ kPa} + 101 \text{ kPa} = 401 \text{ kPa}$$
$$D = ?$$

BASIC EQUATION:

$$\frac{D}{D'} = \frac{P}{P'}$$

WORKING EQUATION:

$$D' = \frac{DP'}{P}$$

SUBSTITUTION:

$$D' = \frac{(2.00 \text{ kg/m}^3)(401 \text{ kPa})}{261 \text{ kPa}}$$
$$= 3.07 \text{ kg/m}^3$$

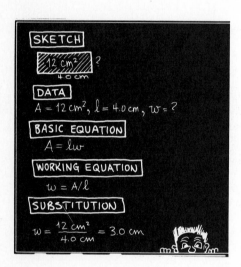

## PROBLEMS

*Use $V/V' = P'/P$ or $D/D' = P/P'$ to find each of the following. (All pressures are absolute unless otherwise stated.)*

1. $V' = 30\overline{0}$ cm³, $P = 101$ kPa, $P' = 85.0$ kPa; find $V$.
2. $V = 45\overline{0}$ L, $V' = 70\overline{0}$ L, $P = 75\overline{0}$ kPa; find $P'$.
3. $V = 76.0$ m³, $V' = 139$ m³, $P' = 41.0$ kPa; find $P$.
4. $V = 439$ in³, $P' = 38.7$ psi, $P = 47.1$ psi; find $V'$.
5. $D = 1.80$ kg/m³, $P = 108$ kPa, $P' = 125$ kPa; find $D'$.
6. $D = 1.65$ kg/m³, $P = 87.0$ kPa, $D' = 1.85$ kg/m³; find $P'$.
7. $P = 51.0$ psi, $P' = 65.3$ psi, $D' = 0.231$ lb/ft³; find $D$.
8. Some air at 22.5 psi occupies 140$\overline{0}$ in³. What is its volume at 18.0 psi?
9. Some gas at a pressure of 110.0 kPa occupies 185 m³. What is its pressure if its volume is changed to 225 m³?
10. Some gas at 185.0 kPa occupies 65.0 L. What is its volume at a pressure of 95.0 kPa?
11. Some gas has a density of 3.75 kg/m³ at 81$\overline{0}$ kPa. What is its density if the pressure is decreased to 72$\overline{0}$ kPa?
12. A gas has a density of 1.75 kg/m³ at normal atmospheric pressure. What is its pressure (in kPa) when the density is changed to 1.45 kg/m³?
13. Some methane at 50$\overline{0}$ kPa gauge pressure occupies 75$\overline{0}$ m³. What is its gauge pressure if its volume is 50$\overline{0}$ m³?
14. Some helium at 15.0 psi gauge pressure occupies 20.0 ft³. What is its volume at 20.0 psi gauge pressure?
15. Some nitrogen at 80.0 psi gauge pressure occupies 13.0 ft³. What is its volume at 50.0 psi gauge pressure?
16. A gas at 30$\overline{0}$ kPa occupies 40.0 m³. What is its pressure if its volume is doubled? tripled? halved?

**16-4**
**CHARLES'
AND BOYLE'S LAWS
COMBINED**

Most of the time it is very difficult to keep the pressure constant or the temperature constant. In this case we combine Charles' law and Boyle's law as follows:

$$\frac{VP}{T} = \frac{V'P'}{T'} \quad \text{or} \quad VPT' = V'P'T$$

## EXAMPLE

Assume we have $50\bar{0}$ in³ of acetylene at $4\bar{0}°$F at $200\bar{0}$ psi. What is the pressure if its volume is changed to $80\bar{0}$ in³ at $10\bar{0}°$F?

DATA:

$$V = 50\bar{0} \text{ in}^3$$
$$P = 200\bar{0} \text{ psi}$$
$$T = 4\bar{0}° + 46\bar{0}° = 50\bar{0}°\text{R}$$
$$V' = 80\bar{0} \text{ in}^3$$
$$P' = ?$$
$$T' = 10\bar{0}° + 46\bar{0}° = 56\bar{0}°\text{R}$$

BASIC EQUATION:

$$\frac{VP}{T} = \frac{V'P'}{T'}$$

WORKING EQUATION:

$$P' = \frac{VPT'}{TV'}$$

SUBSTITUTION:

$$P' = \frac{(50\bar{0} \text{ in}^3)(200\bar{0} \text{ psi})(56\bar{0}°\text{R})}{(50\bar{0}°\text{R})(80\bar{0} \text{ in}^3)}$$
$$= 140\bar{0} \text{ psi}$$

A commonly used reference in gas laws is called *standard temperature and pressure* (STP). Standard temperature is the freezing point of water, 0°C or 32°F. Standard pressure is equivalent to atmospheric pressure, 101.32 kPa or 14.7 lb/in².

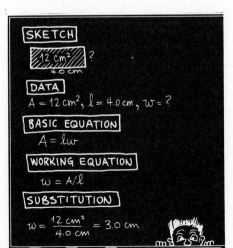

## PROBLEMS

*Use VP/T = V' P' /T' to find each of the following. (All pressures are absolute unless otherwise stated.)*

1. $P = 80\bar{0}$ psi, $T_R = 57\bar{0}°$, $V' = 150\bar{0}$ in³, $P' = 60\bar{0}$ psi, $T'_R = 50\bar{0}°$; find $V$.
2. $V = 50\bar{0}$ in³, $T_R = 50\bar{0}°$, $V' = 80\bar{0}$ in³, $P' = 80\bar{0}$ psi, $T'_R = 45\bar{0}°$; find $P$.
3. $V = 90\bar{0}$ m³, $P = 105$ kPa, $T = 30\bar{0}$ K, $P' = 165$ kPa, $T' = 265$ K; find $V'$.
4. $V = 18.0$ m³, $P = 112$ kPa, $V' = 15.0$ m³, $P' = 135$ kPa, $T' = 235$ K; find $T$.
5. $V = 532$ m³, $P = 135$ kPa, $T = 87°$C, $V' = 379$ m³, $P' = 123$ kPa; find $T'$.
6. We have $60\bar{0}$ in³ of oxygen at $150\bar{0}$ psi at 65°F. What is the volume at $120\bar{0}$ psi at 9\bar{0}°F?
7. We have $80\bar{0}$ m³ of natural gas at 235 kPa at $3\bar{0}°$C. What is the temperature if the volume is changed to $120\bar{0}$ m³ at 215 kPa?

8. We have $140\overline{0}$ L of nitrogen at 135 kPa at 54°C. What is the temperature if the volume changes to $80\overline{0}$ L at 275 kPa?

9. An acetylene welding tank has a pressure of $20\overline{0}0$ psi at $4\overline{0}$°F. If the temperature rises to $9\overline{0}$°F, what is the new pressure?

10. What is the new pressure in Problem 9 if the temperature falls to $-3\overline{0}$°F?

11. An ideal gas occupies a volume of 5.00 L at STP. What is its gauge pressure (in kPa) if the volume is halved and its temperature is increased to $40\overline{0}$°C?

12. An ideal gas occupies a volume of 5.00 L at STP.
    (a) What is its temperature if its volume is halved and its absolute pressure is doubled?
    (b) What is its temperature if its volume is doubled and its absolute pressure is tripled?

# CHAPTER 17

## Change of State

**17-1**
**FUSION**
Many industries are concerned with a change of state in the materials they use. In foundries the principal activity is to change the state of solid metals to liquid, pour the liquid metal into molds, and allow it to become solid again.

This first change of state from solid to liquid is called *melting* or *fusion*. The change from the liquid to solid state is called *freezing* or *solidification*. Most solids have a crystal structure and a definite melting point at any given pressure. Fusion and solidification of these substances occur at the same temperature. For example, water at 32° Fahrenheit (0° Celsius) changes to ice and ice changes to water at the same temperature. There is no temperature change during change of state. Ice at 32°F changes to water at 32°F. Only a few substances, such as butter and glass, have no particular melting temperature but change state gradually.

Although there is no temperature change during a change of state, *there is a transfer of heat.* A melting solid *absorbs* heat and a solidifying liquid *gives off* heat. When 1 g of ice at 0°C melts, it absorbs $8\bar{0}$ cal of heat. Similarly, when 1 g of water freezes at 0°C, ice at 0°C is produced and $8\bar{0}$ cal of heat is released.

When 1 kg of ice at 0°C melts, it absorbs $8\bar{0}$ kcal of heat. And similarly, when 1 kg of water freezes at 0°C, ice at 0°C is produced and $8\bar{0}$ kcal of heat is released.

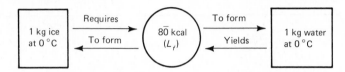

Or, when 1 kg of ice at 0°C melts, it absorbs 335 kilojoules (kJ) of heat. Then, when 1 kg of water freezes at 0°C, ice at 0°C is produced and 335 kJ of heat is released.

When 1 lb of ice at 32°F melts, it absorbs 144 Btu of heat. Similarly, when 1 lb of water freezes at 32°F, ice at 32°F is produced and 144 Btu of heat is released.

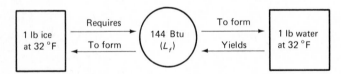

The amount of heat required to melt 1 g or 1 kg or 1 lb of a liquid is called its *heat of fusion,* designated $L_f$.

$$L_f = \frac{Q}{w} \quad \text{(English system)} \qquad L_f = \frac{Q}{m} \quad \text{(metric system)}$$

where:   $L_f$ = heat of fusion
   $Q$ = quantity of heat
   $w$ = weight of substance (English system)
   $m$ = mass of substance (metric system)

### EXAMPLE

If 864 Btu of heat is required to melt 6.00 lb of ice at 32°F into water at 32°F, what is the heat of fusion of water?

DATA:

$$Q = 864 \text{ Btu}$$
$$w = 6.00 \text{ lb}$$
$$L_f = ?$$

BASIC EQUATION:

$$L_f = \frac{Q}{w}$$

WORKING EQUATION: Same

SUBSTITUTION:

$$L_f = \frac{864 \text{ Btu}}{6.00 \text{ lb}}$$

$$= 144 \text{ Btu/lb}$$

| heat of fusion (water) = $8\bar{0}$ cal/g or $8\bar{0}$ kcal/kg or 335 kJ/kg or 144 Btu/lb |
| --- |

**17-2**
**VAPORIZATION**

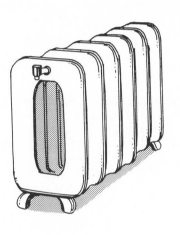

Steam heating systems in homes and factories are important applications of the principles of change of state from liquid to the gaseous or vapor state. This change of state is called *vaporization*. Boiling water shows this change of state. The reverse process (change from gas to liquid) is called *condensation*. As steam condenses in radiators, large amounts of heat are released.

While a liquid is boiling, the temperature of the liquid does not change. There is, however, a transfer of heat. A liquid being vaporized (boiling) *absorbs* heat. As a vapor condenses, heat is given off.

The amount of heat required to vaporize 1 g or 1 kg or 1 lb of a liquid is called its *heat of vaporization,* designated $L_v$. So, when 1 g of water at $10\bar{0}°$C changes to steam at $10\bar{0}°$C, it absorbs $54\bar{0}$ cal; when 1 g of steam at $10\bar{0}°$C condenses to water at $10\bar{0}°$C, $54\bar{0}$ cal of heat is given off.

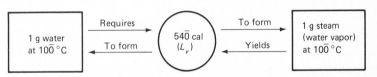

When 1 kg of water at $10\bar{0}°$C changes to steam at $10\bar{0}°$C, it absorbs $54\bar{0}$ kcal of heat. Similarly, when 1 kg of steam at $10\bar{0}°$C condenses to water at $10\bar{0}°$C, $54\bar{0}$ kcal of heat is given off.

Or, when 1 kg of water at $10\bar{0}°$C changes to steam at $10\bar{0}°$C, it absorbs 2.26 MJ ($2.26 \times 10^6$J) of heat. Then, when 1 kg of steam at $10\bar{0}°$C condenses to water at $10\bar{0}°$C, 2.26 MJ of heat is given off.

When 1 lb of water at 212°F changes to steam at 212°F, $97\bar{0}$ Btu of heat is absorbed; when 1 lb of steam at 212°F condenses to water at 212°F, $97\bar{0}$ Btu of heat is given off.

$$L_v = \frac{Q}{w} \quad \text{(English system)} \qquad L_v = \frac{Q}{m} \quad \text{(metric system)}$$

where: $L_v$ = heat of vaporization
$Q$ = quantity of heat
$w$ = weight of substance (English system)
$m$ = mass of substance (metric system)

## EXAMPLE 1

If 135,000 cal of heat is required to vaporize $25\bar{0}$ g of water at $10\bar{0}°C$, what is the heat of vaporization of water?

DATA:

$$Q = 135,000 \text{ cal}$$
$$m = 25\bar{0} \text{ g}$$
$$L_v = ?$$

BASIC EQUATION:

$$L_v = \frac{Q}{m}$$

WORKING EQUATION: Same

SUBSTITUTION:

$$L_v = \frac{135,000 \text{ cal}}{25\bar{0} \text{ g}}$$
$$= 54\bar{0} \text{ cal/g}$$

heat of vaporization (water) = $54\bar{0}$ cal/g or $54\bar{0}$ kcal/kg or 2.26 MJ/kg or $97\bar{0}$ Btu/lb

We may now calculate the amount of heat released when a given quantity of steam is cooled through the change of state. We will use the heat of vaporization and the method of mixtures studied in Chapter 14. From Chapter 14, heat lost during cooling is given by $Q = cw \, \Delta T$. The total amount of heat released by changing steam to water equals the amount of heat lost by the steam during cooling. ($c_{steam} w_{steam} \Delta T_{steam}$), plus the quantity of heat lost during condensation ($L_v w_{steam}$), plus the heat lost by the water during cooling ($c_{water} w_{water} \Delta T_{water}$). Therefore,

$$Q = (c_{steam})(w_{steam})(\Delta T_{steam}) + L_v(w_{steam}) + (c_{water})(w_{water})(\Delta T_{water})$$

## EXAMPLE 2

How many Btu of heat is released when 4.00 lb of steam at 222°F is cooled to water at 82°F?

DATA:

$$w = 4.00 \text{ lb}$$
$$T_i \text{ of steam} = 222°F$$
$$T_f \text{ of water} = 82°F$$
$$Q = ?$$

BASIC EQUATION:

$$Q = (c_{steam})(w_{steam})(\Delta T_{steam}) + L_v(w_{steam}) + (c_{water})(w_{water})(\Delta T_{water})$$

WORKING EQUATION:   Same

SUBSTITUTION:

$$Q = \left(0.48 \frac{Btu}{lb\cdot{}^\circ F}\right)(4.00 \text{ lb})(1\bar{0}\,{}^\circ F) + \left(97\bar{0} \frac{Btu}{lb}\right)(4.00 \text{ lb})$$

$$+ \left(1 \frac{Btu}{lb\cdot{}^\circ F}\right)(4.00 \text{ lb})(13\bar{0}\,{}^\circ F)$$

$$= 4420 \text{ Btu}$$

Note that during changes of state there are no temperature changes.

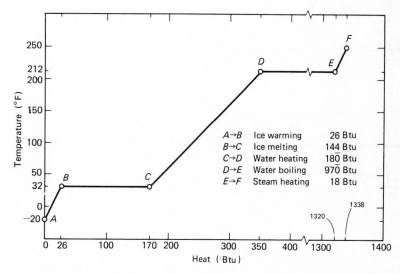

Heat gained by one pound of ice at $-2\bar{0}\,{}^\circ F$ as it is converted to steam at $25\bar{0}\,{}^\circ F$.

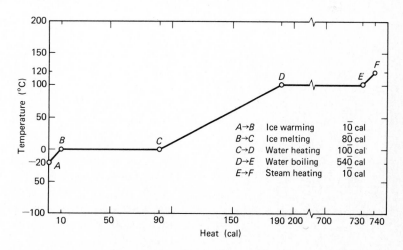

Heat gained by one gram of ice at $-2\bar{0}\,{}^\circ C$ as it is converted to steam at $12\bar{0}\,{}^\circ C$.

The table at the top of page 236 shows some common substances and their heat characteristics.

| | Melting point (°C) | Boiling point (°C) | Specific heat | | Heat of fusion | | Heat of vaporization | |
|---|---|---|---|---|---|---|---|---|
| | | | (cal/g °C or kcal/kg °C or Btu/lb °F) | (J/kg °C) | (cal/g or kcal/kg) | (J/kg) | (cal/g or kcal/kg) | (J/kg) |
| Alcohol, ethyl | −117 | 78.5 | 0.58 | 2400 | 24.9 | $1.04 \times 10^5$ | 204 | $8.54 \times 10^5$ |
| Aluminum | 660 | 2057 | 0.22 | 920 | 76.8 | $3.21 \times 10^5$ | | |
| Brass | 840 | | 0.091 | 390 | | | | |
| Copper | 1083 | 2330 | 0.093 | 390 | 49.0 | $2.05 \times 10^5$ | | |
| Glass | | | 0.21 | 880 | | | | |
| Ice | 0 | | 0.51 | 2100 | $8\bar{0}$ | $3.35 \times 10^5$ | | |
| Iron (steel) | 1540 | 3000 | 0.115 | 481 | 7.89 | $3.30 \times 10^4$ | | |
| Lead | 327 | 1620 | 0.031 | 130 | 5.86 | $2.45 \times 10^4$ | | |
| Mercury | −38.9 | 357 | 0.033 | 140 | 2.82 | $1.18 \times 10^4$ | 65.0 | $2.72 \times 10^5$ |
| Silver | 961 | 1950 | 0.056 | 230 | 26.0 | $1.09 \times 10^5$ | | |
| Steam | | | 0.48 | $2\bar{0}00$ | | | | |
| Water (liquid) | 0 | $10\bar{0}$ | 1.00 | 4190 | | | $54\bar{0}$ | $2.26 \times 10^6$ |
| Zinc | 419 | 907 | 0.092 | 390 | 23.0 | $9.63 \times 10^4$ | | |

**17-3**
EFFECTS OF PRESSURE
AND IMPURITIES
ON CHANGE OF STATE

Automobile cooling systems present important problems concerning change of state. Most substances contract on solidifying. Water and a few other substances, however, expand. The tremendous force exerted by this expansion is shown by the number of cracked automobile blocks and burst radiators suffered by careless motorists every winter.

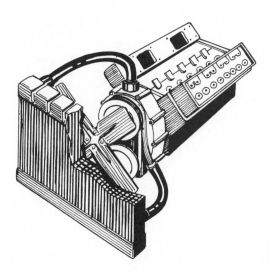

Impurities in water tend to *lower* the freezing point. Alcohol has a lower freezing point than water and is used in some types of antifreeze. By mixing antifreeze with water in the cooling system, the freezing point of the water may be lowered to avoid freezing in winter. Automobile engines may also be ruined in winter by overheating if the water in the radiator is frozen, preventing the engine from being cooled by circulation in the system.

An increase in the pressure on a liquid *raises* the boiling point. Automobile manufacturers utilize this fact by pressurizing their cooling systems and thereby raising the boiling point of the coolant used.

A decrease in the pressure on a liquid *lowers* the boiling point. Frozen concentrated orange juice is produced by subjecting the pure juice to very low pressures at which the water in the juice is evaporated. Then the consumer must restore the lost water before serving the juice.

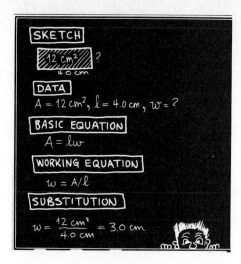

## PROBLEMS

1. How many calories of heat are required to melt 14.0 g of ice at 0°C?

2. How many pounds of ice at 32°F can be melted by the addition of 635 Btu of heat?

3. How many Btu of heat are required to vaporize 11.0 lb of water at 212°F?

4. How many grams of steam in a boiler at $10\bar{0}$°C can be condensed to water at $10\bar{0}$°C by the removal of 1520 cal of heat?

5. How many calories of heat are required to melt $32\bar{0}$ g of ice at 0°C?

6. How many calories of heat are given off when 3250 g of steam is condensed to water at $10\bar{0}$°C?

7. How many Btu of heat are required to melt 33.0 lb of ice and to raise the temperature of the melted ice to 72°F?

8. How many Btu of heat are liberated when 20.0 lb of water at $8\bar{0}$°F is cooled to 32°F and then frozen in an ice plant?

9. How many Btu of heat are required to change 9.00 lb of ice at $1\bar{0}$°F to steam at 232°F?

10. How many calories of heat are liberated when $20\bar{0}$ g of steam at $12\bar{0}$°C is changed to ice at −12°C?

11. How many joules of heat are required to melt 20.0 kg of ice at 0°C?

12. How many kilocalories of heat are required to melt 20.0 kg of ice at 0°C?

13. How many kilocalories of heat are required to melt 50.0 kg of ice at 0°C and to raise the temperature of the melted ice to $2\bar{0}$°C?

14. How many joules of heat are required to melt 15.0 kg of ice and to raise the temperature of the melted ice to 75°C?

15. How many joules of heat need to be removed to condense 1.50 kg of steam at $10\bar{0}$°C?

16. How many litres of water at $10\bar{0}$°C are vaporized by the addition of 5.00 MJ of heat?

17. How many joules of heat need to be removed from 1.25 kg of steam at 115°C to condense it to water and cool the water to 50°C?

18. How many kcal of heat are needed to vaporize 5.00 kg of water at $10\bar{0}$°C and raise the temperature of the steam to 145°C?

# CHAPTER
# *18*

# *Wave Motion*

Energy may be transferred by the motion of particles. Electricity, for example, is conducted along a wire by the motion of electrons. Heat is conducted by the motion of atoms and molecules. Tides and winds are examples of transfer of energy by the motion of fluids.

Another means by which energy transfer may occur is *wave motion*. The sun's energy is transported to the earth by light waves. Radio waves are an illustration of energy transfer for communications. Light waves produced by lasers are being used for voice and data transmission in optical fibers. Sound waves are yet another method by which energy may be transferred. Note that in each case no medium is being transferred by the wave.

A *wave* is a disturbance that moves through a medium or through space. This disturbance may be a displacement of atoms away from their equilibrium positions in an elastic medium, a pulse in a spring, a change in pressure of a gas, or a variation in light intensity. There is a transfer of energy in the direction of propagation of the disturbance for each type of wave.

The elastic medium through which a wave travels or propagates is in many respects similar to a chain of particles connected by a series of springs. If particle $A$ is pulled to the left away from its equilibrium position the neighboring spring exerts a force that tends to return $A$ to its equilibrium position. The same spring exerts an equal but opposite force on particle $B$, which also tends to displace $B$ to the left. As particle $B$ moves to the left, the next particle experiences a force to the left, and so on until each particle experiences a displacement. If particle $A$ is returned to its equilibrium position, the other particles will return to their equilibrium positions at a later time. If particle $A$ is forced to oscillate about its equilibrium position, all the other particles will also oscillate about their equilibrium positions. The kinetic energy given to the first particle is transmitted to each successive particle in the system. Although energy is transferred through the connecting springs, there is no transfer of particles from position *A* to *E*. This energy transfer without matter transfer is typical of all types of wave motion.

Another type of wave motion is shown in the following figure. In this case the elastic medium is a long spring. If the left end of the spring is rapidly lifted up and then returned to its starting position, a crest is formed which travels to the right as shown. If the left end is displaced downward and rapidly returned to its original position, a trough is formed which travels to the right. These nonrepeated disturbances are called *pulses*. If they are

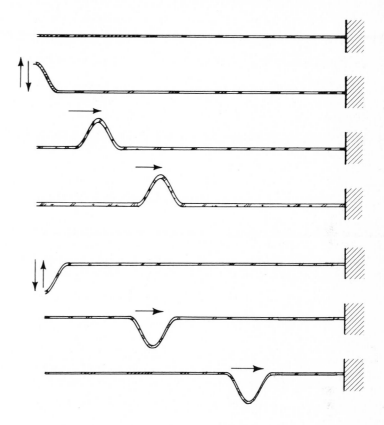

repeated periodically, then a series of crests and troughs will travel through the medium, creating a traveling wave. The displacement of the particles is perpendicular to the direction of wave motion. This is referred to as a *transverse wave*. Water waves are another example of transverse waves.

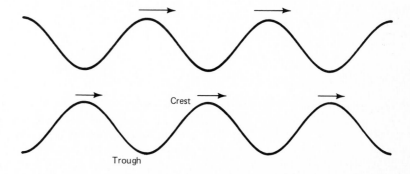

If a spring is compressed at the left end as shown on page 240 and then released, the compression will travel to the right. Similarly, if the spring is stretched, a rarefaction is formed which will propagate to the right. In

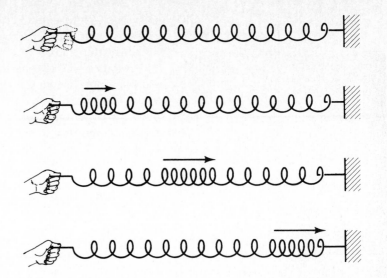

this case the particle motion is along the direction of the wave travel and is referred to as *longitudinal wave* motion. Sound is another example of a longitudinal wave.

The wavelength λ is the minimum distance between particles which have the same displacement and are moving in the same direction. The period

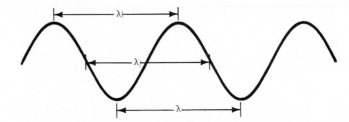

$T$ is the time required for a single wave to pass a given point. The frequency $f$ is the number of complete waves passing a given point per unit time. It is common to use the unit hertz (Hz) for frequency, where one oscillation per second is equal to 1 hertz (1 Hz = 1/s). The period and frequency are related by

$$f = \frac{1}{T}$$

The propagation velocity $v$ of a wave is given by the distance traveled by the wave in one period divided by the period, or

$$v = \frac{\lambda}{T} = \lambda f$$

## EXAMPLE

What is the velocity of a wave that has a wavelength of 2.5 m and a frequency of $4\overline{0}$ Hz?

DATA:

$$\lambda = 2.5 \text{ m}$$
$$f = 4\overline{0} \text{ Hz} = 4\overline{0}/\text{s}$$
$$v = ?$$

BASIC EQUATION:

$$v = \lambda f$$

WORKING EQUATION:   Same

SUBSTITUTION:

$$v = (2.5 \text{ m})(4\bar{0}/\text{s})$$
$$= 1\bar{0}0 \text{ m/s}$$

## PROBLEMS

1. What is the period of a wave whose frequency is $50\bar{0}$ Hz?
2. What is the frequency of a wave whose period is 0.55 s?
3. What is the velocity of a wave that has a wavelength of 2.00 m and a frequency of $40\bar{0}$ Hz?
4. What is the frequency of a light wave that has a wavelength of $5.00 \times 10^{-7}$ m and a velocity of $2.99 \times 10^8$ m/s?
5. What is the period of the wave in Problem 4?

**18-2**
**SUPERPOSITION**

Two waves of similar type can pass through the same medium. These waves will *superimpose* on each other in such a way that their amplitudes add to form a new wave. The shape of the resultant wave is given by the sum of the amplitudes. Where the waves add together to form a larger disturbance, as at point *A,* constructive interference occurs. If the waves oppose each other, producing a smaller disturbance as at point *B,* destructive interference occurs.

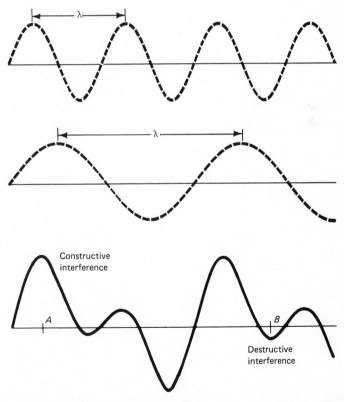

**18-3**
**STANDING WAVES**

When a transverse pulse reaches the end of a spring that is fastened to a rigid support as shown on page 242, the pulse is reflected back along the spring. A traveling wave is also reflected at the rigid end of a spring, producing two waves moving in opposite directions. In one special case it

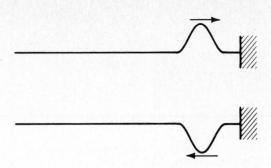

is possible for two waves to combine in such a way that there is no propagation of energy along the wave. This is called a *standing wave*. It can be considered as two waves of equal amplitude and wavelength which are moving in opposite directions. The following figure shows an example of a standing wave on a string which could produce sound on a musical instrument. Note that there is no motion of the string at the end points. Although there is no propagation of energy along the string, there may be energy transfer from the string to the air surrounding it, producing sound waves at the same frequency of oscillation as that of the vibrating string.

**18-4
INTERFERENCE
AND DIFFRACTION**

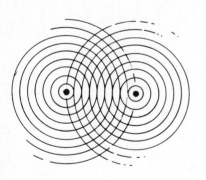

If two rocks are simultaneously dropped into a pool, each will produce a set of waves or ripples. Wherever two wave crests cross each other, the water height is higher than for either crest alone. Where two troughs cross, the water level is lower than for one alone. If a trough crosses a crest, the water level is nearly undisturbed. This is an example of wave *interference*. Constructive interference occurs when two crests or troughs meet, giving a larger disturbance than for either wave alone. Destructive interference occurs when a wave and a trough meet and cancel each other out.

Interference of waves can occur for any type of wave, including sound, light, or radio waves. Some directional radio broadcast antennas rely upon constructive interference between waves from different parts of the antenna to direct the signal in the desired direction. Destructive interference prevents the signal from propagating in undesired directions. "Dead spots" in an auditorium are caused by destructive interference of waves direct from the sound source, with sound waves reflected from the walls, ceiling, or floor of the room. Proper choice of room shape and proper placement of materials that absorb sound well can lead to a room with good acoustical characteristics.

Another interesting property of waves is that they need not travel in straight lines. Waves can bend around obstacles in their path. This is called *diffraction*. Water waves bend around the supports of a pier. Sound waves pass from one room through a door and spread into a second room. Wave

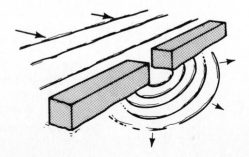

diffraction is commonly observed only when the obstacle or opening is nearly the same size as the wavelength. Water waves and sound waves often have wavelengths in an easily observed range. Light, however, can have a wavelength approximately $5 \times 10^{-7}$ m. For this reason diffraction of light is not as easily recognized as for some other waves. The blue appearance of the sky and the reddish appearance of the setting sun, however, may be accounted for by the diffraction of light off molecules and particles in the atmosphere that have diameters smaller than the longest wavelength of visible light. Those colors with the smallest wavelengths (blue and violet) are diffracted or scattered. The largest wavelengths (red and yellow) are not affected as much by the molecules and small particles and are therefore transmitted without being scattered. In a heavily polluted area, the density of larger particles may be sufficiently high to scatter the red and yellow colors, producing a different hue to the sky.

# CHAPTER
# *19*

## *Sound*

**19-1**
**NATURE**
**AND PRODUCTION**
**OF SOUND**

Each one of us is exposed to sounds of many kinds each day. We communicate with other people using the medium of sound. Whether the sound is pleasant music, the voice of another person, or a loud siren, there are three requirements for the production of sound in a physiological sense. There must be a source of sound, a medium (such as air) for transmitting it, and an ear to receive it. In a physical sense, sound is a vibratory disturbance in an elastic medium which may produce the sensation of sound. The frequency range over which the human ear responds is approximately 20 to 20,000 Hz. Ultrasonic waves have a frequency higher than 20,000 Hz.

Sound is produced by the vibration of a source. A ringing bell, the vibrating head of a drum, and a tuning fork are easily recognized examples of vibrating sources. Other vibrating sources are not as easily recognized. Vibrating vocal cords produce speech. The notes of a clarinet originate with a vibrating reed. An auto horn uses an electrically driven vibrating diaphragm.

**19-2**
**PROPAGATION**
**AND SPEED**
**OF SOUND**

The most common medium for the propagation of sound is air. Sound will also propagate in solids or liquids. A vacuum will not propagate sound. A mechanic may listen to the sounds of a running engine by placing one end of a metal rod against the engine and the other end against his ear. Sounds transmitted through water are utilized by passive sonar receivers aboard ships or submarines to identify other ships nearby.

Sound waves transmitted through the earth may be detected by an instrument called a *seismograph,* which can detect small motion of the earth's crust. An earthquake produces both longitudinal and transverse waves, which propagate with different velocities through the earth's crust. The distance to the source may be determined by measuring the time interval between the arrival of the two types of waves. Comparison of such data from seismographs at several points on the surface of the earth allows the location of the epicenter of the earthquake. Waves may be intentionally set off by exploding buried dynamite. Reflections of these sound waves off different rock formations are recorded by seismographs. These recordings allow geologists to determine the underlying structure of the earth and predict the location of possible oil- or gas-producing regions.

You may have watched distant lightning and noticed the considerable time before the sound of thunder reaches you. This is an example of the relatively slow speed of sound compared to the speed of light ($2.99 \times 10^8$ m/s). The speed of sound in dry air at 1 atm pressure and 0°C is 331 m/s. Changes in humidity, pressure, and temperature cause a variation in the speed of sound. It is found that the increase in the speed of sound with increasing temperature is 0.61 m/s °C. The speed of sound is then given by

$$v = 331 \text{ m/s} + (0.61 \text{ m/s °C})T$$

where: $v$ = speed of sound in air
$T$ = air temperature

## EXAMPLE 1

What is the speed of sound in dry air at 1 atm pressure if the temperature is 23°C?

DATA:

$$T = 23°C$$
$$v = ?$$

BASIC EQUATION:

$$v = 331 \text{ m/s} + (0.61 \text{ m/s °C})T$$

WORKING EQUATION:   Same

SUBSTITUTION:

$$v = 331 \text{ m/s} + (0.61 \text{ m/s °C})(23°C)$$
$$= 345 \text{ m/s}$$

## EXAMPLE 2

What is the time required for the sound from an explosion to reach an observer $190\overline{0}$ m away for the conditions of Example 1?

DATA:

$$v = 345 \text{ m/s}$$
$$s = 190\overline{0} \text{ m}$$
$$t = ?$$

BASIC EQUATION:

$$s = vt \quad \text{(from Chapter 3, Section 3-5)}$$

WORKING EQUATION:

$$t = \frac{s}{v}$$

SUBSTITUTION:

$$t = \frac{190\overline{0} \text{ m}}{345 \text{ m/s}}$$

$$= 5.51 \text{ s} \qquad \boxed{\frac{\text{m}}{\text{m/s}} = \text{m} \div \frac{\text{m}}{\text{s}} = \cancel{\text{m}} \cdot \frac{\text{s}}{\cancel{\text{m}}} = \text{s}}$$

Sound propagates faster in a dense medium such as water than it does in a less dense medium such as air. A list of the speed of sound in various media is given in the following table.

| Medium | Speed | |
|---|---|---|
| | m/s | ft/s |
| Aluminum | 6,420 | 21,100 |
| Brass | 4,7$\overline{0}$0 | 15,400 |
| Steel | 5,960 | 19,500 |
| Granite | 6,0$\overline{0}$0 | 19,700 |
| Alcohol | 1,210 | 3,970 |
| Water (25°C) | 1,5$\overline{0}$0 | 4,920 |
| Air, dry (0°C) | 331 | 1,090 |
| Vacuum | 0 | 0 |

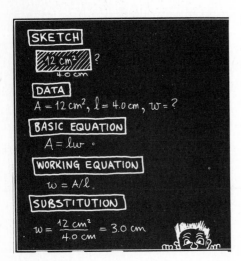

## PROBLEMS

1. What is the speed of sound in m/s at 1$\overline{0}$°C at 1 atm pressure in dry air?

2. What is the speed of sound in m/s at 35°C at 1 atm pressure in dry air?

3. What is the speed of sound in m/s at −23°C at 1 atm pressure in dry air?

4. How long will it take a sound wave to propagate 21 m for the conditions of Problem 1?

5. How long will it take a sound wave to propagate through 75$\overline{0}$0 m of water at 25°C?

6. A sound wave is transmitted through water from one submarine, is reflected off another submarine 15 km away, and returns to the sonar receiver on the first sub. What is the round-trip transit time for the sound wave? Assume that the water temperature is 25°C.

7. A sonar receiver detects a reflected sound wave from another ship 3.52 s after the wave was transmitted. How far away is the other ship? Assume that the water temperature is 25°C.

8. A person is swimming when she hears the underwater sound wave from an exploding ship across the harbor. She immediately lifts her head out of the water. The sound wave from the explosion propagating through the air reaches her 4.00 s later. How far away is the ship? Assume that the water temperature is 25°C and the air temperature is 23°C.

**19-3**
**PROPERTIES**
**OF SOUND**

All sounds have characteristics which we associate only with that sound. A siren is loud. A whisper is soft. Music can be loud or soft. The physical properties that differ for these sounds are intensity and frequency. The physiological characteristics of these sounds are loudness and pitch. Sound quality is related to the number of frequencies present. Pitch and frequency are closely related terms.

*Intensity* is the energy transferred by sound per unit time through unit area. *Loudness* describes how strong or faint the sensation of sound seems to an observer. The ear does not respond equally to all frequencies. A high *frequency* must have a higher intensity to sound as loud as a low-frequency sound. Sound must reach a certain intensity before it can be heard. This threshold of hearing is approximately $10^{-16}$ W/cm². The intensity of a normal conversation is approximately $10^{-9}$ W/cm² or nearly $10^7$ times the hearing threshold. Thunder is nearly $10^{11}$ times more intense than the threshold value, or about $10^{-5}$ W/cm². Sound waves with an intensity greater than $10^{-4}$ W/cm² will produce a painful sensation. This is called the *threshold of pain*.

It is found that an increase in the intensity of a sound by a factor of 2 does not produce the sensation of a sound that is twice as loud. Instead, it is found necessary to increase the intensity by a factor of 10 to produce a sound that is twice as loud. To produce a sound that is three times as loud, the intensity must be increased by 100 times. This type of relationship is called *logarithmic*.

**19-4**
**BEATS**

Sound waves from two sources arriving at an observer at the same time will constructively or destructively interfere with each other (see Section 18-4). Identical waves from two separate sources such as stereo loudspeakers will enhance or reduce each other at different points in space. So sound waves of slightly different frequencies will enhance or reduce each other at different points in time. The alternating periods of increasing and decreasing volume are called *beats*. The number of times per second the sound reaches a maximum is known as the *beat frequency* and is equal to the difference in frequency of the two sound waves. This phenomenon is used for tuning of musical instruments against a set of standard-frequency tuning forks. When the beat frequency is small, the instrument is well matched with the tuning fork.

**19-5**
**THE DOPPLER EFFECT**

As an automobile goes by at a high speed sounding its horn, the frequency or pitch of the sound drops noticeably as the auto passes the observer. This variation in pitch is called the *Doppler effect*. Radar units employed by police to monitor highway speeds employ this principle. A high-frequency radio signal is transmitted by the radar unit toward an oncoming car. Because of the motion of the auto, the reflected signal has a different frequency. This signal is received by the radar unit, which electronically measures the frequency shift to determine the auto speed.

Motion of a source of sound toward an observer increases the rate at which he or she receives the vibrations. The velocity of each vibration is the speed of sound whether the source is moving or not. Each vibration from an approaching source has a smaller distance to travel. The vibrations are therefore received at a higher frequency than they are sent. Similarly, sound waves from a receding source are received at a lower frequency than that at which they are sent.

The apparent Doppler shifted frequency for sound is given by the equation

$$f' = f\left(\frac{v}{v \pm v_s}\right)$$

where   $f'$ = Doppler shifted frequency
   $f$ = actual source frequency
   $v$ = speed of sound
   $v_s$ = speed of the source

The + sign in the denominator is used when the source is moving away (receding) from the observer. The − sign is used when the source is moving toward the observer.

EXAMPLE

---

An automobile sounds its horn while passing an observer at 25 m/s. The actual horn frequency is $40\bar{0}$ Hz.
(a) What is the frequency heard by the observer while the car is approaching?
(b) What is the frequency heard when the car is receding?
   Assume that the speed of sound is 345 m/s.

(a) DATA:

$$f = 40\overline{0} \text{ Hz}$$
$$v = 345 \text{ m/s}$$
$$v_s = 25 \text{ m/s toward observer}$$
$$f' = ?$$

BASIC EQUATION:

$$f' = f\left(\frac{v}{v - v_s}\right)$$

WORKING EQUATION: Same

SUBSTITUTION:

$$f' = (40\overline{0} \text{ Hz})\left(\frac{345 \text{ m/s}}{345 \text{ m/s} - 25 \text{ m/s}}\right)$$
$$= 431 \text{ Hz}$$

(b) We simply change the sign from $-$ to $+$ in the basic equation of part (a). All other data remain the same. We then find

$$f' = (40\overline{0} \text{ Hz})\left(\frac{345 \text{ m/s}}{345 \text{ m/s} + 25 \text{ m/s}}\right)$$
$$= 373 \text{ Hz}$$

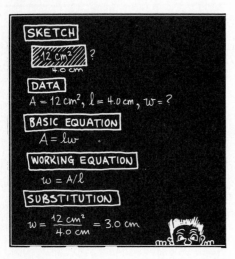

## PROBLEMS

1. A train traveling at a speed of $4\overline{0}$ m/s approaches an observer at a station and sounds a $55\overline{0}$-Hz whistle. What frequency will be heard by the observer? Assume that the sound velocity in air is 345 m/s.

2. What frequency is heard by an observer who hears the $45\overline{0}$-Hz siren on a police car traveling at 35 m/s away from her? Assume that the velocity of sound in air is 345 m/s.

3. A car is traveling toward you at 40.0 mi/h. The car horn produces a sound at a frequency of $48\overline{0}0$ Hz. What frequency do you hear? Assume that the sound velocity in air is 1090 ft/s.

4. A car is traveling away from you at 40.0 mi/h. The car horn produces a sound at a frequency of $48\overline{0}0$ Hz. What frequency do you hear? Assume that the sound velocity in air is 1090 ft/s.

**19-6**
RESONANCE

When a tuning fork is struck with a rubber hammer, it vibrates at its *natural frequency*. This frequency depends on the length, thickness, and the material from which the tuning fork is made. Strings on a guitar also vibrate at a natural frequency. The sounding boards of a guitar are forced to vibrate at the same frequency as the strings because of energy transfer from the strings to the sounding board. This is an example of *forced vibration*. The natural frequency of the board is typically different from that of the strings or tuning fork. Because the area of the sounding board is large, energy transfer into sound waves is very efficient. Therefore, the vibrating string or tuning fork loses its energy or dies out more rapidly if in contact with a sounding board.

Consider two objects such as tuning forks with the same natural frequency which are set close together. One is set into vibration and then stopped after a few seconds. It is found that the other tuning fork is weakly vibrating. The sound waves of the first fork cause the second to vibrate. This is called *sympathetic vibration* or resonance. Energy transfer into vibrations of the second fork is found to be much more efficient when both forks have the same frequency than when they have different frequencies. Large vibrations can be set up if the driving force is at the natural frequency of a system. Auto body rattles sometimes occur at certain speeds and disappear for small speed changes. Radio receivers operate on the principle of resonance. The natural frequency of vibration of electrical currents in a circuit may be tuned to that of an incoming radio signal, which is then amplified and converted into sound.

# CHAPTER
# *20*

## *Static Electricity*

**20-1**
ELECTRIFICATION

The discovery of electrification was made when an amber rod was rubbed with a wool cloth and the rod attracted small bits of paper. When two objects are rubbed together, they become electrified.

When you slide rubber-soled shoes on a wool rug on a dry day, you become electrified. We say you have acquired a static charge. This static charge is usually lost when you touch a metal object or other ground, such as a person.

Trucks that carry flammable liquids prevent a buildup of static charge by dragging a chain on the pavement. Otherwise, a spark caused by a discharge could cause an explosion.

**20-2**
ELECTRICAL CHARGES

There are two types of electrical charges that can be produced. They can be observed indirectly by using an electroscope. A very simple electroscope is a ball of wood pith on a silk thread (Fig. 1).

We can produce a charge on a hard rubber rod by rubbing it with a wool cloth. The rubber rod acquires a negative charge and the wool, a positive charge. Now transfer some of this negative charge from the rubber rod to the pith ball (Fig. 2).

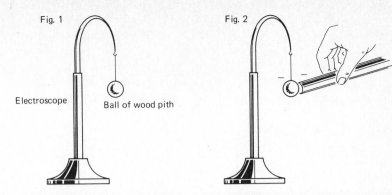

Fig. 1

Electroscope          Ball of wood pith

Fig. 2

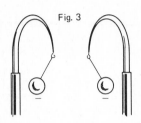

Fig. 3

Another pith ball charged in the same way is repelled by the other pith ball (Fig. 3). This charge is called a *negative charge* (−).

Now rub a glass rod with silk. The glass rod acquires a positive charge and the silk acquires a negative charge. Now transfer some of the positive charge from the glass rod to the pith ball (Fig. 4). This pith ball is attracted to the negatively charged pith ball (Fig. 5).

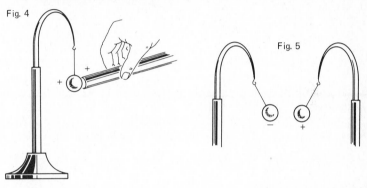

Fig. 4

Fig. 5

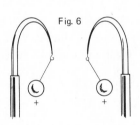

Fig. 6

The charge produced by glass and silk is called a *positive charge* (+). Two pith balls that are positively charged will repel each other (Fig. 6).

In summary,

---

*Like charges repel each other, and unlike charges attract each other.*

---

To understand electricity we need to know more about the structure of matter. We have seen that all matter is made up of atoms. These atoms are made of electrons, protons, and neutrons. Each proton has one unit of positive charge, and each electron has one unit of negative charge. The neutron has no charge. The protons and neutrons are tightly packed into what is called the nucleus. Electrons may be thought of as small "planets" which orbit the nucleus.

An atom normally has the same number of electrons as protons and thus is uncharged.

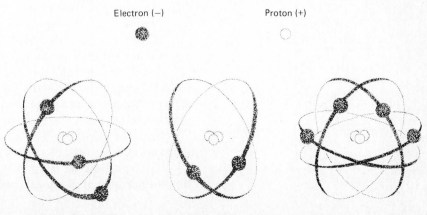

Electron (−)          Proton (+)

Normal atom (uncharged)          Atom with a positive charge          Atom with a negative charge

If an electron is removed, the atom is left with a positive charge. If an extra electron is added, the atom has a negative charge.

When two materials are rubbed together, the atoms on the two surfaces move across each other and brush off electrons. The electrons are transferred from one surface to the other. One surface is then left with a positive charge and the other is negative. This is the process we have called *electrification*.

**20-3**
## COULOMB'S LAW

In 1789, Charles Coulomb made a scientific study of the forces of attraction and repulsion between charged objects. He determined the existence of an "inverse square law" for charged particles which can be used to measure the forces of attraction or repulsion between charged objects.

---

*COULOMB'S LAW OF ELECTROSTATICS*

The force between two point charges $q$ and $q'$ is directly proportional to the product of their magnitudes and inversely proportional to the square of the distance between them.

---

We must use a "proportionality constant" $k$ in writing Coulomb's law as an equation to take into account the air or other medium between the charges. Written in equation form, Coulomb's law becomes

$$F = \frac{kqq'}{r^2}$$

where:     $F =$ force of attraction or repulsion, measured in newtons

$k = 9 \times 10^9 \dfrac{\text{newton-metre}^2}{\text{coulomb}^2}$     (coulomb is a unit of measure of charge; $k$ was found by experiment)

$q,\ q' =$ size of charges, measured in coulombs
$r =$ distance between the charges

The force between the charges is a vector quantity that acts on each charge.

**20-4**
## ELECTRIC FIELDS

So far, we have discussed electrification due to the brushing of electrons from a surface. The concept of the electric field is also an important part of the study of static electricity.

Two magnets may either attract or repel each other even though they may not be touching each other. This illustrates the idea of the "field" in that even though they are not touching each other, there is an invisible region around each one where the other is affected if placed in the region.

In terms of static electricity, an electric field exists where an electric force (of attraction or repulsion) acts on a charge brought into the area.

A balloon clinging to a wall illustrates this principle. The balloon has acquired a negative charge through friction, but the area of wall acquires a positive charge, becoming a positively charged electric field. Such an invisible electric field is present around every charged object.

Static electricity can be a real hazard in industry as well as a curiosity and sometimes a nuisance in daily life.

The electrical spark from static electricity, particularly in synthetic fiber textile mills, is extremely dangerous. Also, some workers in cosmetic factories where aerosol (spray) products are made with hydrocarbon (petroleum type) propellants are required to wear cotton clothes rather than those made with synthetic fibers. Your own calculator may make mistakes in very dry weather due to static electricity introduced by your hands.

Lightning is simply a huge static electricity spark produced in the atmosphere by moving air masses; it is a tremendous discharge.

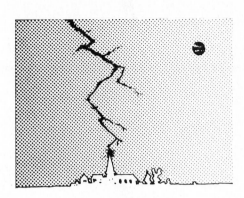

**20-5
ELECTROSTATIC
GENERATORS**

There are two common electrostatic generators in use in physics laboratories. They are the van de Graff generator and the Wimshurst machine. The van de Graff consists of an electron source, a rubber belt driven by a motor, and a metal sphere supported by an insulating stand. The electron source charges the rubber belt as it passes by and carries the electrons up to the sphere and deposits them there. This builds up a high voltage (several hundred thousand volts) between the sphere and the ground.

The Wimshurst machine produces charge by a more complicated method which will not be discussed here. The voltage built up by a Wimshurst machine is not as large as that produced by a van de Graff. However, the current delivered by a Wimshurst machine can be very large.

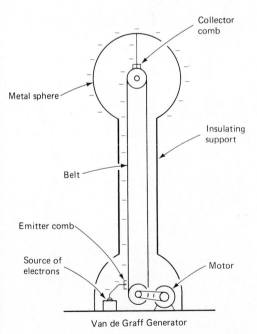

Van de Graff Generator

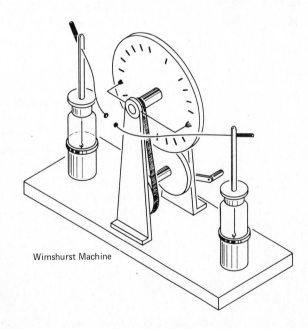

Wimshurst Machine

# CHAPTER
# *21*

## Direct-Current Electricity

**21-1**
**MOTION**
**OF ELECTRONS**

Electrons moving in a wire make up a current in the wire. When the electron current flows in only one direction, it is called *direct current* (dc). Current that changes direction is called *alternating current* (ac). Alternating current will be considered in later chapters.

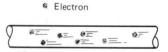
Electron

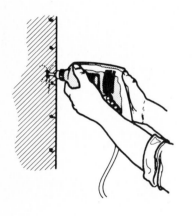

An electric current is a convenient means of transmitting energy. Technicians face many situations every day that require energy to do a particular task. To drill a hole in a metal block, energy must be supplied and transformed into mechanical energy to turn the drill bit. The problem the technician faces is how to supply energy to the machine being used in a form that the machine can turn into useful work. Electricity is often the most satisfactory means of transmitting energy. We begin our study of the use of electricity in transferring energy by looking at an example.

**21-2**
**SIMPLE CIRCUITS**

The following figure shows a circuit such as that of a simple flashlight. An electric circuit is a conducting loop in which electrons carrying electric energy may be transferred from a suitable source to do useful work and back to the source. Energy is stored in the battery. When the switch is closed, energy is transmitted to the light and the light glows.

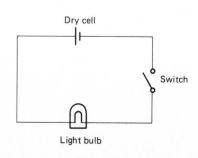
Dry cell

Switch

Light bulb

Compared with static electricity, current electricity is the flow of energized electrons through an electron carrier called a *conductor*. The electrons move from the energy *source* (the battery, here) to the *load* (the place where the transmitted energy is turned into useful work). There they lose energy picked up in the source.

Let us consider each part of the circuit and determine its function.

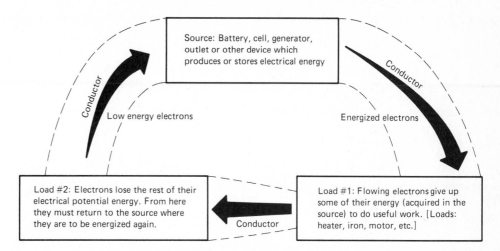

**21-3**
**THE SOURCE**
**OF ENERGY**

The dry cell is a device that converts chemical energy to electrical energy. How the cell does this will be studied in Chapter 23. Here, we simply state that, by chemical action, electrons are given energy in the cell. When energy is given to electrons in this manner, their electrical potential energy is raised.

What does "electrical potential energy" mean? Remember, when a stone is lifted off the ground, it is given gravitational potential energy. (It has the ability to do work in falling back to the ground under the pull of gravity—stored energy.) There is a difference in potential due to its position or condition. Work has been done in lifting it against gravity to the higher position.

In a source of electrical energy something similar happens. In the source (the battery), work is done on electrons which gives them potential energy. This potential difference between the energized electrons [at the negative (−) pole] and the low-potential-energy electrons [at the positive (+) pole] causes the electrons to flow from one point (−) to the other (+) when connected.

It may be helpful to think of this potential difference between the two points in terms of there being an electric field set up between the poles which drives the electrons from the negative pole to the positive pole. The energized electrons collect at the negative pole, repel each other, and flow through the circuit to the positive pole. They lose their potential electrical energy to the load.

**21-4**
**THE CONDUCTOR**

A conductor carries or transfers the electrical charge to the load (here, the light bulb). *Conductors* are substances (such as copper) that have large numbers of free electrons (electrons that are free to move throughout the

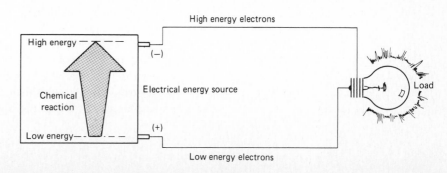

Good conductor

Poor conductor

conductor). As high-energy electrons from the dry cell pass through the conductor, they collide with other electrons in the conductor. These electrons then carry the energy farther along the wire until they collide again and transfer energy on through the wire.

Silver, copper, and aluminum are metals that allow electrons to pass freely through and thus are good conductors. Other metals offer more opposition to the flow of electrons and are poorer conductors.

Substances that do not allow electrons to pass readily are called *insulators*. Common insulators are rubber, wool, silk, glass, wood, distilled water, and dry air.

A small number of materials, called *semiconductors*, fall between conductors and insulators. Their importance is due to the fact that these materials under certain conditions allow current to flow in one direction only. Silicon and germanium are examples of semiconductors and are used in transistors.

**21-5**
**THE LOAD**

In the load, electrons lose their energy. The load converts the electrical energy to other useful forms. In a light bulb, electrical energy is changed to light and heat. An electric motor changes electrical energy to mechanical energy. The load may be a complex motor or only a simple resistor with heat the only new form of energy. The electrons do not collect and remain in the load, but continue back to the low-energy side of the battery (+). There they may be energized again for another trip through the circuit.

**21-6**
**CURRENT**

The flow of electrons through a conductor is called *current*. We could count the electrons passing a point during a certain time to get the rate of flow. This is impractical because the number of electrons is so large (about $10^{18}$/s).

To have a workable unit of electric charge, we will define the charge on $6.25 \times 10^{18}$ electrons as 1 *coulomb*. The *ampere* (A) is the rate of flow of 1 coulomb of charge passing a point in 1 second. Now we can define a unit for the rate of flow of charge.

$$1 \text{ ampere (A)} = \frac{1 \text{ coulomb (C)}}{1 \text{ second (s)}}$$

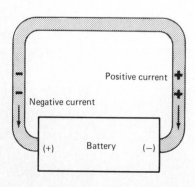

Positive current

Negative current

(+) Battery (−)

As mentioned above, the charge carriers in metals are electrons. In some other conductors, such as electrolytes (conducting liquid solutions), the charge carriers may be positive or negative or both. An agreement must be made to determine which charge carriers should be assumed in our following discussions.

Note that positive charges would flow in the opposite direction (toward the negative terminal) from that followed by negative charges (toward the positive terminal) when a battery is connected to a circuit. A positive current moving in one direction is equivalent for almost all measurements to a current of negative charges flowing in the opposite direction.

We will assume in all the following discussions that the charge carriers are positive, and we will draw our current arrows in the direction that a positive charge would flow. This is the practice of the majority of engineers and technicians. If the student should encounter the negative current convention at a later date, he or she should remember that a negative current flows in the opposite direction from that of a positive current. Regardless of the method used, the analysis of a situation by either method will give the correct result. Some of the rules discussed later, such as the right-hand rule for finding the direction of the magnetic field, will be different if the negative-current convention is used.

**21-7**
**VOLTAGE**

We have seen that current flows in a circuit because of the difference in potential of the different points in the circuit. In *sources,* the raising of the potential energy of electrons which results in a potential difference is called *emf (E).* In *circuits,* the lowering of the potential energy of electrons in a load is called *voltage drop.*

The *volt* (V) is the unit of both emf and voltage drop. We define the volt as the potential difference between two points if 1 joule of work is produced or used in moving 1 coulomb of charge from one point to another.

$$1 \text{ volt (V)} = \frac{1 \text{ joule (J)}}{1 \text{ coulomb (C)}}$$

**21-8**
**RESISTANCE**

Not all substances and not even all metals are good conductors of electricity. Those with few free electrons tend to have greater opposition to the flow of charge. This opposition is called *resistance.* The unit of resistance is the ohm ($\Omega$). It is not a fundamental unit and will be discussed in a later section.

The resistance of a wire is determined by several factors. Among these are:

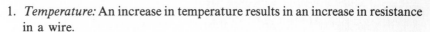
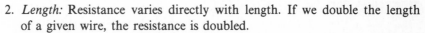

1.  *Temperature:* An increase in temperature results in an increase in resistance in a wire.
2.  *Length:* Resistance varies directly with length. If we double the length of a given wire, the resistance is doubled.
3.  *Cross-sectional area:* Resistance varies inversely with cross-sectional area. If we double the cross section of a wire, the resistance is *halved.* This is similar to water flowing through two pipes. It flows more easily through the larger pipe. (Note that doubling the radius of a wire *more than* doubles the cross-sectional area: $A = \pi r^2$.)
4.  *Material:* Resistance depends on the nature of the material. For example, copper is a better conductor than steel. The conducting characteristic of various materials is described by resistivity. *Resistivity* ($\rho$) is the resistance of a specified amount of the material. These factors are related by the equation

$$R = \frac{\rho l}{A}$$

where:   $R$ = resistance ($\Omega$)
$\rho$ = resistivity ($\Omega$ cm)
$l$ = length (cm)
$A$ = cross-sectional area (cm$^2$)

What is the resistance of a copper wire 20.0 m long with cross-sectional area of $6.56 \times 10^{-3}$ cm² at 20°C? The resistivity of copper at 20°C is $1.72 \times 10^{-6}$ Ω cm.

DATA:

$$l = 20.0 \text{ m} = 2.00 \times 10^3 \text{ cm}$$
$$A = 6.56 \times 10^{-3} \text{ cm}^2$$
$$\rho = 1.72 \times 10^{-6} \text{ Ω cm}$$
$$R = ?$$

BASIC EQUATION:

$$R = \frac{\rho l}{A}$$

WORKING EQUATION:  Same

SUBSTITUTION:

$$R = \frac{(1.72 \times 10^{-6} \text{ Ω cm})(2.00 \times 10^3 \text{ cm})}{6.56 \times 10^{-3} \text{ cm}^2}$$
$$= 0.524 \text{ Ω}$$

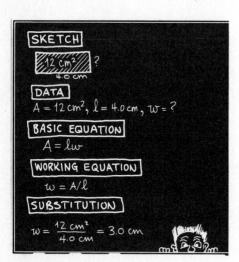

## PROBLEMS

1. What is the resistance of 78.0 m of No. 20 aluminum wire at 20°C? ($\rho = 2.83 \times 10^{-6}$ Ω cm, $A = 2.07 \times 10^{-2}$ cm².)

2. What is the resistance of 375 ft of No. 24 copper wire if it has a resistance of 0.0262 Ω/ft?

3. What is the resistance per foot of No. 22 copper wire if $58\overline{0}$ ft has a resistance of 9.57 Ω?

4. At 77°F, $10\overline{0}$ ft of No. 18 copper wire has a resistance of 0.651 Ω. What is the resistance of $50\overline{0}$ ft of this wire?

5. What is the resistance of 125 m of No. 20 copper wire at 20°C? ($\rho = 1.72 \times 10^{-6}$ Ω cm, $A = 2.07 \times 10^{-2}$ cm².)

6. What is the resistance of $10\overline{0}$ m of No. 20 copper wire at 20°C? ($\rho = 1.72 \times 10^{-6}$ Ω cm, $A = 2.07 \times 10^{-2}$ cm².)

7. What is the resistance of 50.0 m of No. 20 aluminum wire at 20°C? ($\rho = 2.83 \times 10^{-6}$ Ω cm, $A = 2.07 \times 10^{-2}$ cm².)

8. Find the length of a quantity of copper wire with resistance 0.0262 Ω/ft and total resistance 3.00 Ω.

9. Find the cross-sectional area of copper wire at 20°C which is 40.0 m long and has resistivity $\rho = 1.72 \times 10^{-6}$ Ω cm and resistance 0.524 Ω.

# Ohm's Law and dc Circuits

**22-1
OHM'S LAW**

When a current flows through a conductor, there is a definite relationship between current, voltage drop, and resistance. A German physicist, Georg Simon Ohm, studied this relationship and formulated

| | |
|---|---|
| *Ohm's law:* | *Ohm's law can also be written:* |
| $$I = \frac{V}{R}$$ | $$I = \frac{E}{R}$$ |
| where: $I$ = current<br>$V$ = voltage drop<br>$R$ = resistance | where: $E$ = emf of the source<br>of electrical energy |

### EXAMPLE 1

A soldering iron operating on a 115-V outlet has a resistance of 15.0 Ω. What current does it draw?

DATA:

$$E = 115 \text{ V}$$
$$R = 15.0 \text{ }\Omega$$
$$I = ?$$

BASIC EQUATION:

$$I = \frac{E}{R}$$

WORKING EQUATION:  Same

SUBSTITUTION:

$$I = \frac{115 \text{ V}}{15.0 \text{ }\Omega}$$
$$= 7.67 \text{ V/}\Omega$$

$$= 7.67 \text{ A} \qquad \boxed{A = \frac{V}{\Omega}}$$

Ohm's law applies to all dc circuits and those ac circuits containing only resistance. It may be applied to the whole circuit or to any part of it.

Ohm's law should aid us now in understanding resistance. As we mentioned earlier, the ohm ($\Omega$) is a derived unit. From Ohm's law,

$$I = \frac{V}{R}$$

Solving for $R$,

$$R = \frac{V}{I}$$

Substituting units,

$$\boxed{\Omega = \frac{V}{A}}$$

### EXAMPLE 2

A flashlight bulb is connected to a 1.50-V dry cell. If it draws 0.250 A, what is its resistance?

SKETCH:

$$R = ?$$
$$E = 1.50 \text{ V}$$
$$I = 0.250 \text{ A}$$

DATA:

$$E = 1.50 \text{ V}$$
$$I = 0.250 \text{ A}$$
$$R = ?$$

BASIC EQUATION:

$$I = \frac{E}{R}$$

WORKING EQUATION:

$$R = \frac{E}{I}$$

SUBSTITUTION:

$$R = \frac{1.50 \text{ V}}{0.250 \text{ A}}$$
$$= 6.00 \text{ V/A}$$
$$= 6.00 \ \Omega$$

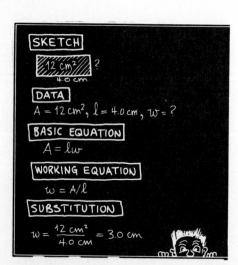

## PROBLEMS

1. A heating element operates on a 115-V line. If it has a resistance of 12.0 $\Omega$, what current does it draw?

2. A given coffeepot operates on 12.0 V. If it draws 2.50 A, what is its resistance?

3. An electric heater draws a maximum of 14.0 A. If its resistance is 15.7 $\Omega$, to what voltage line should it be connected?

4. A heating coil operates on a 220-V line. If it draws 11.0 A, what is its resistance?

5. If an appliance draws 0.750 A on a 115-V line, what is its resistance?

6. What current does a 75.0-$\Omega$ resistance draw on a 115-V line?

In order to communicate about problems in electricity, technicians have developed a "language" of their own. It is a picture language using symbols and diagrams. Some of the most often used symbols appear in the following figure.

**ELECTRICAL SYMBOLS**

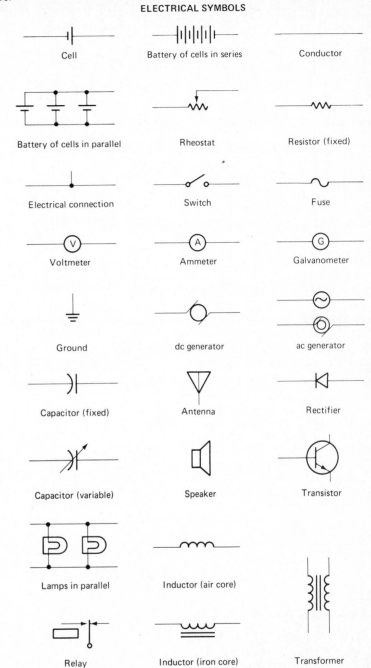

| | | |
|---|---|---|
| Cell | Battery of cells in series | Conductor |
| Battery of cells in parallel | Rheostat | Resistor (fixed) |
| Electrical connection | Switch | Fuse |
| Voltmeter | Ammeter | Galvanometer |
| Ground | dc generator | ac generator |
| Capacitor (fixed) | Antenna | Rectifier |
| Capacitor (variable) | Speaker | Transistor |
| Lamps in parallel | Inductor (air core) | Transformer |
| Relay | Inductor (iron core) | |

The circuit diagram is the most common and useful way to show a circuit. Note how each component (part) of the picture is represented by its symbol in the circuit diagram in its relative position.

The light bulb may be represented as a resistance. Then the following circuit diagram would appear as at the top of page 262.

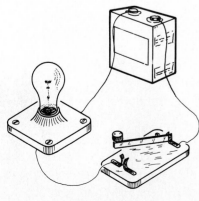

Picture diagram

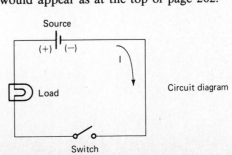

Source

(+)   (−)

Load

Circuit diagram

Switch

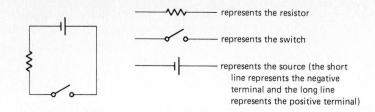

There are two basic types of circuits: series and parallel. A fuse in a house is wired in series with its outlets. The outlets, themselves, are wired in parallel. A study of series and parallel circuits is basic to a study of electricity.

An electrical circuit with only one path for the current to flow is called a *series* circuit.

The current in a series circuit is the same throughout. That is, the total current is the same as the current flowing through each resistance in the circuit.

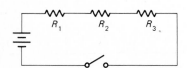

Series

$$I = I_1 = I_2 = I_3 = \cdots$$

where: $I$ = total current
$I_1$ = current through $R_1$
$I_2$ = current through $R_2$
$I_3$ = current through $R_3$

In a series circuit the emf of the source equals the sum of the separate voltage drops in the circuit.

Series

$$E = V_1 + V_2 + V_3 + \cdots$$

where: $E$ = emf of the source
$V_1$ = voltage drop across $R_1$
$V_2$ = voltage drop across $R_2$
$V_3$ = voltage drop across $R_3$

The resistance of the conducting wires is very small and will be neglected here. The total resistance of a series circuit is equal to the sum of all the resistances in the circuit.

Series

$$R = R_1 + R_2 + R_3 + \cdots$$

where: $R$ = total resistance of the circuit
$R_1$ = resistance of first load
$R_2$ = resistance of second load
$R_3$ = resistance of third load

EXAMPLE 1

What is the total resistance of the circuit shown?

SKETCH:

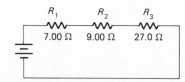

262

$$R_1 = 7.00 \ \Omega$$
$$R_2 = 9.00 \ \Omega$$
$$R_3 = 27.0 \ \Omega$$
$$R = ?$$

BASIC EQUATION:

$$R = R_1 + R_2 + R_3$$

WORKING EQUATION: Same

SUBSTITUTION:

$$R = 7.00 \ \Omega + 9.00 \ \Omega + 27.0 \ \Omega$$
$$= 43.0 \ \Omega$$

## EXAMPLE 2

What is the current in the circuit shown?

SKETCH:

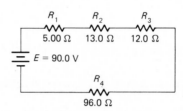

DATA:

$$R_1 = 5.00 \ \Omega$$
$$R_2 = 13.0 \ \Omega$$
$$R_3 = 12.0 \ \Omega$$
$$R_4 = 96.0 \ \Omega$$
$$E = 90.0 \ V$$
$$I = ?$$

BASIC EQUATIONS:

$$R = R_1 + R_2 + R_3 + R_4 \quad \text{and} \quad I = \frac{E}{R}$$

WORKING EQUATIONS: Same

SUBSTITUTIONS:

$$R = 5.00 \ \Omega + 13.0 \ \Omega + 12.0 \ \Omega + 96.0 \ \Omega$$
$$= 126.0 \ \Omega$$

$$I = \frac{90.0 \ V}{126.0 \ \Omega}$$
$$= 0.714 \ A$$

## EXAMPLE 3

What is the value of $R_3$ in the following circuit?

SKETCH:

$$I = 3.00 \text{ A}$$
$$E = 115 \text{ V}$$
$$R_1 = 23.0 \ \Omega$$
$$R_2 = 14.0 \ \Omega$$
$$R_3 = ?$$

BASIC EQUATIONS:

$$I = \frac{E}{R} \quad \text{and} \quad R = R_1 + R_2 + R_3$$

WORKING EQUATIONS:

$$R = \frac{E}{I} \quad \text{and} \quad R_3 = R - R_1 - R_2$$

SUBSTITUTIONS:

$$R = \frac{115 \text{ V}}{3.00 \text{ A}}$$
$$= 38.3 \ \Omega$$

$$R_3 = 38.3 \ \Omega - 23.0 \ \Omega - 14.0 \ \Omega$$
$$= 1.3 \ \Omega$$

## EXAMPLE 4

What is the voltage drop across $R_3$ in Example 3?

DATA:

$$I = I_3 = 3.00 \text{ A}$$
$$R_3 = 1.3 \ \Omega$$
$$V_3 = ?$$

BASIC EQUATION:

$$I_3 = \frac{V_3}{R_3}$$

WORKING EQUATION:

$$V_3 = I_3 R_3$$

SUBSTITUTION:

$$V_3 = (3.00 \text{ A})(1.3 \ \Omega)$$
$$= 3.9 \text{ V}$$

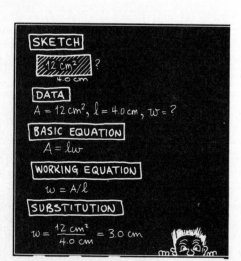

## CHARACTERISTICS OF SERIES CIRCUITS

| | Series |
|---|---|
| Current | $I = I_1 = I_2 = I_3 = \cdots$ |
| Resistance | $R = R_1 + R_2 + R_3 + \cdots$ |
| Voltage | $E = V_1 + V_2 + V_3 + \cdots$ |

## PROBLEMS

1. Three resistors of 2.00 $\Omega$, 5.00 $\Omega$, and 7.00 $\Omega$ are connected in series with a 24.0-V battery. What is the total resistance of the circuit?
2. What is the current in Problem 1?

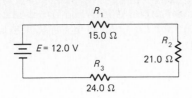

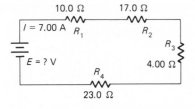

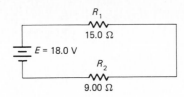

3. What is the total resistance in the circuit shown at left?
4. Find the current through $R_2$ in Problem 3.
5. Find the current in the following circuit.

6. What is the voltage drop across $R_1$ in Problem 5?
7. What emf is needed for the circuit shown at left?
8. What is the voltage drop across $R_3$ in Problem 7?
9. Find the total resistance in the circuit shown below.

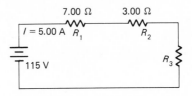

10. Find $R_3$ in the circuit in Problem 9.

**22-4**
**PARALLEL CIRCUITS**

An electrical circuit with more than one path for the current to flow is called a *parallel* circuit.

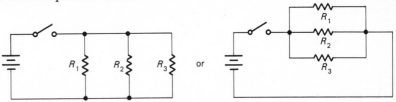

The current in a parallel circuit is divided among the branches of the circuit. How it is divided depends on the resistance of each branch. Since the current divides, the current from the source is equal to the sum of the currents through each of the branches.

Parallel

$$I = I_1 + I_2 + I_3 + \cdots$$

where:   $I =$ total current in the circuit
$I_1 =$ current through $R_1$
$I_2 =$ current through $R_2$
$I_3 =$ current through $R_3$

The emf of the source is the same as the voltage drop across each resistance in the circuit. Therefore, several different loads requiring the same voltage may be connected in parallel.

Parallel

$$E = V_1 = V_2 = V_3 = \cdots$$

where:   $E =$ emf of the source
$V_1 =$ voltage drop across $R_1$
$V_2 =$ voltage drop across $R_2$
$V_3 =$ voltage drop across $R_3$

The equivalent resistance of a parallel circuit is less than the resistance of any single branch of the circuit. To find the equivalent resistance, use the formula

Parallel

$$\frac{1}{R} = \frac{1}{R_1} + \frac{1}{R_2} + \frac{1}{R_3} + \cdots$$

where:  $R$ = equivalent resistance
$R_1$ = resistance of $R_1$
$R_2$ = resistance of $R_2$
$R_3$ = resistance of $R_3$

For comparison to parallel circuits, consider the water system shown.

1. The total amount of water flowing through $R_1 + R_2 + R_3$ is equal to the amount flowing through $A$ or $B$.

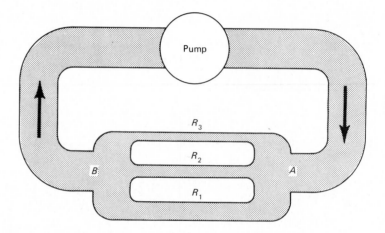

2. The water flowing past point $A$ divides into the three branches $R_1$, $R_2$, and $R_3$.

3. The larger pipes have *less* opposition to water flow than do the smaller pipes. Because $R_1$ has a larger cross-sectional area than $R_2$ or $R_3$, it has less opposition to the flow of water and, therefore, carries more water than $R_2$ or $R_3$.

Similarly, in a parallel circuit:

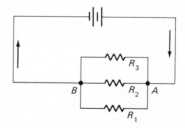

1. The total amount of current flowing through $R_1 + R_2 + R_3$ is equal to the amount flowing through $A$ or $B$.

2. The current flowing past point $A$ divides into the three branches $R_1$, $R_2$, and $R_3$.

3. The smaller resistors have *less* opposition to current flow and therefore carry larger currents.

EXAMPLE 1

What is the equivalent resistance of the circuit shown?

SKETCH:

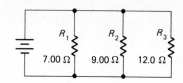

DATA:

$$R_1 = 7.00\ \Omega$$
$$R_2 = 9.00\ \Omega$$
$$R_3 = 12.0\ \Omega$$
$$R = ?$$

BASIC EQUATION:

$$\frac{1}{R} = \frac{1}{R_1} + \frac{1}{R_2} + \frac{1}{R_3}$$

WORKING EQUATION:  When using this formula, it is advisable to substitute directly and then solve for the unknown.

SUBSTITUTION:

$$\frac{1}{R} = \frac{1}{7.00\ \Omega} + \frac{1}{9.00\ \Omega} + \frac{1}{12.0\ \Omega}$$

$$\frac{1}{R} = \frac{0.143}{\Omega} + \frac{0.111}{\Omega} + \frac{0.0833}{\Omega}$$

$$\frac{1}{R} = \frac{0.337}{\Omega}$$

$$R = \frac{1\ \Omega}{0.337}$$

$$= 2.97\ \Omega$$

EXAMPLE 2

What is the total current in the circuit shown?

SKETCH:

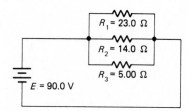

DATA:

$$R_1 = 23.0\ \Omega$$
$$R_2 = 14.0\ \Omega$$
$$R_3 = 5.00\ \Omega$$
$$E = 90.0\ V$$
$$I = ?$$

First, find the equivalent resistance, $R$. Second, find the total current, $I$. To find $R$:

BASIC EQUATION:

$$\frac{1}{R} = \frac{1}{R_1} + \frac{1}{R_2} + \frac{1}{R_3}$$

SUBSTITUTION:

$$\frac{1}{R} = \frac{1}{23.0\ \Omega} + \frac{1}{14.0\ \Omega} + \frac{1}{5.00\ \Omega}$$

$$\frac{1}{R} = \frac{0.0435}{\Omega} + \frac{0.0714}{\Omega} + \frac{0.200}{\Omega}$$

$$\frac{1}{R} = \frac{0.315}{\Omega}$$

$$R = \frac{1\ \Omega}{0.315}$$

$$= 3.17\ \Omega$$

To find *I*:

BASIC EQUATION:

$$I = \frac{E}{R}$$

WORKING EQUATION:  Same

SUBSTITUTION:

$$I = \frac{90.0\ V}{3.17\ \Omega}$$

$$= 28.4\ A$$

## EXAMPLE 3

What is the equivalent resistance and the value of $R_3$ in the circuit?

SKETCH:

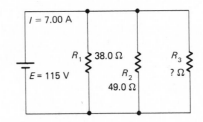

DATA:

$$E = 115\ V$$
$$I = 7.00\ A$$
$$R_1 = 38.0\ \Omega$$
$$R_2 = 49.0\ \Omega$$
$$R_3 = ?$$

First find *R*:

BASIC EQUATION:

$$I = \frac{E}{R}$$

WORKING EQUATION:

$$R = \frac{E}{I}$$

SUBSTITUTION:

$$R = \frac{115\ V}{7.00\ A}$$

$$= 16.4\ \Omega$$

To find $R_3$:

$$\frac{1}{R} = \frac{1}{R_1} + \frac{1}{R_2} + \frac{1}{R_3}$$

SUBSTITUTION:

$$\frac{1}{16.4\ \Omega} = \frac{1}{38.0\ \Omega} + \frac{1}{49.0\ \Omega} + \frac{1}{R_3}$$

$$\frac{1}{R_3} = \frac{1}{16.4\ \Omega} - \frac{1}{38.0\ \Omega} - \frac{1}{49.0\ \Omega}$$

$$\frac{1}{R_3} = \frac{0.0610}{\Omega} - \frac{0.0263}{\Omega} - \frac{0.0204}{\Omega}$$

$$\frac{1}{R_3} = \frac{0.0143}{\Omega}$$

$$R_3 = \frac{1\ \Omega}{0.0143}$$

$$= 69.9\ \Omega$$

## CHARACTERISTICS OF PARALLEL CIRCUITS

| | Parallel |
|---|---|
| Current | $I = I_1 + I_2 + I_3 + \cdots$ |
| Resistance | $\frac{1}{R} = \frac{1}{R_1} + \frac{1}{R_2} + \frac{1}{R_3} + \cdots$ |
| Voltage | $E = V_1 = V_2 = V_3 = \cdots$ |

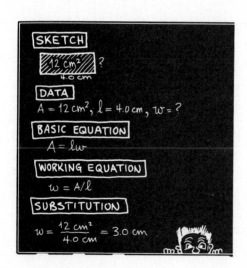

## PROBLEMS

**1.** Find the equivalent resistance in the following circuit.

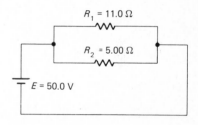

**2.** (a) What is the total current in the circuit in Problem 1?
(b) What is the current through $R_1$?
(c) What is the current through $R_2$?

**3.** (a) Find $I_2$ (current through $R_2$) in the following circuit.

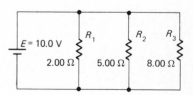

(b) Find $I_3$.
(c) Find $I_1$.

**4.** Find the total current in the circuit in Problem 3.

**5.** Find the equivalent resistance in the circuit in Problem 3.

**6.** Find the resistance of $R_3$ in the circuit at left.

**7.** (a) What is the current through $R_1$ in Problem 6?
(b) What is the current through $R_3$?

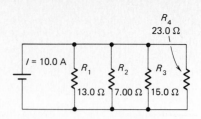

8. What is the equivalent resistance in the above circuit?

9. (a) What emf is required for the circuit in Problem 8?
   (b) What is the voltage drop across each resistance?

10. What is the current through each resistance in Problem 8?

**22-5**
**SERIES–PARALLEL**
**CIRCUITS**

Many circuits cannot be solved directly because of the number and arrangement of the resistances. To simplify this kind of circuit, we usually apply the rules for series and parallel circuits to find an equivalent circuit which reduces to a circuit with one resistance.

### EXAMPLE 1

Circuit $A$ is equivalent to circuit $B$, where $R_4 = R_1 + R_2$. Then, circuit $B$ is equivalent to circuit $C$, where $\dfrac{1}{R_5} = \dfrac{1}{R_3} + \dfrac{1}{R_4}$.

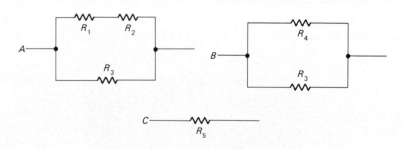

### EXAMPLE 2

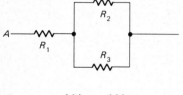

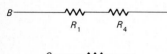

Circuit $A$ (at the left) is equivalent to circuit $B$, where $\dfrac{1}{R_4} = \dfrac{1}{R_2} + \dfrac{1}{R_3}$.

Then, circuit $B$ is equivalent to circuit $C$, where $R_5 = R_1 + R_4$.

### EXAMPLE 3

Find the total current in the circuit shown.

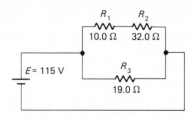

SOLUTION:

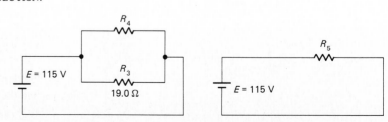

$$E = 115 \text{ V}$$
$$R_1 = 10.0 \ \Omega$$
$$R_2 = 32.0 \ \Omega$$
$$R_3 = 19.0 \ \Omega$$
$$R_4 = R_1 + R_2 = 10.0 \ \Omega + 32.0 \ \Omega = 42.0 \ \Omega$$
$$I = ?$$

First, find the equivalent resistance, $R_5$. Second, find the total current, $I$. To find $R_5$:

BASIC EQUATION:

$$\frac{1}{R_5} = \frac{1}{R_3} + \frac{1}{R_4}$$

SUBSTITUTION:

$$\frac{1}{R_5} = \frac{1}{19.0 \ \Omega} + \frac{1}{42.0 \ \Omega}$$

$$\frac{1}{R_5} = \frac{0.0526}{1 \ \Omega} + \frac{0.0238}{1 \ \Omega}$$

$$\frac{1}{R_5} = \frac{0.0764}{1 \ \Omega}$$

$$R_5 = \frac{1 \ \Omega}{0.0764}$$

$$= 13.1 \ \Omega$$

To find $I$:

BASIC EQUATION:

$$I = \frac{E}{R_5}$$

WORKING EQUATION:   Same

SUBSTITUTION:

$$I = \frac{115 \text{ V}}{13.1 \ \Omega}$$

$$= 8.78 \text{ A}$$

## EXAMPLE 4

Find the equivalent resistance in the circuit shown at left.

SOLUTION:

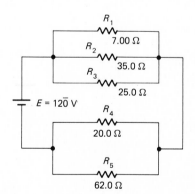

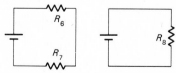

DATA:

$$R_1 = 7.00 \ \Omega$$
$$R_2 = 35.0 \ \Omega$$
$$R_3 = 25.0 \ \Omega$$
$$R_4 = 20.0 \ \Omega$$
$$R_5 = 62.0 \ \Omega$$
$$E = 12\overline{0} \text{ V}$$
$$R_8 = ?$$

First, find $R_6$. Second, find $R_7$. Third, find the equivalent resistance, $R_8$. To find $R_6$:

BASIC EQUATION:

$$\frac{1}{R_6} = \frac{1}{R_1} + \frac{1}{R_2} + \frac{1}{R_3}$$

SUBSTITUTION:

$$\frac{1}{R_6} = \frac{1}{7.00\ \Omega} + \frac{1}{35.0\ \Omega} + \frac{1}{25.0\ \Omega}$$

$$\frac{1}{R_6} = \frac{0.143}{1\ \Omega} + \frac{0.0286}{1\ \Omega} + \frac{0.0400}{1\ \Omega}$$

$$\frac{1}{R_6} = \frac{0.212}{1\ \Omega}$$

$$R_6 = \frac{1\ \Omega}{0.212}$$

$$= 4.72\ \Omega$$

To find $R_7$:

BASIC EQUATION:

$$\frac{1}{R_7} = \frac{1}{R_4} + \frac{1}{R_5}$$

$$\frac{1}{R_7} = \frac{1}{20.0\ \Omega} + \frac{1}{62.0\ \Omega}$$

$$\frac{1}{R_7} = \frac{0.0500}{1\ \Omega} + \frac{0.0161}{1\ \Omega}$$

$$\frac{1}{R_7} = \frac{0.0661}{1\ \Omega}$$

$$R_7 = \frac{1\ \Omega}{0.0661}$$

$$= 15.1\ \Omega$$

To find $R_8$:

BASIC EQUATION:

$$R_8 = R_6 + R_7$$

WORKING EQUATION:    Same

SUBSTITUTION:

$$R_8 = 4.72\ \Omega + 15.1\ \Omega$$

$$= 19.82\ \Omega \quad \text{or} \quad 19.8\ \Omega$$

## EXAMPLE 5

Find the total current in Example 4.

DATA:

$$E = 12\overline{0}\ \text{V}$$

$$R_8 = 19.8\ \Omega$$

$$I = ?$$

BASIC EQUATION:

$$I = \frac{E}{R_8}$$

WORKING EQUATION:    Same

SUBSTITUTION:

$$I = \frac{12\overline{0}\ \text{V}}{19.8\ \Omega}$$

$$= 6.06\ \text{A}$$

| | Series | Parallel |
|---|---|---|
| Current | $I = I_1 = I_2 = I_3 = \cdots$ | $I = I_1 + I_2 + I_3 + \cdots$ |
| Resistance | $R = R_1 + R_2 + R_3 + \cdots$ | $\dfrac{1}{R} = \dfrac{1}{R_1} + \dfrac{1}{R_2} + \dfrac{1}{R_3} + \cdots$ |
| Voltage | $E = V_1 + V_2 + V_3 + \cdots$ | $E = V_1 = V_2 = V_3 = \cdots$ |

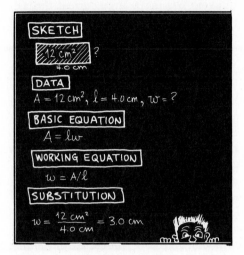

## PROBLEMS

*Circuit A: For use in Problems 1 through 5.*

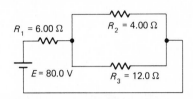

1. (a) Which resistances are connected in parallel?
   (b) What is the equivalent resistance of the resistances connected in parallel?
2. What is the equivalent resistance of the entire circuit?
3. What is the current in $R_1$?
4. What is the voltage drop across $R_1$?
5. (a) Find the current through $R_3$.
   (b) Find the current through $R_2$.

*Circuit B: For use in Problems 6 through 12.*

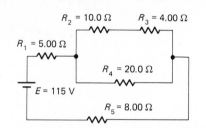

6. What is the equivalent resistance of the resistances connected in parallel?
7. What is the equivalent resistance of the circuit?
8. What is the current in $R_1$?
9. What is the voltage drop across the parallel part of the circuit?
10. What is the current through $R_3$?
11. What is the current through $R_5$?
12. What is the voltage drop across $R_3$?

*Circuit C: For use in Problems 13 through 20.*

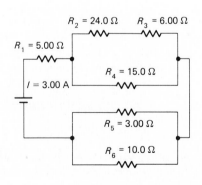

13. What is the equivalent resistance of the parallel arrangement in the upper branch?
14. What is the equivalent resistance of the parallel arrangement in the lower branch?
15. What is the equivalent resistance of the entire circuit?
16. What emf is required for the given current flow in the circuit?
17. What is the voltage drop across the parallel arrangement in the upper branch?
18. What is the voltage drop across $R_4$?
19. What is the voltage drop across $R_6$?
20. What is the current through $R_6$?

*Circuit D: For use in Problems 21 through 25.*

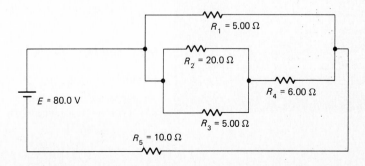

21. What is the equivalent resistance in the circuit?
22. What is the current in $R_5$?
23. What is the voltage drop across $R_5$?
24. What is the voltage drop across $R_4$?
25. What is the current through $R_2$?

## 22-6
## ELECTRICAL INSTRUMENTS

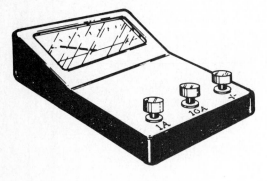

In the laboratory we will use several kinds of electric meters for measurements. Great care must be taken to avoid passing a large current through the meter. Meters are fragile instruments and abuse will ruin them. A large current will burn out the meter.

Measurements in ac circuits and dc circuits require different kinds of meters. One type of meter may *not* be used in the other type of circuit.

Many instruments have more than one range on which readings are to be made. If you connect the lead wires to the (+) and 10-A posts, each scale division represents 1 A. If you connect the lead wires to the (+) and 1-A posts, each scale division represents 0.1 A.

Because of the construction of the meter, the most accurate readings are obtained from the middle of the scale. You should *start with the highest range* and then adjust downward until you get a midscale reading. The reading and use of the different kinds of meters will be studied in the laboratory.

*The voltmeter*　This instrument measures the difference in potential between two points in a circuit. It should *always* be connected in *parallel* with the part of the circuit over which we wish to measure the voltage drop. The voltmeter is a high-resistance instrument and draws very little current.

*The ammeter*　The ammeter measures the current flowing in a circuit. Therefore, it is connected in *series* in the circuit. Since all the current flows through the meter, it has very low resistance so that its effect on the circuit will be as small as possible.

*The galvanometer*　The galvanometer is a very sensitive instrument which is used to detect the presence and direction of *very small* currents.

*The ohmmeter*　The ohmmeter is used to measure the resistance of a circuit component. It should only be used when there is *no current* flowing in the circuit. The ohmmeter has a small battery as a built-in source of energy. Voltmeters and ammeters do not need any source of energy unless they are electronic in nature.

## PROBLEMS

*Using the formulas for series and parallel circuits, fill in the blanks in the tables shown opposite each circuit. In the blanks across from Battery under*

*V:*　Write the emf of the battery.
*I:*　Write the total current in the circuit.
*R:*　Write the equivalent or total resistance of the entire circuit.

*In the blanks across from $R_1$ under*

*V:*　Write the voltage drop across $R_1$.
*I:*　Write the current flowing through $R_1$.
*R:*　Write the resistance of $R_1$.

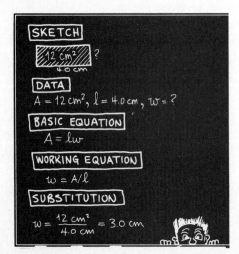

*In the blanks across from $R_2$, $R_3$, . . . , fill in the appropriate numbers under V, I, and R. (Begin by looking for key information given in the table and work from there.)*

1.

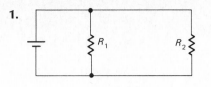

| | V | I | R |
|---|---|---|---|
| Battery | 12.0 V | A | Ω |
| $R_1$ | V | A | 2.00 Ω |
| $R_2$ | V | A | 4.00 Ω |

2.

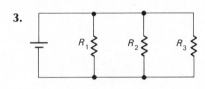

| | V | I | R |
|---|---|---|---|
| Battery | V | A | Ω |
| $R_1$ | V | 2.00 A | 4.00 Ω |
| $R_2$ | V | A | 6.00 Ω |
| $R_3$ | V | A | 8.00 Ω |

3.

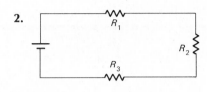

| | V | I | R |
|---|---|---|---|
| Battery | V | A | Ω |
| $R_1$ | V | 2.00 A | Ω |
| $R_2$ | V | 3.00 A | 12.0 Ω |
| $R_3$ | V | 1.00 A | Ω |

4.

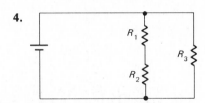

| | V | I | R |
|---|---|---|---|
| Battery | 12.0 V | 2.00 A | Ω |
| $R_1$ | V | A | 6.00 Ω |
| $R_2$ | V | A | 4.00 Ω |
| $R_3$ | V | A | 15.0 Ω |

5.

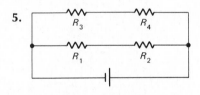

| | V | I | R |
|---|---|---|---|
| Battery | 50.0 V | 5.00 A | Ω |
| $R_1$ | V | 2.00 A | Ω |
| $R_2$ | 25.0 V | A | Ω |
| $R_3$ | 10.0 V | A | Ω |
| $R_4$ | V | 3.00 A | Ω |

6.

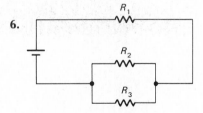

| | V | I | R |
|---|---|---|---|
| Battery | 24.0 V | A | Ω |
| $R_1$ | 8.00 V | A | Ω |
| $R_2$ | V | 4.00 A | Ω |
| $R_3$ | V | 2.00 A | Ω |

7.

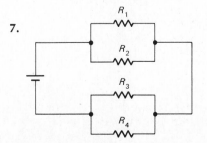

| | V | I | R |
|---|---|---|---|
| Battery | V | A | Ω |
| $R_1$ | 12.0 V | A | 2.00 Ω |
| $R_2$ | V | A | 4.00 Ω |
| $R_3$ | 24.0 V | A | 4.00 Ω |
| $R_4$ | V | A | 8.00 Ω |

**8.**

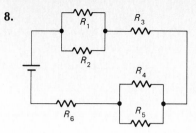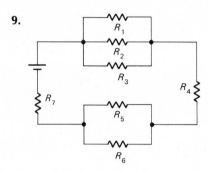

|         | V       | I      | R      |
|---------|---------|--------|--------|
| Battery | 30.0 V  | A      | Ω      |
| $R_1$   | 6.00 V  | 3.00 A | Ω      |
| $R_2$   | V       | 2.00 A | Ω      |
| $R_3$   | V       | A      | 3.00 Ω |
| $R_4$   | V       | 1.00 A | Ω      |
| $R_5$   | 8.00 V  | A      | Ω      |
| $R_6$   | V       | A      | Ω      |

**9.**

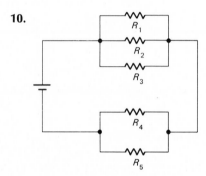

|         | V       | I      | R      |
|---------|---------|--------|--------|
| Battery | V       | 12.0 A | Ω      |
| $R_1$   | V       | A      | Ω      |
| $R_2$   | 18.0 V  | 2.00 A | Ω      |
| $R_3$   | V       | A      | 3.00 Ω |
| $R_4$   | V       | A      | 4.00 Ω |
| $R_5$   | V       | A      | 2.00 Ω |
| $R_6$   | V       | 8.00 A | Ω      |
| $R_7$   | 6.00 V  | A      | Ω      |

**10.**

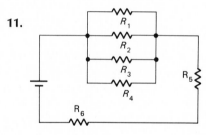

|         | V       | I      | R      |
|---------|---------|--------|--------|
| Battery | 46.0 V  | A      | Ω      |
| $R_1$   | V       | 3.00 A | Ω      |
| $R_2$   | V       | 4.00 A | Ω      |
| $R_3$   | V       | A      | 6.00 Ω |
| $R_4$   | V       | 3.00 A | Ω      |
| $R_5$   | V       | 7.00 A | Ω      |

**11.**

|         | V       | I      | R       |
|---------|---------|--------|---------|
| Battery | V       | A      | Ω       |
| $R_1$   | V       | A      | 20.0 Ω  |
| $R_2$   | 10.0 V  | A      | Ω       |
| $R_3$   | V       | A      | 4.00 Ω  |
| $R_4$   | V       | 1.00 A | Ω       |
| $R_5$   | V       | 5.00 A | 5.00 Ω  |
| $R_6$   | V       | A      | 6.00 Ω  |

# dc Sources

**23-1
THE LEAD
STORAGE CELL**

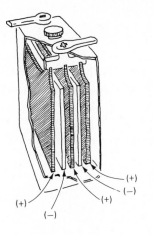

Lead storage batteries are used in automobiles and many other types of vehicles and machinery. A battery is a group of cells connected together. Each cell consists of a positive plate and a negative plate in a conducting solution. These lead cells are *secondary cells*, which means that they are rechargeable. The passing of an electric current through the cell to restore the original chemicals is called *recharging*. Cells, such as the dry cell, which cannot be efficiently recharged are called *primary cells*.

*Note:* The positive plates are all connected together and the negative plates are all connected together with a common terminal for each (+) and (−).

Six storage cells of 2.0 V each connected in series gives 12.0 V for an automobile storage battery.

Lead storage cells are made up of two kinds of lead plates (lead and lead oxide) submerged in a solution of distilled water and sulfuric acid. This acid solution is called an *electrolyte*.

Acid solution

The chemical action between the lead plates and the acid solution produces large numbers of free electrons at the negative (−) pole of the battery. These electrons have a large amount of electrical potential energy which is used in the load in the circuit (for instance, to operate headlights or to turn a starter motor).

As the electrical energy is used in the load, the battery must be recharged. This is done by a generator or an alternator. Such devices provide an electric current to reverse the chemical reaction taking place in the battery. The recharging process extends the life of the battery, which would otherwise be very short.

**23-2**
**THE DRY CELL**

The dry cell is the most widely used primary cell. This is the kind of cell we use in flashlights and portable radios. A dry cell is made of a carbon rod, which is the positive (+) terminal or pole, and a zinc can, which acts as the negative (−) terminal. In between is a paste of chemicals and water which reacts with the terminals to provide energized electrons. These cells are available in a wide range of sizes. Common sizes are in the range 1.5 to 9 V.

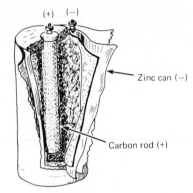

The dry cell, as well as the lead cell, has resistance within the cell itself which opposes the movement of the electrons. This is called the *internal resistance (r)* of the cell. Because current flows in the cell, the emf of the cell is reduced by the voltage drop across the internal resistance. The voltage applied to the external circuit is then

$$V = E - Ir$$

where:   $V$ = voltage applied to circuit
   $E$ = emf of the cell
   $I$ = current through cell
   $r$ = internal resistance of cell

Many times, the current or voltage available from a single cell is inadequate to do a particular job. Then we usually connect two or more cells in a series or parallel arrangement.

Alkaline cells resemble carbon–zinc cells but are five to eight times longer lasting. An alkaline cell has a highly porous zinc anode that oxidizes more readily than does the carbon–zinc cell's anode. Its electrolyte is a strong alkali solution called potassium hydroxide. This compound conducts electricity inside the cell very well and enables the alkaline cell to deliver relatively high currents with greater efficiency than that of carbon–zinc cells.

**23-3**
**CELLS IN SERIES**

To connect cells in series, the positive terminal of one is connected to the negative terminal of the next cell. This procedure is continued until the desired number of cells are all connected (see top of p. 279).

The rules for cells connected in series and parallel are similar to those for simple resistances.

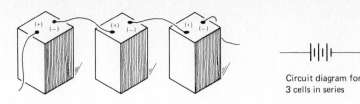

Series connected dry cells

Circuit diagram for
3 cells in series

---

*CELLS IN SERIES*

**1.** The current in the circuit is the same as the current in any single cell:

$$I = I_1 = I_2 = I_3 = \cdots$$

**2.** The internal resistance of the battery is equal to the sum of the individual internal resistances of the cells:

$$r = r_1 + r_2 + r_3 + \cdots$$

**3.** The emf of the battery is equal to the sum of the emfs of the individual cells:

$$E = E_1 + E_2 + E_3 + \cdots$$

---

## EXAMPLE

Two 6.00-V cells with internal resistance of 0.100 Ω each are connected in series to form a battery with a current of 0.750 A in each cell.
(a) What is the emf of the battery?
(b) What is the internal resistance of the battery?
(c) What is the current in the external circuit?

Series

(a) $E = E_1 + E_2 = 6.00 \text{ V} + 6.00 \text{ V} = 12.00 \text{ V}$ (rule 3)
(b) $r = r_1 + r_2 = 0.100 \ \Omega + 0.100 \ \Omega = 0.200 \ \Omega$ (rule 2)
(c) $I = 0.750 \text{ A}$ (rule 1)

**23-4**
**CELLS**
**IN PARALLEL**

To connect cells in parallel, the positive terminals of all the cells are connected together and the negative terminals are all connected together. The leads from the external circuit may be connected to any positive and negative terminals. (The external circuit is all of the circuit *outside* the battery or cell.)

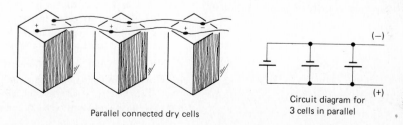

Parallel connected dry cells

Circuit diagram for
3 cells in parallel

CELLS IN PARALLEL

1. The total current is the sum of the individual currents in each cell:

$$I = I_1 + I_2 + I_3 + \cdots$$

2. The internal resistance is equal to the resistance of one cell divided by the number of cells:

$$r = \frac{r \text{ of one cell}}{\text{number of cells}} \quad *$$

3. The emf of the battery is equal to the emf of any single cell:

$$E = E_1 = E_2 = E_3 = \cdots$$

## EXAMPLE

Four cells, each 1.50 V and internal resistance of 0.0500 $\Omega$, are connected in parallel to form a battery with a current output of 0.250 A in each cell.
(a) What is the emf of the battery?
(b) What is the internal resistance of the battery?
(c) What is the current in the external circuit?

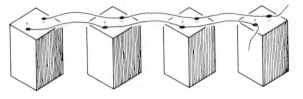

Parallel cells

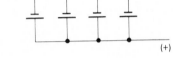

Circuit diagram

(a) $E = 1.50$ V   (rule 3)

(b) $r = \dfrac{r \text{ of one cell}}{\text{number of cells}} = \dfrac{0.0500 \ \Omega}{4} = 0.0125 \ \Omega$   (rule 2)

(c) $I = I_1 + I_2 + I_3 + I_4 = 0.250$ A $+ 0.250$ A $+ 0.250$ A $+ 0.250$ A
   $= 1.000$ A   (rule 1)

## PROBLEMS

1. A cell has an emf of 1.50 V and an internal resistance of 0.0500 $\Omega$. If there is 0.250 A in the cell, what voltage is applied to the external circuit?

2. The voltage applied to a circuit is 11.8 V when the current through the battery is 0.500 A. If the internal resistance of the battery is 0.300 $\Omega$, what is the emf of the battery?

3. The emf of a battery is 12.0 V. If the internal resistance is 0.300 $\Omega$ and the voltage applied to the circuit is 11.6 V, what is current through the battery?

4. Three 1.50-V cells, each with internal resistance of 0.0500 $\Omega$, are connected in series to form a battery with a current of 0.850 A in each cell.
   (a) What is the current in the external circuit?
   (b) What is the emf of the battery?
   (c) What is the internal resistance of the battery?

5. Five 9.00-V cells, each with internal resistance of 0.100 $\Omega$ and current output of 0.750 A, are connected in parallel to form a battery in a certain circuit.
   (a) What is the current in the external circuit?
   (b) What is the emf of the battery?
   (c) What is the internal resistance of the battery?

6. What is the current in the circuit at left?

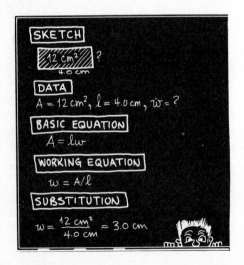

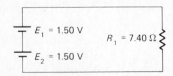

* This formula works only when all the cells have the same internal resistance.

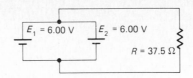

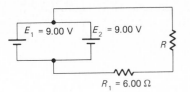

7. Find the current in the circuit shown above.

8. If the current in the circuit at left is 1.20 A, what is the value of *R*?

9. Find the current in the circuit shown below.

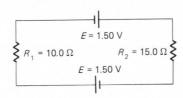

**23-5**

**ELECTRICAL POWER**

Tremendous quantities of energy are used by industry. This energy is, of course, not free but is sold by power companies. The rate of consuming energy is called *power*. The unit of power is the watt. One watt is the power generated by a current of 1 A flowing because of a potential difference of 1 V.

$$P = VI$$

where: $P$ = power (watts)
$V$ = voltage drop
$I$ = current

Recalling Ohm's law: $I = V/R$, we can find two other equations for power:

Given: $P = VI$

Substitute: $V = IR$ $\quad P = (IR)I$

Result: $$P = I^2R$$

Also given: $P = I^2R$

Substitute: $I = \dfrac{V}{R}$ $\quad P = \left(\dfrac{V}{R}\right)^2 R = \dfrac{V^2}{R^2} \cdot R$

Result: $$P = \dfrac{V^2}{R}$$

EXAMPLE 1

A soldering iron draws 7.50 A on a 115 V circuit. What is its wattage rating?

DATA:

$I = 7.50$ A
$V = 115$ V
$P = ?$

BASIC EQUATION:

$$P = VI$$

WORKING EQUATION: Same

Substitution:

$$P = (115 \text{ V})(7.50 \text{ A})$$
$$= 863 \text{ W}$$

## EXAMPLE 2

A hand drill draws 4.00 A and has a resistance of 14.6 $\Omega$. What power does it use?

Data:

$$I = 4.00 \text{ A}$$
$$R = 14.6 \ \Omega$$
$$P = ?$$

Basic Equation:

$$P = I^2 R$$

Working Equation:   Same

Substitution:

$$P = (4.00 \text{ A})^2 (14.6 \ \Omega)$$
$$= 234 \text{ W}$$

Since the watt is a relatively small unit, the kilowatt (1 kW = 1000 watts) is commonly used in industry.

Although we speak of "paying our power bill," what power companies actually sell is energy. Energy is sold in kilowatt-hours (kWh). The amount of energy consumed is equal to the power used times the time it is used. Therefore,

$$\text{energy} = \text{power} \times \text{time}$$

or

$$\text{energy (in kWh)} = (VI)t$$

$$\text{number of kWh} = VIt$$

This equation is useful in finding the cost of electrical energy. Cost is measured in cents per kilowatt-hour.

The cost of operating an electrical device may be found as follows:

$$\text{cost} = (\text{kWh})\left(\frac{\text{cents}}{\text{kWh}}\right)$$

$$\boxed{\text{cost} = \text{power} \times \text{hours} \times \frac{\text{cents}}{\text{kWh}} \times \frac{\text{kW}}{1000 \text{ W}}}$$

$$\text{conversion factor}$$

## EXAMPLE 3

An iron is rated at 550 W. How much would it cost to operate it for 1.50 h at \$0.05/kWh?

Data:

$$P = 550 \text{ W}$$
$$t = 1.50 \text{ h}$$
$$\text{rate} = \$0.05/\text{kWh}$$
$$\text{cost} = ?$$

BASIC EQUATION:

$$\text{cost} = Pt \left(\frac{\text{cents}}{\text{kWh}}\right)\left(\frac{\text{kW}}{1000 \text{ W}}\right)$$

WORKING EQUATION:   Same

SUBSTITUTION:

$$\text{cost} = (550 \text{ W})(1.50 \text{ h}) \left(\frac{\$0.05}{\text{kWh}}\right)\left(\frac{\text{kW}}{1000 \text{ W}}\right)$$
$$= \$0.04$$

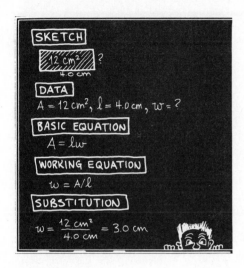

## PROBLEMS

1. A heater draws 8.70 A on a $11\overline{0}$-V line. What is its wattage rating?

2. What power is needed for a sander that draws 3.50 A and has a resistance of 6.70 $\Omega$?

3. How many amperes will a 50.0-W lamp on a $11\overline{0}$-V line draw?

4. What is the resistance of the lamp in Problem 3?

5. How many amperes will a $75\overline{0}$-W lamp draw on a $11\overline{0}$-V circuit?

6. What would it cost to operate the lamp in Problem 5 for 40.0 h if the cost of energy is $0.04 per kWh?

7. Six 50.0-W bulbs are operated on a 115-V circuit. They are in use for 25.0 h in a certain month. If energy costs $0.045 per kWh, what is the cost of operating them for the month?

8. A small furnace expends 2.00 kW of power. If the cost of operation of the furnace is $2.40 for a 24.0-h period, what is the cost of energy per kWh?

9. Will a 20.0-A fuse blow if a $10\overline{0}0$-W hair dryer, a $12\overline{0}0$-W electric skillet, and a $11\overline{0}0$-W toaster are all used at once on a $11\overline{0}$-V line?

10. How long could you operate a $10\overline{0}0$-W soldering iron for $0.50 if the cost of energy is $0.045/kWh?

11. What would be the cost of operating a 1.50-A motor on a $11\overline{0}$-V circuit for 2.00 h at $0.05/kWh?

# CHAPTER
# 24

## Magnetism

**24-1**
**INTRODUCTION TO MAGNETISM**

Many devices that use or produce electrical energy depend on the relation of magnetism and electric currents. Motors and meters are designed to use the fact that electric currents in wires behave like magnets. Generators produce electrical current due to the movement of wires near very large magnets.

We will investigate the basic properties of magnets and the relation between currents and magnetism in this chapter. Later chapters on generators, motors, and transformers will use the basic principles of magnetism that are developed here.

**24-2**
**MAGNETIC MATERIALS**

Certain kinds of metals have been found to have the ability to attract pieces of iron, steel, and some other metals. Metals that have this ability are said to be magnetic. Deposits of iron ore that are naturally magnetic have been found. This ore is called lodestone.

Artificial magnets can be made from iron, steel, and several special alloys such as permalloy and alnico. We will discuss the process of creating artificial magnets later. Materials which can be made into magnets are called magnetic materials. Most materials are nonmagnetic (examples are wood, aluminum, copper, and zinc).

**24-3**
**FORCES BETWEEN MAGNETS**

Suppose that a bar magnet is suspended by a string so that it is free to rotate. It will rotate until one end points north and the other south. The end that points north is called the north-seeking pole, or just north (N) pole. The other end is the south-seeking pole or south (S) pole.

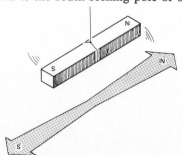

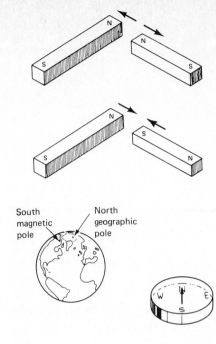

South magnetic pole

North geographic pole

If the north pole of another bar magnet is brought near the north pole of this magnet, the two like poles will repel. The south pole of one magnet will attract the north pole of the other.

In summary:

> *Like magnetic poles repel each other, and unlike magnetic poles attract each other.*

A magnet lines up along a north-south line because of the attraction of the magnetic poles of the earth. The earth has a south magnetic pole which attracts the north pole of a magnet and a north magnetic pole which attracts the south end of a magnet. The earth's south magnetic pole is near its north geographic pole, and vice versa.

A compass is simply a small magnetic needle that is free to rotate on a bearing. The needle's north pole always points to the south magnetic pole of the earth.

**24-4**
**MAGNETIC FIELDS OF FORCE**

There is a magnetic field near a magnetic pole. The existence of this field can be detected by using another magnet. We can represent this field of force by drawing lines that indicate the direction of the force exerted on a north pole placed there.

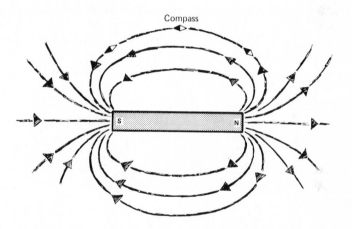

Compass

The field of a bar magnet can be mapped by moving a small compass around the magnet as shown above. These resulting lines are called *flux lines* (lines of force).

The flux lines can also be found by sprinkling iron filings on a sheet of paper laid over a magnet. (See figure at left below.) The fields of combinations of magnets can also be found in this way.

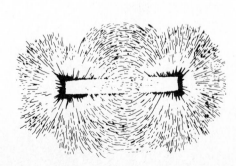

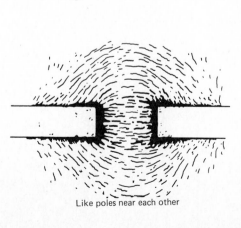

Like poles near each other

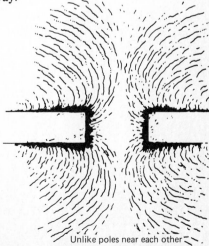

Unlike poles near each other

The magnetic field around a magnet is a three-dimensional field as shown above.

**24-5**
MAGNETIC EFFECTS
OF CURRENTS

When a current passes through a conductor, it sets up a magnetic field. A compass placed near the current shows the direction of this magnetic field. We can show this by connecting a battery to a wire (refer to Figs. 1 to 6 for this discussion). A compass needle is placed under the wire (Fig. 1). When the switch is closed, the needle deflects (Fig. 2). If the terminals of the battery are reversed, the needle deflects in the opposite direction (Fig. 3).

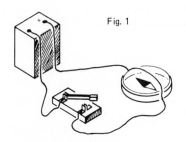

Fig. 1

Compass below conductor, SW open

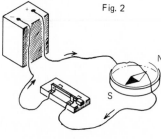

Fig. 2

Compass below conductor, SW closed

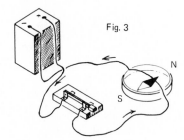

Fig. 3

Compass below conductor, SW closed
Battery terminals reversed

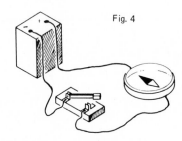

Fig. 4

Compass above conductor, SW open

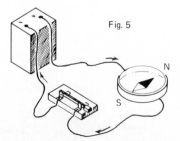

Fig. 5

Compass above conductor, SW closed

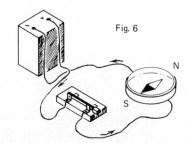

Fig. 6

Compass above conductor, SW closed
Battery terminals reversed

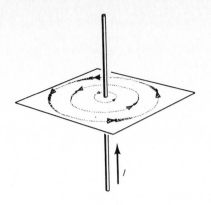

When the compass needle is placed on top of the conductor, the direction of deflection is reversed in each case. (See Figs. 4 through 6.) When the current in a wire flows in a given direction, the flux lines point in one direction below the wire and point in the opposite direction above the wire.

The direction of the flux lines actually curves around the straight current-carrying wire. Iron filings on a sheet of paper perpendicular to a current-carrying wire show the direction of the field. The magnetic field is stronger for large currents than for small currents. The direction of the field near a current in a straight wire is given by the following rule:

---

AMPÈRE'S RULE

*Hold the conductor in your right hand, with your thumb extended in the direction of the current. Your fingers circle the wire in the direction of the flux lines.*

---

The magnetic field near a long current-carrying wire, measured in units of teslas, is circular about the wire and given by Ampère's law:

$$B = \frac{\mu_0 I}{2\pi R}$$

where: $B$ = magnetic field in teslas
$I$ = current through the wire
$R$ = perpendicular distance from the center of the wire
$\mu_0 = 4\pi \times 10^{-7}$ T m/A

The magnetic field, $B$, has the unit tesla (T) and is defined in terms of electric current by the constant $\mu_0$, the permeability constant. The value of $\mu_0$ is not experimentally determined but is an assigned value which explicitly defines magnetic field in terms of electric current.

EXAMPLE

---

A power line carrying $40\overline{0}$ A is 9.00 m above a transit used by a surveying student.
(a) What is the magnetic field because of the power-line current above the transit?
(b) If the earth's horizontal component of magnetic field is $5.20 \times 10^{-5}$ T at that location, what error could be introduced in the angular measurement? (Assume that the power line runs north-south.)
(a) DATA:

$$I = 40\overline{0} \text{ A}$$
$$R = 9.00 \text{ m}$$
$$\mu_0 = 4\pi \times 10^{-7} \text{ T m/A}$$
$$B = ?$$

BASIC EQUATION:

$$B = \frac{\mu_0 I}{2\pi R}$$

WORKING EQUATION: Same

SUBSTITUTION:

$$B = \frac{(4\pi \times 10^{-7} \text{ T m/A})(40\overline{0} \text{ A})}{2\pi(9.00 \text{ m})}$$
$$= 8.89 \times 10^{-6} \text{ T}$$

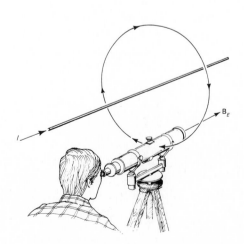

With the current from south to north, the rule of thumb shows the direction of $B$ to be east to west.

(b) The angle that the resultant vector [the earth's field plus the wire's field $(B_E + B)$] makes with $B_E$ would be the angular error.

DATA:

$$B_x = B_E = 5.20 \times 10^{-5} \text{ T} \qquad \text{earth's component}$$
$$B_y = B = 8.89 \times 10^{-6} \text{ T} \qquad \text{[part (a)]}$$
$$\theta = ?$$

BASIC EQUATION:

$$\tan \theta = \frac{B_y}{B_x}$$

WORKING EQUATION:  Same

SUBSTITUTION:

$$\tan \theta = \frac{8.89 \times 10^{-6} \text{ T}}{5.20 \times 10^{-5} \text{ T}} = 0.171$$
$$\theta = 9.70°$$

The bearing on the surveying student's transit could be in error by 9.70° because of the power line.

## PROBLEMS

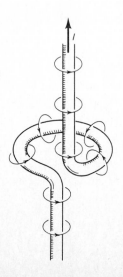

**1.** What is the magnetic field at 0.100 m from a long wire carrying a current of 15.0 A?

**2.** What is the magnetic field at 0.500 m from a long wire carrying a current of 7.50 A?

**3.** What is the current in a wire if the magnetic field is $4.0 \times 10^{-6}$ T at a distance of 2.0 m from the wire?

**4.** A power line runs north-south carrying 675 A and is 5.00 m above a transit used by a surveyor.
(a) What is the magnetic field at the transit because of the power-line current?
(b) If the earth's horizontal component of magnetic field is $5.20 \times 10^{-5}$ T, what is the error introduced in the surveyor's angular measurement?

**24-6**
**MAGNETIC FIELD**
**OF A LOOP**

To determine the direction of the flux lines of a current in a loop, use Ampère's rule as shown at left. If several loops are made into a tight spiral as shown below, the flux lines add to form the field shown. A coil of tightly wrapped wire is called a *solenoid*.

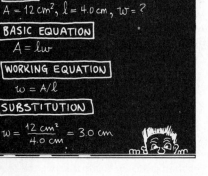

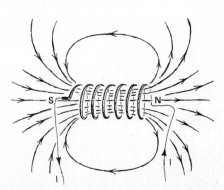

The left side of this solenoid acts like a south magnetic pole. The right side acts like a north magnetic pole. This polarity could be found by using a compass. The rule for finding the polarity of a solenoid is:

*Hold the solenoid in your right hand so that your fingers circle it in the same direction as the current. Your thumb points to the north pole of the solenoid.*

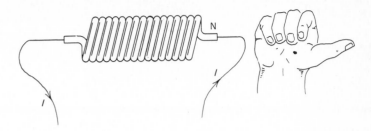

For a long coil that is tightly turned, the field strength at its center is

$$B = \mu_0 In$$

where: $B$ = magnetic field in the region at the center of the solenoid
$\mu_0$ = permeability constant
$I$ = current through the solenoid
$n$ = turn per unit length of solenoid

The longer the solenoid is with respect to its radius, the more uniform the magnetic field is inside the solenoid; and for an infinitely long solenoid, the value of $B$ is uniform throughout.

### EXAMPLE

What is the magnetic field at the center of a solenoid that is 0.425 m long, 0.075 m in diameter, and has three layers of $85\bar{0}$ turns each, when 0.25 A flows throughout?

DATA:

$$I = 0.25 \text{ A}$$

$$n = \frac{3 \times 85\bar{0} \text{ turns}}{0.425 \text{ m}} = 60\bar{0}0 \text{ turns/m}$$

$$\mu_0 = 4\pi \times 10^{-7} \text{ T m/A}$$

$$B = ?$$

BASIC EQUATION:

$$B = \mu_0 In$$

WORKING EQUATION: Same

SUBSTITUTION:

$$B = (4\pi \times 10^{-7} \text{ T m/A})(0.25 \text{ A})(60\bar{0}0/\text{m})$$
$$= 1.9 \times 10^{-3} \text{ T}$$

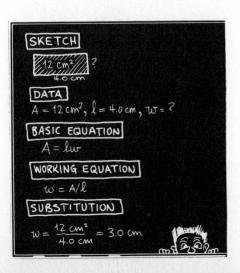

### PROBLEMS

1. A solenoid has $100\bar{0}$ turns of wire, is 0.32 m long, and carries a current of 5.0 A. What is the magnetic field at the center of the solenoid?

2. A solenoid has $300\bar{0}$ turns of wire and is 0.35 m long. What current is required to produce a magnetic field of 0.10 T at the center of the solenoid?

3. A small solenoid is 0.15 m in length, 0.015 m in diameter, and has $60\bar{0}$ turns of wire. What current is required to produce a magnetic field of $1.25 \times 10^{-3}$ T at the center of the solenoid?

When a magnetic material such as iron is placed in the core of a current-carrying solenoid, the material becomes very strongly magnetized. This is called *induced magnetism*. The solenoid and magnetic core are called an *electromagnet*.

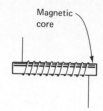

Magnetic core

When the current through the coil is turned off, the strength of the induced magnet decreases, but some remains. When the core is removed, a magnetic field remains in the core. In materials such as soft iron, very little magnetic field remains in the core after the current flow stops. In other materials, such as steel, alnico, and permalloy, a much stronger field remains. These latter materials are used for permanent magnets. However, they are undesirable for use as a core in an induction motor. Soft-iron cores are often used for this application because less energy is required to reverse the polarity of the induced magnetic field.

A magnet can be thought to consist of many atoms, each behaving like a small magnet. In each atom, the electrons orbit about the nucleus and each electron spins about its own axis. These motions produce a magnetic field in the atom.

Usually, the atoms in a material are arranged so that their magnetic fields point in different directions. The result is that the magnetism of one atom is canceled out by its neighbors.

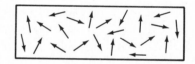

When a magnetic material is magnetized, many of the atoms' magnetic fields line up to point in one direction. The magnetic fields of these atoms add together to give the material a magnetic field.

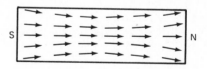

# CHAPTER
## 25

# *Generators and Motors*

**25-1**
**INDUCED CURRENT**

When a magnet is moved so that its flux lines cut across a wire, an emf is induced in the wire. The strength of this induced emf depends on the strength of the magnetic field and on the rate at which the flux lines are cut by moving the magnet or wire. Increasing the strength of the field or increasing the rate at which the flux lines are cut also causes the current to increase.

While the magnet in the diagram is moved downward, the galvanometer indicates that a current flows through the wire.

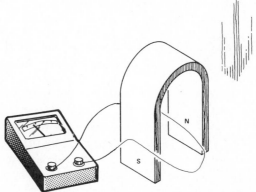

Magnet moving downward.

If the magnet is moved upward, the induced current would be in the opposite direction.

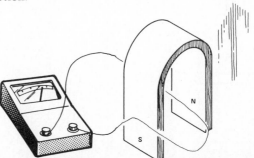

Magnet moving upward.

A current would also flow in the wire if the wire is moved and the magnet is stationary. The current is produced by the relative motion of the magnet and wire. In commercial generators, magnets are spun inside a set of coils of wire.

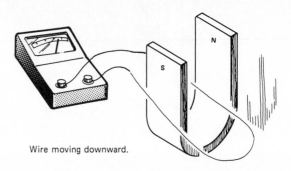

Wire moving downward.

The induced emf is increased by replacing the single wire with a coil of many turns. For example, doubling the number of turns doubles the induced emf.

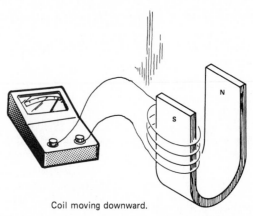

Coil moving downward.

**25-2**
**ac GENERATORS**

The induction of an emf in a coil of wire can be used to supply electrical power. A coil of wire rotating in a magnetic field produces a fluctuating (changing) emf in the wire. This is the simplest kind of generator to build in the laboratory, so we will study its operation here and compare it to commercial generators where the magnets (electromagnets) are rotated.

The current produced by rotating the wire through the magnetic field is called an *alternating current* (ac). As side *A* of the current loop passes downward by the north pole, the induced current is in one direction.

As side *A* (same side of the rotating loop) passes upward by the south pole, the induced current is in the opposite direction. The result is an alternating current induced in the rotating wire. As side *B* (the other side of the rotating loop) passes upward and then downward, the current in it also alternates.

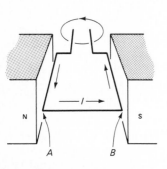

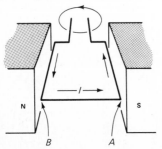

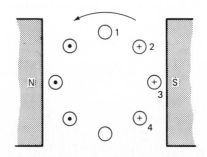

• Current out of page

+ Current into page

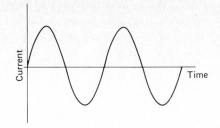

The direction of current flow in a wire as it moves between the north and south magnet poles is illustrated in the figure at the bottom of page 292.

A graph of the induced current is shown at left. One cycle is produced by one revolution of the wire. The time required for one cycle depends on the rotational speed of the coil. If the coil rotates 60 times each second, an alternating current of frequency 60 hertz (cycles per second) is produced.

The current produced in the coil is conducted by brushes on slip rings to the external circuit as shown. The rotating coil is called the *rotor* or *armature,* and the field magnets are called the *stator.*

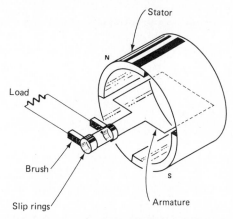

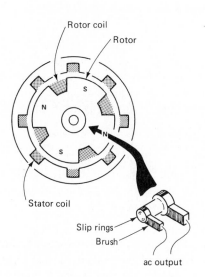

The generator does not actually create electrical energy; it changes the mechanical energy of rotation into electrical energy.

The energy to turn the rotor may be supplied by water falling down a waterfall, a diesel engine, or a steam turbine.

Power companies use large commercial ac generators to produce the current they need to supply to their customers. These generators work in the same manner as the generator discussed here; but they have many coils, and electromagnets are used instead of permanent magnets.

The large generators used by electrical power companies can produce voltages as large as 13,000 V and currents up to 10 A. The alternator used in automobiles is an ac generator that produces about 10 V and up to 40 A.

**25-3**
**dc GENERATORS**

By the use of a special device called a *commutator,* the ac generator can be used to produce direct current. The commutator is a split ring that replaces the slip rings as shown.

When side *A* of the coil passes upward along the N pole, the induced current flows in the direction shown and is picked up by brush number 1. The current in the external circuit is also shown.

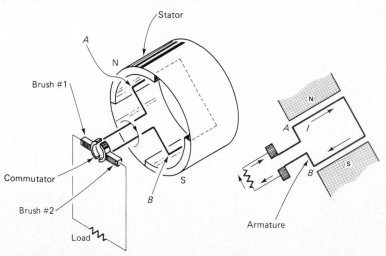

When side *B* of the coil passes upward along the N pole, the induced current flows in the direction shown and is picked up by brush number 1. The current in the external circuit is in the same direction as it was when *A* passed along the N pole. Thus, this is a direct current.

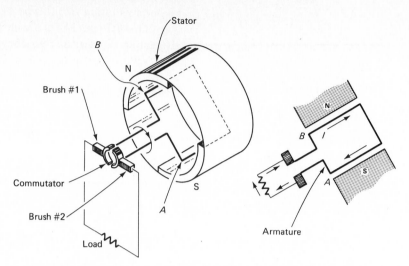

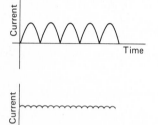

The current produced by this dc generator does not have the same value at all times. A graph of the induced current is shown at top left.

Commercial dc generators that are used for industrial purposes contain many coils. The output current has almost the same value at all times due to the use of the large number of coils.

**25-4**

**THE MOTOR PRINCIPLE**

We have seen that like poles of magnets repel each other. A magnet that is pivoted will spin due to the repulsion of another magnet nearby (Fig. 1).

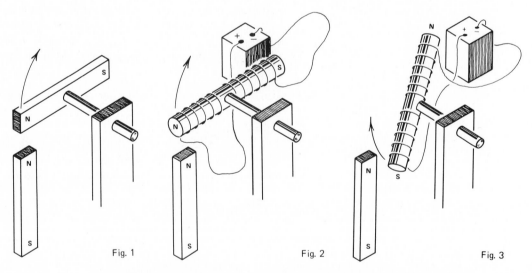

Fig. 1          Fig. 2          Fig. 3

We can construct an electromagnet by wrapping wire around an iron core and running a current through the wire (Fig. 2). The N pole of the electromagnet will be repelled by a north pole of another magnet.

The electromagnet will turn until its S pole is next to the N pole of the permanent magnet (Fig. 3).

If we could suddenly change the polarity of the electromagnet (often called the *armature*), the magnet would repel the N pole, and the electromagnet would continue to spin (Fig. 4).

If a dc current supply is used, this change can be made by using a

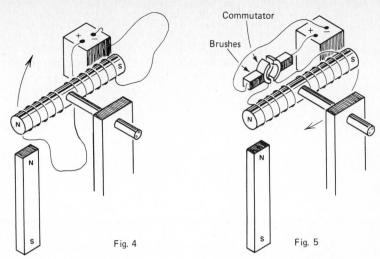

Fig. 4          Fig. 5

commutator (split ring) to change the direction of the current in the electromagnet. Changing the direction of the current flowing through the coil of the electromagnet changes the poles (Fig. 5).

As the current changes direction, the electromagnet spins due to the repulsion of like poles. A shaft may be connected to the electromagnet so that the rotational motion can be used to do work. This device is called a *motor*. A motor converts electrical energy to mechanical energy and thus performs the reverse function of a generator (Fig. 6).

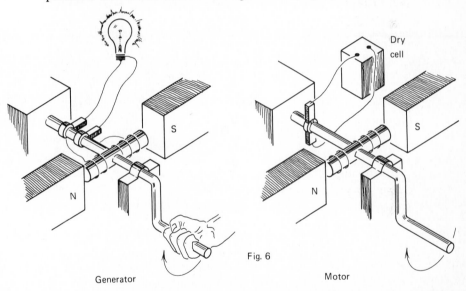

Fig. 6

Generator          Motor

If ac current is supplied to the electromagnet, slip rings are used instead of a commutator. The use of alternating current makes the commutator unnecessary. The changes in direction are supplied by the ac current itself (Fig. 7).

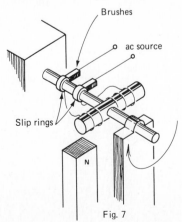

Fig. 7

Commercial motors operate in the same way as the motors discussed above. However, they usually use electromagnets in place of the permanent magnets and are much more complex. Slip rings are not necessary in ac motors. The current in the rotating electromagnet can be induced in the same way a current is induced in a generator.

Motors can be designed for many different purposes. Heavily loaded motors need certain types of starters. The torque and power outputs can be greatly varied by differences in design.

Several types of ac motors are discussed below.

1. The *universal motor* can be run on either ac or dc power. Slip rings are used in this type of motor as in dc motors. This motor is often used in small hand tools and appliances (see the figure).

2. The *induction motor* (see the following figure) is the most widely used ac motor. The rotating electromagnet is not connected to a power source by slip rings. Instead, the current in the electromagnet is induced by a moving magnetic field caused by the ac current.

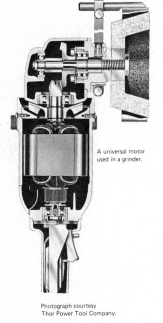

A universal motor used in a grinder.

Photograph courtesy Thor Power Tool Company.

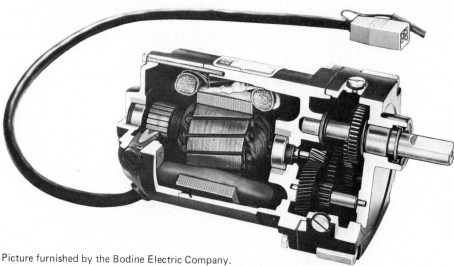

Picture furnished by the Bodine Electric Company.

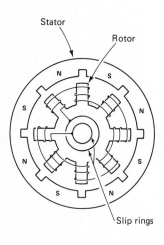

Stator

Rotor

Slip rings

3. The *synchronous motor* is very similar to the slip ring ac motor discussed earlier. The rotating electromagnet is supplied with current through slip rings. The speed of rotation of a synchronous motor is constant and depends on the number of coils and on the frequency of the power supply. The synchronous motor will work only when operated with an ac power source of the frequency for which it is designed. The word "synchronous" is derived from *syn*, meaning same, and *chrona*, or time. Synchronous motors are used to operate clocks and other devices needing accurate speed control.

# Alternating-Current Electricity

**26-1
WHAT IS
ALTERNATING CURRENT?**

Alternating current has many more applications in industry and everyday experience than does direct current. We first studied direct current to learn some of the basic ideas of all electricity. Now we will turn our attention to the more common alternating current.

As its name implies, *alternating current* is current that flows in one direction in a conductor, changes direction, and then flows in the other direction. The direction of flow changes many times in one second. Ordinary household current is 60-Hz current.

Every time the current repeats itself—flows, changes direction, flows, and changes direction—it goes through one *cycle*. The reason for this alternation is that this is how current comes from electric generators. The emf and current produced by a generator do not alternate instantly between maximum values in each direction, but they build up to maximum values and then decrease, change direction, and build to maximum values in the other direction.

1 cycle

3 cycles

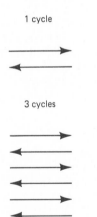

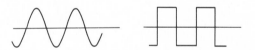

Direct current is usually a steady flow at a fairly constant value. Graphically, it can be represented as shown.

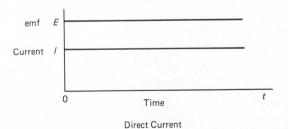

Direct Current

Alternating current, however, is constantly changing. To graphically represent ac, we must show that it builds up and drops off. This can be demonstrated by the curve shown on page 298, called a *sine curve*. We form

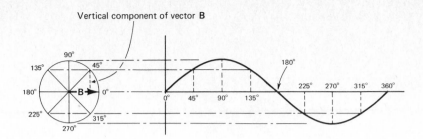

Vertical component of vector **B**

the curve by rotating a vector **B** about a point and plotting the vertical components of **B**. Rotating **B** through 360° is graphing one cycle.

The graph above shows the ac-current curve. A graph of ac voltage is also a sine curve.

Ac current and voltage are constantly changing. We can find the value of current or voltage at any instant by using the fact that each makes a sine curve.

$$i = I_{max} \sin \theta$$
$$e = E_{max} \sin \theta$$

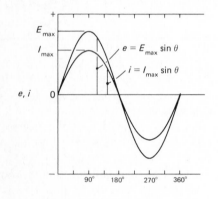

where:   $i =$ instantaneous current (current at any instant)
  $I_{max} =$ maximum instantaneous current
  $\theta =$ angle measured from the beginning of cycle (see preceding graph),
  $e =$ instantaneous voltage
  $E_{max} =$ maximum instantaneous voltage

Both of the curves at the left show $e$ and $i$ reaching a maximum at the same time and falling to zero at the same time. When this occurs, they are said to be "in phase." $e$ and $i$ are in phase in electrical circuits when there is only resistance in the circuit. In a later chapter we will study some other things about ac that will shift the phase.

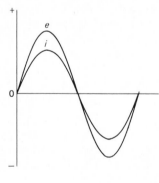

$e$ and $i$ in phase

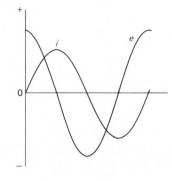

$e$ and $i$ out of phase

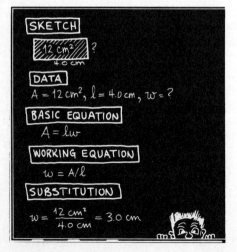

## PROBLEMS

1. What is the maximum value of the voltage in a circuit where the instantaneous value of the voltage at $\theta = 35.0°$ is 17.0 V?

2. The instantaneous voltage at $\theta = 50.0°$ is 82.0 V. What is the maximum voltage?

3. If the maximum ac voltage on a line is $16\overline{0}$ V, what is the instantaneous voltage at $\theta = 45.0°$?

4. If the maximum ac voltage on a line is $220\overline{0}$ V, what is the instantaneous voltage at $\theta = 60.0°$?

A direct measurement of ac is difficult because it is constantly changing. The most useful value of ac is based on its heating effect and is called its *effective value*.

The effective value of an alternating current is the number of amperes that produce the same amount of heat in a resistance as an equal number of amperes of a steady direct current.

$$I = \frac{1}{\sqrt{2}} \, I_{max} = 0.707 I_{max}$$

$$I_{max} = \sqrt{2} \, I = 1.41 I$$

where:  $I$ = effective value of current (sometimes called rms value)
$I_{max}$ = maximum instantaneous current

The numerical factors in the foregoing equations are derived from an average of the sine-wave time variation of the ac current. When this time average is taken, the factors $\sqrt{2} = 1.414$ and $1/\sqrt{2} = 0.707$ are found.

### EXAMPLE

The current supplied to a woodworking shop is rated at 10.0 A. What is the maximum value of the current supplied?

DATA:

$$I = 10.0 \text{ A}$$
$$I_{max} = ?$$

BASIC EQUATION:

$$I_{max} = 1.41 \, I$$

WORKING EQUATION:   Same

SUBSTITUTION:

$$I_{max} = 1.41(10.0 \text{ A})$$
$$= 14.1 \text{ A}$$

The effective value for ac voltage may be expressed similarly:

$$E = \frac{1}{\sqrt{2}} \, E_{max} = 0.707 E_{max}$$

$$E_{max} = \sqrt{2} \, E = 1.41 E$$

where:  $E$ = effective value of voltage
$E_{max}$ = maximum instantaneous voltage

When we say a house is wired for 120 volts, we are using the effective value of the voltage. Actually, the voltage varies between +170 V and −170 V during each cycle.

Unless otherwise stated, ac voltage and current are *always* expressed in terms of effective values.

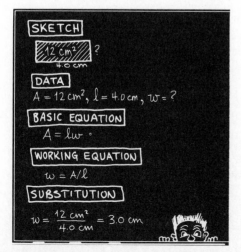

## PROBLEMS

1. What is the effective value of an ac voltage whose maximum voltage is $220\bar{0}$ V?
2. What is the maximum current in a circuit that has a current of 6.00 A in it?
3. What is the effective value of an ac voltage whose maximum voltage is 165 V?
4. What is the maximum current in a circuit that has a current of 4.00 A in it?
5. What is the effective value of a current that reaches a maximum of 17.0 A?

**26-3**
**ac POWER**

When the load has only resistance, power in ac circuits is determined in the same way as in dc circuits.

$$P = I^2R = VI = \frac{V^2}{R}$$

## EXAMPLE 1

What power is expended in a resistance of 37.0 Ω if it has a current of 0.480 A flowing through it?

DATA:

$$R = 37.0 \ \Omega$$
$$I = 0.480 \ A$$
$$P = ?$$

BASIC EQUATION:

$$P = I^2R$$

WORKING EQUATION: Same

SUBSTITUTION:

$$P = (0.480 \ A)^2(37.0 \ \Omega)$$
$$= 8.52 \ W$$

## EXAMPLE 2

What power is expended in a load of 12.0 Ω resistance if the voltage drop across it is $11\overline{0}$ V?

DATA:

$$R = 12.0 \ \Omega$$
$$V = 11\overline{0} \ V$$
$$P = ?$$

BASIC EQUATION:

$$P = \frac{V^2}{R}$$

WORKING EQUATION: Same

SUBSTITUTION:

$$P = \frac{(11\overline{0} \ V)^2}{12.0 \Omega}$$
$$= 1010 \ W$$

The preceding relationships are true only when $e$ and $i$ are in phase. Phase differences are produced by capacitance and inductance in ac circuits. These new effects are called *impedance*. Capacitance, inductance, and impedance will be studied in Chapter 28.

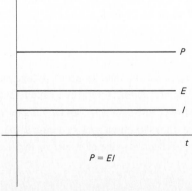

dc circuit

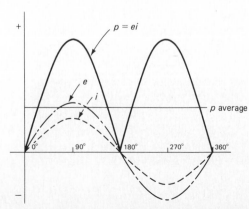

ac circuit

Note that, in the graph comparing dc and ac power on page 300, ac power varies but is always positive (+). The sign indicates only the direction of the current. Even so, $P$ is positive in calculations because the product of $-e$ and $-i$ is positive: $P = (-e)(-i) = ei$.

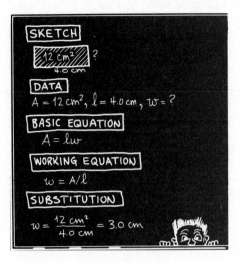

## PROBLEMS

1. A soldering iron is rated at $35\overline{0}$ W. If the current in the iron is 4.00 A, what is the resistance of the iron?

2. What power is developed by a device that draws 6.00 A and has a resistance of 12.0 Ω?

3. What is the output power of a transformer that has an output voltage of $50\overline{0}$ V and current 7.00 A?

4. A heater operates on a $11\overline{0}$ V line and is rated at $45\overline{0}$ W. What is the resistance of the element?

5. A heating element draws 6.00 A on a $22\overline{0}$-V line. What power is expended in the element?

6. A resistance coil has a resistance of 32.0 Ω. If it expends 375 W of power, what is the current in the coil?

7. What power is used by a heater that has a resistance of 12.0 Ω and draws a current of 7.00 A?

**26-4**
**COMMERCIAL GENERATOR**
**POWER OUTPUT**

The power output of the generator is the product of voltage and current. The ac generator converts mechanical energy to electrical energy by performing three functions:

1. The production of voltage—electrical pressure which pushes the current through the loads.
2. The production of power current which is converted into heat, light, and mechanical power.
3. The production of magnetizing current—current transferred back and forth for magnetizing purposes in the generation of electrical power—it is called reactive kVA (kilovolt-amperes).

*Apparent power and reactive kVA* If the current and voltage are not in phase, the resultant product of current and voltage is *apparent power* instead of actual power. Apparent power is measured in kVA (kilovolt-amperes). Actual power is the product of apparent power and the *power factor*.

$$\text{Power factor} = \frac{\text{actual power}}{\text{apparent power}}$$

where the actual power is measured in kW, the apparent power is measured in kVA and is called reactive kVA, and the power factor is a unitless ratio less than 1. Note that 1 VA = 1 W.

Mathematically, the power factor is equal to the cosine of the angle by which the current lags behind (or in rare cases leads) the voltage.

The power factor is really a correction factor which must be applied to determine actual power produced. The situation is very similar to finding the amount of work done when a force and the motion are not in the same direction.

What is the actual power produced by a generating system that produces 13,600 kVA with a power factor of 0.900?

DATA:

$$\text{apparent power} = 13,600 \text{ kVA}$$
$$\text{power factor} = 0.900$$
$$\text{actual power} = ?$$

BASIC EQUATION:

$$\text{power factor} = \frac{\text{actual power}}{\text{apparent power}}$$

WORKING EQUATION:

$$\text{actual power} = (\text{apparent power})(\text{power factor})$$

SUBSTITUTION:

$$\text{actual power} = (13,600 \text{ kVA})(0.900)$$
$$= 12,200 \text{ kVA}$$
$$= 12,200 \text{ kW}$$

SKETCH

DATA
$A = 12 \text{ cm}^2$, $l = 4.0 \text{ cm}$, $w = ?$

BASIC EQUATION
$A = lw$

WORKING EQUATION
$w = A/l$

SUBSTITUTION
$w = \dfrac{12 \text{ cm}^2}{4.0 \text{ cm}} = 3.0 \text{ cm}$

## PROBLEMS

1. What is the actual power produced by a generating station that produces 14,800 kVA with a power factor of 0.85?
2. A generating station operates with a power factor of 0.87. What actual power is available on the transmission lines if the apparent power is 12,800 kVA?
3. What is the apparent power produced by a generating station whose actual power is 120,000 kW and whose power factor is 0.90?
4. What is the apparent power produced by a generating station whose actual power is 1,900,000 kW and whose power factor is 0.80?

**26-5
USES
OF ac AND dc**

The uses of direct current in industry are somewhat limited. Primary applications are in charging storage batteries, electroplating, the generation of alternating current, electrolysis, electromagnets, and automobile ignition systems. In all cases, however, ac can be changed to dc by a simple device called a *rectifier*.

Much more can be done with alternating current. From Mom's mixer to the largest industrial motors, ac finds wide application. There are very practical reasons for this. The voltage of ac can be easily and efficiently changed in transformers to give almost any desired values. Transformers will be studied in Chapter 27. Ac can be used for most purposes just as efficiently as dc. A big advantage of ac is that it can be transmitted over large distances with very little heat loss. [Heat lost in any electrical device is found by the formula: heat loss (power) $P = I^2R$. The energy wasted as heat can be reduced by making the current smaller. Transformers reduce the current by increasing the voltage.]

# CHAPTER
# 27

# Transformers

**27-1**
**CHANGING VOLTAGE WITH TRANSFORMERS**

The major advantage of ac over dc is that ac voltage can be easily changed to meet our needs. It can also be transmitted with very little waste of power due to heat loss $(I^2R)$. The amount of the loss is determined by the square of the current.

### EXAMPLE 1

A plant generates 50.0 kW (50,$\bar{0}$00 W) of power to be sent to a substation on a line with a resistance of 3.00 $\Omega$. We know that some power will be lost as heat during the transmission. The power lost is $P_{lost} = I^2R$.

(a) How much power is lost if the transmission is at 11$\bar{0}$0 V?
(b) What percent of the power generated is lost in transmission at 11$\bar{0}$0 V?
(c) How much power is lost if the transmission is at 11,$\bar{0}$00 V?
(d) What percent of the power generated is lost in transmission at 11,$\bar{0}$00 V?

 Now compare the power losses at the two different transmission voltages.
(a) At 11$\bar{0}$0 V:

$$P = VI$$

$$I = \frac{P}{V}$$

$$I = \frac{50,\bar{0}00 \text{ W}}{11\bar{0}0 \text{ V}}$$

$$= 45.5 \text{ A}$$

$$P_{lost} = I^2R$$

$$P_{lost} = (45.5 \text{ A})^2(3.00 \text{ }\Omega)$$

$$= 6210 \text{ W}$$

$$= 6.21 \text{ kW}$$

(b)

$$\%_{lost} = \frac{\text{power lost}}{\text{power generated}} \times 100\%$$

$$\%_{lost} = \frac{6.21 \text{ kW}}{50.0 \text{ kW}} \times 100\%$$

$$= 0.124 \times 100\%$$

$$= 12.4\%$$

(c) At $11,\overline{0}00$ V:

$$P = VI$$

$$I = \frac{P}{V}$$

$$I = \frac{50,\overline{0}00 \text{ W}}{11,\overline{0}00 \text{ V}}$$

$$= 4.55 \text{ A}$$

$$P_{\text{lost}} = I^2 R$$

$$P_{\text{lost}} = (4.55 \text{ A})^2 (3.00 \ \Omega)$$

$$= 62.1 \text{ W}$$

$$= 0.0621 \text{ kW}$$

(d)

$$\%_{\text{lost}} = \frac{\text{power lost}}{\text{power generated}} \times 100\%$$

$$\%_{\text{lost}} = \frac{0.0621 \text{ kW}}{50.0 \text{ kW}} \times 100\%$$

$$= 0.00124 \times 100\%$$

$$= 0.124\%$$

This example shows that where 12.4% of the power would be lost during transmission at $11\overline{0}0$ V, only 0.124% would be lost at $11,\overline{0}00$ V. So by increasing the voltage, the current is correspondingly lowered, and the power wasted in transmission is greatly reduced. The transformer is a device that can be used to change the voltage to reduce the current and thereby lessen the power loss.

A transformer consists of two coils of wire wrapped on an iron core. It is used to change the voltage of electricity.

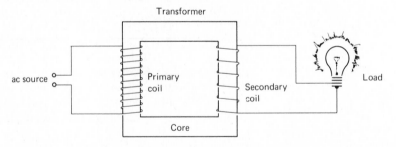

Transformer

When an alternating current passes through the primary coil, it induces an alternating magnetic field in the core. This magnetic field in turn induces an alternating voltage in the secondary coil. The magnitude of the voltage induced in the secondary coil depends on:

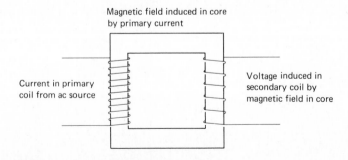

1. The voltage applied to the primary coil.
2. The number of turns on the primary coil.
3. The number of turns on the secondary coil.
4. The power loss between primary and secondary coils.

If we assume that there is no power loss between the primary and secondary coils, we have the following equation:

$$\frac{V_P}{V_S} = \frac{N_P}{N_S}$$

where: $V_P$ = primary voltage
$V_S$ = secondary voltage
$N_P$ = number of primary turns
$N_S$ = number of secondary turns

## EXAMPLE 2

A transformer on a neon sign has $10\bar{0}$ turns on its primary coil and $15,\bar{0}00$ turns on its secondary coil. If the voltage applied to the primary is $11\bar{0}$ V, what is the secondary voltage?

DATA:

$$V_P = 11\bar{0} \text{ V}$$
$$N_P = 10\bar{0} \text{ turns}$$
$$N_S = 15,\bar{0}00 \text{ turns}$$
$$V_S = ?$$

BASIC EQUATION:

$$\frac{V_P}{V_S} = \frac{N_P}{N_S}$$

WORKING EQUATION:

$$V_S = \frac{V_P N_S}{N_P}$$

SUBSTITUTION:

$$V_S = \frac{(11\bar{0} \text{ V})(15,\bar{0}00 \text{ turns})}{10\bar{0} \text{ turns}}$$
$$= 16,500 \text{ V}$$

Since transformers are used to raise or lower voltage, they are called either step-up or step-down transformers.

*Step-up transformers* are used when a high voltage is needed to operate X-ray tubes, neon signs, and for transmission of electric power over long distances. A step-up transformer raises the voltage by having more turns on the secondary than on the primary coil.

*Step-down transformers* are used to lower the voltage from high-voltage transmission lines to regular 110 V and 220 V for home and industrial use. Voltage is lowered in the step-down transformer because it has more turns in the primary coil than in the secondary coil.

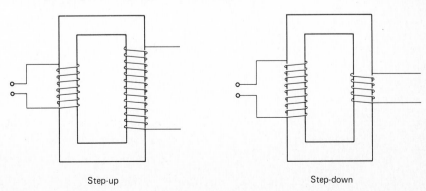

Step-up                         Step-down

Auto transformers are used when a variable output voltage is needed. In this type of transformer, contact can be made across a variable number of the secondary coils using a brush contact. The output voltage is therefore variable from nearly zero to some maximum value. This type of transformer is often used to supply ac power to resistive heater elements to control the heating output.

**27-2**
**TRANSFORMERS AND POWER**

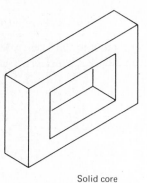

Solid core

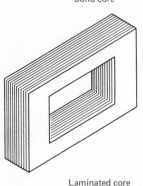

Laminated core

Transformers do not create energy. Some energy is lost, however, during the change of voltage. Energy losses in transformers are of three types:

1. *Copper losses*—these result from the resistance of the copper wires in the coils and are unavoidable.
2. *Magnetic losses* (called *hysteresis losses*)—some energy is lost (turned into heat) by reversing the magnetism in the core.
3. *Eddy currents*—when a mass of metal (the core) is subjected to a changing magnetic field, currents are set up in the metal which do no useful work, waste energy, and produce heat. These losses can be lessened by *laminating* the core. Instead of using a solid block of metal for the core, thin sheets of metal with insulated surfaces are used, reducing these induced currents.

When a transformer steps up the voltage applied to its primary, it reduces the current. Energy is conserved—we cannot get any more electrical energy out of a transformer than we put into it. The relationship between primary and secondary currents is

$$\frac{I_S}{I_P} = \frac{N_P}{N_S}$$

where:   $I_S$ = current in secondary coil
$I_P$ = current in primary coil
$N_P$ = number of turns in primary
$N_S$ = number of turns in secondary

## EXAMPLE 1

The primary current in a transformer is 10.0 A. If the primary has $55\bar{0}$ turns and the secondary has $250\bar{0}$ turns, what current flows in the secondary coil?

DATA:

$$N_P = 55\bar{0} \text{ turns}$$
$$I_P = 10.0 \text{ A}$$
$$N_S = 250\bar{0} \text{ turns}$$
$$I_S = ?$$

BASIC EQUATION:

$$\frac{I_S}{I_P} = \frac{N_P}{N_S}$$

WORKING EQUATION:

$$I_S = \frac{I_P N_P}{N_S}$$

SUBSTITUTION:

$$I_S = \frac{(10.0 \text{ A})(55\bar{0} \text{ turns})}{250\bar{0} \text{ turns}}$$

$$= 2.20 \text{ A}$$

Another way of showing power is conserved is:

$$\boxed{I_P V_P = I_S V_S \quad \text{or} \quad P_P = P_S}$$

where: $I_P$ = current in primary coil
$I_S$ = current in secondary coil
$V_P$ = voltage in primary coil
$V_S$ = voltage in secondary coil

where: $P_P$ = power (watts) in primary coil
$P_S$ = power (watts) in secondary coil

## EXAMPLE 2

The power in the primary coil of a transformer is 375 W. If the current in the secondary is 11.4 A, what is the voltage in the secondary?

DATA:

$$P_P = 375 \text{ W}$$
$$I_S = 11.4 \text{ A}$$
$$V_S = ?$$

BASIC EQUATIONS:

$$P_P = P_S \quad \text{and} \quad P_S = V_S I_S$$

WORKING EQUATIONS:

$$P_P = V_S I_S \quad (\textit{Note:} \text{ Substitute for } P_S)$$
$$V_S = \frac{P_P}{I_S}$$

SUBSTITUTION:

$$V_S = \frac{375 \text{ W}}{11.4 \text{ A}}$$
$$= 32.9 \text{ V}$$

Good transformers are more than 98% efficient. This is very important in power transmission. It is impractical to generate electricity at high voltage, but high voltage is desirable for transmission. Therefore, transformers are used to step up the voltage for transmission. High voltage is unsuitable, though, for consumer use, so transformers are used to reduce the voltage.

A simplified diagram of a power distribution system follows.

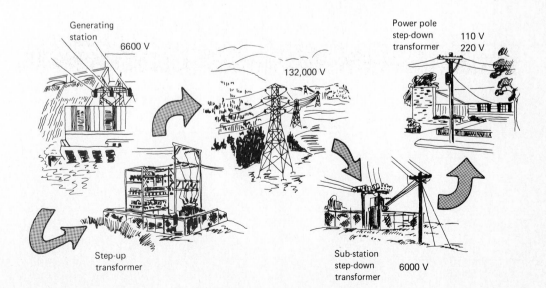

Generating station

6600 V

132,000 V

Power pole step-down transformer   110 V   220 V

Step-up transformer

Sub-station step-down transformer   6000 V

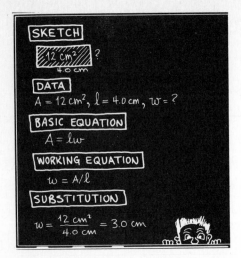

SKETCH

12 cm² ?
4.0 cm

DATA
$A = 12 \text{ cm}^2$, $\ell = 4.0 \text{ cm}$, $w = ?$

BASIC EQUATION
$A = \ell w$

WORKING EQUATION
$w = A/\ell$

SUBSTITUTION
$w = \dfrac{12 \text{ cm}^2}{4.0 \text{ cm}} = 3.0 \text{ cm}$

## PROBLEMS

1. $V_P = 20.0$ V
   $V_S = 30.0$ V
   $N_S = 15.0$ turns
   Find $N_P$.

2. $V_P = 25\overline{0}$ V
   $N_P = 73\overline{0}$ turns
   $N_S = 375$ turns
   Find $V_S$.

3. $I_P = 6.00$ A
   $I_S = 4.00$ A
   $V_P = 39.0$ V
   Find $V_S$.

4. A step-up transformer on a 115-V line provides a voltage of $230\overline{0}$ V. If the primary has 65.0 turns, how many turns does the secondary have?

5. A step-down transformer on a 115-V line provides a voltage of 11.5 V. If the secondary has 35.0 turns, how many turns does the primary have?

6. A transformer has 20.0 turns in the primary coil and $220\overline{0}$ in the secondary. If the primary voltage is 12.0 V, what is the secondary voltage?

7. If there is a current of 7.00 A in the primary in Problem 3, what is the current in the secondary?

8. If the voltage in the secondary coil of a transformer is $11\overline{0}$ V and the current in it is 15.0 A, what power does it supply?

9. A neon sign has a transformer that changes electricity from $11\overline{0}$ V to $15,\overline{0}00$ V. If the primary current is 10.0 A, what is the current in the secondary?

10. What is the power in the primary in Problem 9?

11. A transformer has an output power of 990 W. If the current in the secondary is 0.45 A, what is the voltage in the secondary?

12. The current in the secondary of a transformer is 3.00 A. What is the voltage in the secondary if the power is 775 W?

13. A transformer steps down $66\overline{0}0$ V to $12\overline{0}$ V. If the secondary current is 14.0 A, what is the primary current?

14. What is the power in the primary in Problem 13?

308

# *ac Circuits*

**28-1**
**INDUCTANCE**

Electronic circuitry in televisions, radios, computers, and electronic instruments has many components other than resistors. These components include capacitors, inductors, diodes, and transistors. With these components weak signals can be amplified, noise can be reduced, and signals at certain frequencies can be detected while signals from other frequencies can be rejected (that is, a circuit can be "tuned in" to a frequency). The analysis of the behavior of circuits with these components can become very complex. As a start toward understanding these circuits, we discuss inductors and capacitors. The operation of diodes and transistors will be discussed only briefly. We begin with a discussion of inductors.

A coil of wire in an ac circuit opposes a change in the value of the current. This is due to the emf induced in the coil itself as the magnetic field of the coil changes. This emf opposes a change in the current.

The unit of inductance *(L)* is the henry (H). A coil has an inductance of 1 henry if an emf of 1 volt is induced when the current changes at the rate of 1 A/s. A henry can be expressed as $\Omega$ s.

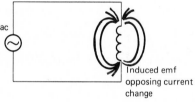

$$\boxed{1 \text{ henry} = 1\ \Omega\ \text{s}}$$

The henry is a large unit. A more practical unit is the millihenry (mH), which is one-thousandth of a henry.

Inductance can be illustrated by connecting a coil with a large number of turns and a lamp in series. When connected to a dc source, the lamp burns brightly. However, when this circuit is connected to an ac power source of the same voltage, the lamp is dimmer because of the inductance of the coil.

Following is the circuit symbol for inductance.

**28-2**
**INDUCTIVE REACTANCE**

The opposition to ac current flow in an inductor is called *inductive reactance* and is measured in ohms. This is usually represented by $X_L$. The inductive reactance of a coil is directly proportional to frequency and is found by the following:

$$X_L = 2\pi fL$$

where: $X_L$ = inductive reactance
$f$ = frequency of the ac voltage, expressed in hertz (cycles per second), such as $6\bar{0}$ Hz or $6\bar{0}$/s
$L$ = inductance, in henries

The current in a circuit that has only an ac voltage source and an inductor is given by

$$I = \frac{E}{X_L}$$

where: $I$ = current
$E$ = voltage
$X_L$ = inductive reactance

## EXAMPLE

A coil with inductance of 0.100 H is connected to a 60.0-Hz ac power source of $11\bar{0}$ V. What is the current in the circuit?

SKETCH:

DATA:

$E = 11\bar{0}$ V
$L = 0.100$ H
$f = 60.0$ Hz $= 60.0$/s
$I = ?$

BASIC EQUATIONS:

$$X_L = 2\pi fL \quad \text{and} \quad I = \frac{E}{X_L}$$

WORKING EQUATIONS: Same

SUBSTITUTIONS:

$$X_L = 2\pi(60.0/\text{s})(0.100 \text{ H})$$
$$= 37.7 \frac{\text{H}}{\text{s}}$$
$$= 37.7 \frac{\text{H}}{\cancel{s}}\left(\frac{1 \,\Omega \,\cancel{s}}{1 \,\cancel{H}}\right) \quad \text{(note conversion factor)}$$
$$= 37.7 \,\Omega$$

$$I = \frac{E}{X_L}$$
$$I = \frac{11\bar{0} \text{ V}}{37.7 \,\Omega}$$
$$= 2.92 \frac{\cancel{V}}{\cancel{\Omega}}\left(\frac{\text{A}\cancel{\Omega}}{\cancel{V}}\right)$$
$$= 2.92 \text{ A}$$

In an inductive circuit the current lags behind the voltage by a quarter of a cycle. The maximum voltage in a 60-Hz circuit thus occurs

$$\frac{1}{4} \times \frac{1}{60} \, \text{s} = \frac{1}{240} \, \text{s}$$

before the maximum current. The current lag is usually measured in degrees. A quarter of a cycle is 90°.

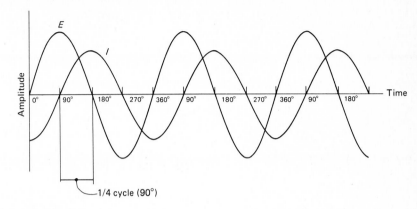

1/4 cycle (90°)

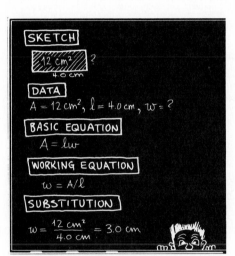

## PROBLEMS

*Calculate the inductive reactance of each inductance at the given frequency.*

1. $L = 5.00 \times 10^{-3}$ H, $f = 6\bar{0}$ Hz, $X_L =$ _____ Ω
2. $L = 2.00 \times 10^{-2}$ H, $f = 9\bar{0}$ Hz, $X_L =$ _____ Ω
3. $L = 7.00 \times 10^{-4}$ H, $f = 1.00 \times 10^4$ Hz, $X_L =$ _____ Ω
4. $L = 8.00 \times 10^{-1}$ H, $f = 1.00 \times 10^1$ Hz, $X_L =$ _____ Ω
5. $L = 4.00 \times 10^{-2}$ H, $f = 1.00 \times 10^3$ Hz, $X_L =$ _____ Ω

*Calculate the current in the inductive circuits with the following characteristics.*

6. $L = 3.00 \times 10^{-4}$ H, $f = 10\bar{0}$ Hz, $E = 17.0$ V, $I =$ _____ A
7. $L = 1.00 \times 10^{-3}$ H, $f = 1.00 \times 10^4$ Hz, $E = 15\bar{0}$ V, $I =$ _____ A
8. $L = 5.00 \times 10^{-3}$ H, $f = 2.00 \times 10^3$ Hz, $E = 50.0$ V, $I =$ _____ A
9. $L = 4.00 \times 10^{-2}$ H, $f = 7.00 \times 10^2$ Hz, $E = 27.0$ V, $I =$ _____ A
10. $L = 7.20 \times 10^{-3}$ H, $f = 1.00 \times 10^6$ Hz, $E = 10\bar{0}$ V, $I =$ _____ A

**28-3**
**INDUCTANCE**
**AND RESISTANCE**
**IN SERIES**

Most ac circuits have resistance in the form of lights or resistors in addition to inductance. The current lags behind the voltage by any amount of time greater than zero and as large as a quarter cycle.

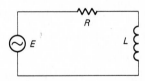

The effect of both the resistance and the inductance on a circuit is called the *impedance*. Ohm's law in an ac circuit can be written as

$$I = \frac{E}{Z}$$

where:  $I =$ current
$E =$ voltage
$Z =$ impedance

The impedance of a series circuit containing a resistance and an inductance is

$$Z = \sqrt{R^2 + X_L{}^2}$$
$$Z = \sqrt{R^2 + (2\pi fL)^2}$$

where:  $Z$ = impedance
$R$ = resistance
$X_L$ = inductive reactance
$f$ = frequency
$L$ = inductance

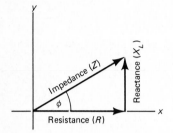

The impedance can be represented vectorially as the hypotenuse of the right triangle shown. The resistance is always drawn as a vector pointing in the positive $x$ direction. The inductive reactance is drawn as a vector pointing in the positive $y$ direction.

The angle $\phi$ shown is the *phase angle* and is equal to the amount by which the current lags behind the voltage. The phase angle is given by

$$\tan \phi = \frac{X_L}{R}$$

EXAMPLE
___

A lamp of resistance 40.0 $\Omega$ is connected in series with an inductance of 0.100 H. This circuit is connected to a $11\bar{0}$-V, 60.0-Hz power supply. What is the current in the circuit? What is the phase angle?

SKETCH:

$R = 40.0 \ \Omega$
$E = 11\bar{0}$ V
$f = 60.0$ Hz
$L = 0.100$ H

DATA:

$E = 11\bar{0}$ V
$f = 60.0$ Hz $= 60.0/\text{s}$
$R = 40.0 \ \Omega$
$L = 0.100$ H
$I = ?$
$\phi = ?$

BASIC EQUATIONS:

$$Z = \sqrt{R^2 + (2\pi fL)^2} \quad \text{and} \quad I = \frac{E}{Z}$$

WORKING EQUATIONS:   First calculate the impedance.

$$Z = \sqrt{(40.0 \ \Omega)^2 + [(2\pi)(60.0/\text{s})(0.100 \ \text{H})]^2}$$
$$= \sqrt{16\bar{0}0 \ \Omega^2 + (37.7 \ \text{H/s})^2}$$
$$= \sqrt{16\bar{0}0 \ \Omega^2 + \left[\left(37.7 \frac{\text{H}}{\text{s}}\right)\left(\frac{1 \ \Omega \ \text{s}}{1 \ \text{H}}\right)\right]^2} \quad \text{(note conversion factor)}$$
$$= \sqrt{16\bar{0}0 \ \Omega^2 + 1420 \ \Omega^2}$$
$$= \sqrt{3020 \ \Omega^2}$$
$$= 55.0 \ \Omega$$

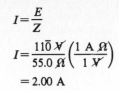

$$I = \frac{E}{Z}$$

$$I = \frac{11\bar{0} \, V}{55.0 \, \Omega} \left( \frac{1 \, A \, \Omega}{1 \, V} \right)$$

$$= 2.00 \, A$$

To find the phase angle ($\phi$), we construct the vector right triangle. Then find the angle whose tangent is given by

$$\tan \phi = \frac{X_L}{R}$$

$$\tan \phi = \frac{37.7 \, \Omega}{40.0 \, \Omega}$$

$$\tan \phi = 0.943$$

$$\phi = 43°$$

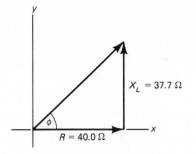

$X_L = 37.7 \, \Omega$

$R = 40.0 \, \Omega$

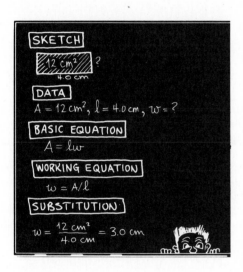

## PROBLEMS

*Find the impedance of the circuits with the following characteristics.*

1. $R = 200 \, \Omega$, $L = 1.00 \times 10^{-2}$ H, $f = 15\bar{0}0$ Hz, $Z =$ _____ $\Omega$
2. $R = 16.0 \, \Omega$, $L = 1.00 \times 10^{-3}$ H, $f = 90\bar{0}$ Hz, $Z =$ _____ $\Omega$
3. $R = 1.00 \times 10^3 \, \Omega$, $L \doteq 5.00 \times 10^{-2}$ H, $f = 1.00 \times 10^4$ Hz, $Z =$ _____ $\Omega$
4. $R = 2.00 \times 10^3 \, \Omega$, $L = 7.00 \times 10^{-2}$ H, $f = 5.00 \times 10^3$ Hz, $Z =$ _____ $\Omega$
5. $R = 3.00 \times 10^2 \, \Omega$, $L = 2.00 \times 10^{-3}$ H, $f = 40\bar{0}0$ Hz, $Z =$ _____ $\Omega$
6. Find the phase angle in Problem 1.
7. Find the phase angle in Problem 2.
8. Find the phase angle in Problem 3.
9. Find the phase angle in Problem 4.
10. Find the phase angle in Problem 5.
11. Find the current in Problem 1 if the voltage is 50.0 V.
12. Find the current in Problem 2 if the voltage is 10.0 V.
13. Find the current in Problem 3 if the voltage is 15.0 V.
14. Find the current in Problem 4 if the voltage is 12.0 V.
15. Find the current in Problem 5 if the voltage is 8.00 V.

**28-4**
**CAPACITANCE**

An important component of many ac circuits is the *capacitor*. A capacitor consists of two conductors which are usually parallel plates separated by a thin insulator. The plates are often made of a metal foil which is rolled to a convenient size. Capacitors are represented in circuit diagrams, as shown.

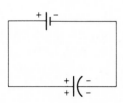

The unique property of a capacitor is that it can build up and store charge. When a capacitor is connected to a battery, electrons flow from the negative terminal to one capacitor plate as shown at left.

When the capacitor is removed from the battery, the charges remain on the capacitor.

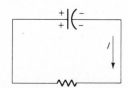

If the capacitor is then connected to a resistor, electrons will flow through the circuit until the capacitor has lost its charge.

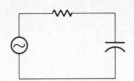

A capacitor in an ac circuit causes a current to flow when the ac voltage is zero. The unit of capacitance is the *farad* (F). A more practical unit is the microfarad ($\mu$F; $10^{-6}$ F). The effect of a capacitor on a circuit is inversely proportional to frequency and is measured as capacitive reactance. It is measured in ohms and given by

$$X_C = \frac{1}{2\pi f C}$$

where:  $X_C$ = capacitive reactance
$f$ = frequency
$C$ = capacitance (farads)

$$1 \text{ F} = 1 \text{ s}/\Omega$$

In a circuit that contains only capacitors, the current *leads* the voltage by 90° (one quarter cycle).

EXAMPLE
_____

Calculate the capacitive reactance of a $1.00 \times 10^{-5}$ F capacitor in a circuit of frequency $10\bar{0}$ Hz.

DATA:

$$C = 1.00 \times 10^{-5} \text{ F}$$
$$f = 10\bar{0} \text{ Hz} = 10\bar{0}/\text{s}$$
$$X_C = ?$$

BASIC EQUATION:

$$X_C = \frac{1}{2\pi f C}$$

WORKING EQUATION:   Same

SUBSTITUTION:

$$X_C = \frac{1}{2\pi(10\bar{0}/\text{s})(1.00 \times 10^{-5} \text{ F})}$$

$$= \frac{1}{\left(6.28 \times 10^{-3} \frac{\text{F}}{\text{s}}\right)\left(\frac{1 \text{ s}}{1 \text{ F } \Omega}\right)} \quad \text{(note conversion factor)}$$

$$= 159 \ \Omega$$

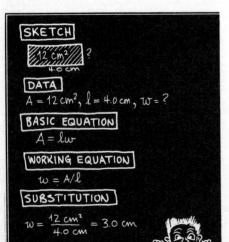

## PROBLEMS

*Find the capacitive reactance of the circuits with the following characteristics.*

1. $C = 4.00 \times 10^{-5}$ F, $f = 1.00 \times 10^3$ Hz, $X_C = $ _____ $\Omega$
2. $C = 7.00 \times 10^{-3}$ F, $f = 1.00 \times 10^2$ Hz, $X_C = $ _____ $\Omega$
3. $C = 6.00 \times 10^{-4}$ F, $f = 1.00 \times 10^2$ Hz, $X_C = $ _____ $\Omega$
4. $C = 3.00 \times 10^{-3}$ F, $f = 1.00 \times 10^4$ Hz, $X_C = $ _____ $\Omega$
5. $C = 8.00 \times 10^{-4}$ F, $f = 1.00 \times 10^5$ Hz, $X_C = $ _____ $\Omega$

**28-5**
CAPACITANCE
AND RESISTANCE
IN SERIES

The combined effect of capacitance and resistance in series is measured by the impedance of the circuit, $Z$.

$$Z = \sqrt{R^2 + X_C^2}$$

$$Z = \sqrt{R^2 + \left(\frac{1}{2\pi f C}\right)^2}$$

where   $Z$ = impedance
$R$ = resistance
$X_C$ = capacitive reactance
$f$ = frequency
$C$ = capacitance

The current is given by *Ohm's law:*

$$I = \frac{E}{Z}$$

where:   $I$ = current
$E$ = voltage
$Z$ = impedance

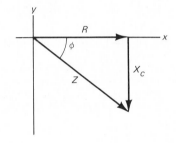

The phase angle can be found by drawing the resistance as a vector in the positive $x$ direction and the capacitive impedance as a vector in the negative $y$ direction as shown. The phase angle gives the amount by which the voltage lags the current.

$$\tan \phi = \frac{X_C}{R}$$

EXAMPLE

What current will flow in a 60.0-Hz ac circuit that includes a $11\overline{0}$-V source, a capacitor of 40.0 $\mu$F, and a 16.0-$\Omega$ resistance in series? Also find the phase angle.

SKETCH:

$R = 16.0\ \Omega$

$E = 11\overline{0}$ V
$f = 60.0$ Hz

$C = 40.0\ \mu$F

DATA:

$E = 11\overline{0}$ V

$f = 60.0$ Hz $= 60.0/$s

$R = 16.0\ \Omega$

$C = 40.0\ \mu$F $= 4.00 \times 10^{-5}$ F

$Z = ?$

$I = ?$

BASIC EQUATIONS:

$$Z = \sqrt{R^2 + \left(\frac{1}{2\pi f C}\right)^2} \quad \text{and} \quad I = \frac{E}{Z}$$

315

WORKING EQUATIONS:   Same

SUBSTITUTIONS:

First, find $Z$:

$$Z = \sqrt{(16.0 \ \Omega)^2 + \left(\frac{1}{2\pi \times 60.0/\text{s} \times 4.00 \times 10^{-5} \ \text{F}}\right)^2}$$

$$= \sqrt{256 \ \Omega^2 + \left(\frac{1}{0.0151 \ \text{F/s}}\right)^2}$$

$$= \sqrt{256 \ \Omega^2 + \left(66.2 \ \frac{\cancel{s}}{\cancel{F}} \times \frac{1 \ \Omega \ \cancel{F}}{1 \ \cancel{s}}\right)^2}$$

$$= \sqrt{256 \ \Omega^2 + 4380 \ \Omega^2}$$

$$= 68.1 \ \Omega$$

Then use $Z$ to find $I$:

$$I = \frac{E}{Z}$$

$$I = \frac{11\overline{0} \ \text{V}}{68.1 \ \Omega}$$

$$= 1.62 \ \frac{\cancel{V}}{\cancel{\Omega}} \left(\frac{1 \ \cancel{\Omega} \ \text{A}}{1 \ \cancel{V}}\right)$$

$$= 1.62 \ \text{A}$$

Then find the phase angle:

$$\tan \phi = \frac{X_C}{R}$$

$$\tan \phi = \frac{66.2 \ \cancel{\Omega}}{16.0 \ \cancel{\Omega}}$$

$$\tan \phi = 4.14$$

so

$$\phi = 76°$$

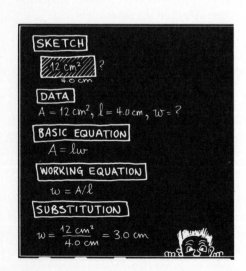

## PROBLEMS

*In Problems 1 through 5, find the impedance in each circuit.*

1. $R = 1.00 \times 10^3 \ \Omega$, $C = 1.00 \times 10^{-6}$ F, $V = 1.0\overline{0}$ V, $f = 20\overline{0}$ Hz, $Z = $ _____ $\Omega$
2. $R = 25\overline{0} \ \Omega$, $C = 5.00 \times 10^{-6}$ F, $V = 20.0$ V, $f = 50\overline{0}$ Hz, $Z = $ _____ $\Omega$
3. $R = 7.80 \ \Omega$, $C = 45.0 \times 10^{-6}$ F, $V = 15.0$ V, $f = 95\overline{0}$ Hz, $Z = $ _____ $\Omega$
4. $R = 145 \ \Omega$, $C = 1.00 \times 10^{-5}$ F, $V = 7.00$ V, $f = 75\overline{0}$ Hz, $Z = $ _____ $\Omega$
5. $R = 10.0 \ \Omega$, $C = 5.00 \times 10^{-5}$ F, $V = 15.0$ V, $f = 1.00 \times 10^2$ Hz, $Z = $ _____ $\Omega$
6. Find the phase angle in Problem 1.
7. Find the phase angle in Problem 2.
8. Find the phase angle in Problem 3.
9. Find the phase angle in Problem 4.
10. Find the phase angle in Problem 5.
11. Find the current in Problem 1.
12. Find the current in Problem 2.
13. Find the current in Problem 3.
14. Find the current in Problem 4.
15. Find the current in Problem 5.

Many circuits that are important in the design of electronic equipment contain all three types of circuit elements discussed in this chapter. The impedance of a circuit containing resistance, capacitance, and inductance in series can be found from the equation

$$Z = \sqrt{R^2 + (X_L - X_C)^2}$$

where: $Z$ = impedance
$R$ = resistance
$X_L$ = inductive reactance
$X_C$ = capacitive reactance

The vector diagram for this type of circuit is shown in the figures. The phase angle is given by

$$\tan \phi = \frac{X_L - X_C}{R}$$

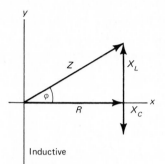

Inductive

In a circuit containing $R$, $L$, and $C$ components, the circuit is said to be *inductive* if $X_L > X_C$ and the current lags the voltage. A circuit is *capacitive* if $X_C > X_L$. In this case the voltage and current are *in phase*. It is desirable to keep the voltage and current in phase if the circuit power is to be maximized.

The current in this type of circuit is given by

$$I = \frac{E}{Z} = \frac{E}{\sqrt{R^2 + \left(2\pi f L - \dfrac{1}{2\pi f C}\right)^2}}$$

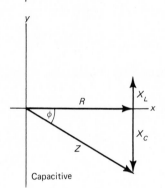

Capacitive

EXAMPLE

A circuit contains a $10\bar{0}$-$\Omega$ resistance, a $1.00 \times 10^{-5}$ F capacitor, and a $1.00 \times 10^{-2}$ H inductance in series with a 25.0-V, $20\bar{0}$-Hz source. Find the impedance and the current.

SKETCH:

$$R = 10\bar{0}\ \Omega$$

$E = 25.0\ \text{V} \quad L = 1.00 \times 10^{-2}\ \text{H}$

$f = 20\bar{0}\ \text{Hz}$

$C = 1.00 \times 10^{-5}\ \text{F}$

DATA:

$$R = 10\bar{0}\ \Omega$$
$$C = 1.00 \times 10^{-5}\ \text{F}$$
$$L = 1.00 \times 10^{-2}\ \text{H}$$
$$E = 25.0\ \text{V}$$
$$f = 20\bar{0}\ \text{Hz} = 20\bar{0}/\text{s}$$
$$Z = ?$$
$$I = ?$$

317

$$X_L = 2\pi f L$$

$$X_C = \frac{1}{2\pi f C}$$

$$Z = \sqrt{R^2 + (X_L - X_C)^2}$$

$$I = \frac{E}{Z}$$

WORKING EQUATIONS: Same

SUBSTITUTIONS:

First find $X_L$:

$$X_L = 2\pi f L$$

$$X_L = 2\pi \ (20\overline{0}/\text{s})(1.00 \times 10^{-2} \ \text{H})$$

$$= 12.6 \ \frac{\cancel{H}}{\cancel{s}}\left(\frac{1 \ \Omega \ \cancel{s}}{1 \ \cancel{H}}\right)$$

$$= 12.6 \ \Omega$$

Then find $X_C$:

$$X_C = \frac{1}{2\pi f C}$$

$$X_C = \frac{1}{2\pi \ (20\overline{0}/\text{s})(1.00 \times 10^{-5} \ \text{F})}$$

$$= 79.6 \ \Omega$$

Then find the impedance:

$$Z = \sqrt{R^2 + (X_L - X_C)^2}$$

$$Z = \sqrt{(10\overline{0} \ \Omega)^2 + (12.6 \ \Omega - 79.6 \ \Omega)^2}$$

$$= \sqrt{1.00 \times 10^4 \ \Omega^2 + (-67.0 \ \Omega)^2}$$

$$= \sqrt{1.00 \times 10^4 \ \Omega^2 + 4490 \ \Omega^2}$$

$$= 12\overline{0} \ \Omega$$

Then find the current:

$$I = \frac{E}{Z}$$

$$I = \frac{25.0 \ \text{V}}{12\overline{0} \ \Omega}$$

$$= 0.208 \ \frac{\cancel{V}}{\cancel{\Omega}}\left(\frac{1 \ \text{A} \ \cancel{\Omega}}{1 \ \cancel{V}}\right)$$

$$= 0.208 \ \text{A}$$

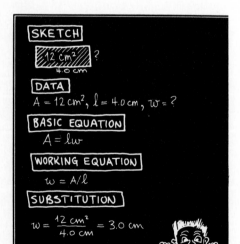

## PROBLEMS

*Find the impedance and current in the following circuits.*

1. $R = 40.0 \ \Omega$, $L = 0.500 \ \text{H}$, $C = 50.0 \times 10^{-6} \ \text{F}$, $f = 60 \ \text{Hz}$, $V = 5.00 \ \text{V}$.
2. $R = 20\overline{0} \ \Omega$, $L = 1.00 \times 10^{-2} \ \text{H}$, $C = 2.00 \times 10^{-7} \ \text{F}$, $f = 1.00 \times 10^3 \ \text{Hz}$, $V = 10.0 \ \text{V}$.
3. $R = 1.00 \times 10^2 \ \Omega$, $L = 1.00 \times 10^{-2} \ \text{H}$, $C = 3.00 \times 10^{-4} \ \text{F}$, $f = 1.00 \times 10^4 \ \text{Hz}$, $V = 15.0 \ \text{V}$.
4. $R = 60.0 \ \Omega$, $L = 0.700 \ \text{H}$, $C = 30.0 \times 10^{-6} \ \text{F}$, $f = 6\overline{0} \ \text{Hz}$, $V = 8.00 \ \text{V}$.

**28-7**
RESONANCE

The current in a circuit containing resistance, capacitance, and inductance is given by the equation

$$I = \frac{E}{\sqrt{R^2 + (X_L - X_C)^2}}$$

When the inductive reactance equals the capacitive reactance, they nullify each other and the current is given by

$$I = \frac{E}{R}$$

which is its maximum possible value. When this condition exists, the circuit is in *resonance* with the applied voltage.

Resonant circuits are used in radios and televisions. The frequency of a certain station is tuned in when a resonant circuit (antenna circuit) is adjusted to that frequency. This is accomplished by changing the capacitance until the capacitive reactance equals the inductive reactance. The applied voltage is the radio signal picked up by the antenna.

The resonant frequency occurs when $X_L = X_C$. To find this frequency, we write

$$X_L = X_C$$
$$2\pi f L = \frac{1}{2\pi f C}$$
$$f^2 = \frac{1}{4\pi^2 LC}$$
$$f = \frac{1}{\sqrt{4\pi^2 LC}}$$

$$f = \frac{1}{2\pi\sqrt{LC}}$$

The circuit can be adjusted to any frequency by varying the capacitance or the inductance.

EXAMPLE

Find the resonant frequency of a circuit containing a $5.00 \times 10^{-9}$ F capacitor in series with a $2.60 \times 10^{-6}$ H inductor.

SKETCH:

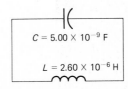

$C = 5.00 \times 10^{-9}$ F

$L = 2.60 \times 10^{-6}$ H

DATA:

$$C = 5.00 \times 10^{-9} \text{ F}$$
$$L = 2.60 \times 10^{-6} \text{ H}$$
$$f = ?$$

BASIC EQUATION:

$$f = \frac{1}{2\pi\sqrt{LC}}$$

WORKING EQUATION:  Same

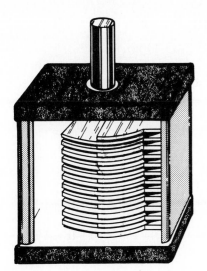

SUBSTITUTION:

$$f=\frac{1}{2\pi\sqrt{2.60\times10^{-6}\text{ H})(5.00}\times10^{-9}\text{ F)}}$$

$$=\frac{1}{2\pi\sqrt{1.30\times10^{-14}\text{ HF}\left(\dfrac{1\ \Omega\ s}{1\ H}\right)\left(\dfrac{1\ s}{1\ F\ \Omega}\right)}}$$

$$=\frac{1}{7.16\times10^{-7}\text{ s}}$$

$$=1.40\times10^{6}\ \frac{\text{cycles}}{\text{s}}\quad\text{or}\quad140\bar{0}\ \frac{\text{kilocycles}}{\text{s}}\quad\text{or}\quad140\bar{0}\text{ kHz}$$

This frequency is in the AM radio band.

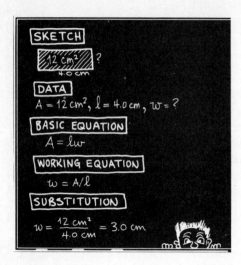

## PROBLEMS

1. Find the resonant frequency of a circuit with $L = 1.00 \times 10^{-5}$ H and $C = 5.00 \times 10^{-6}$ F.

2. Find the resonant frequency of a circuit with $L = 2.00 \times 10^{-4}$ H and $C = 4.00 \times 10^{-5}$ F.

3. Find the resonant frequency of a circuit with $L = 3.00 \times 10^{-2}$ H and $C = 6.00 \times 10^{-4}$ F.

**28-8**
RECTIFICATION

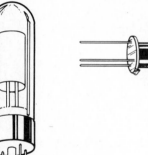

Technicians often find it necessary to change ac into dc. This process is called *rectification*. A device that accomplishes this is called a *diode*. Early diodes were constructed as vacuum tubes.

Many diodes today are constructed out of a small piece of semiconductor material. These diodes are usually less than $\frac{1}{8}$ in. long.

A diode allows current to flow in only one direction. It is similar to a turnstile that revolves in only one direction.

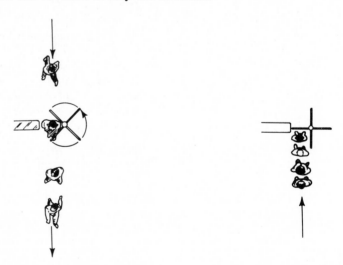

People may pass the turnstile in one direction but are blocked when attempting to pass in the opposite direction.

A diode allows electrons to pass in only one direction and not in the other.

An alternating current is allowed to pass only in one direction and is thus changed to a direct current.

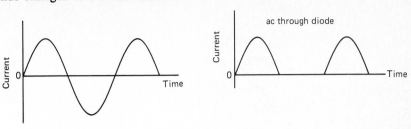

Additional circuit devices can be added to the rectifier which will smooth out the direct current so that it appears as shown.

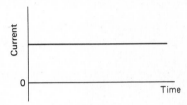

Rectifiers are used in automobiles to change the alternating current produced by the alternator into direct current. They are used in all ac radios and televisions to produce dc power for the tubes and transistors.

**28-9**
AMPLIFICATION

It is often necessary to increase the strength of an electronic signal. This is referred to as *amplification*. Radios, stereos, and many other instruments contain one or more amplifier circuits. Early amplifiers utilized vacuum tubes together with other components. The transistor has replaced the vacuum tube in most circuitry because of its smaller size and power consumption. A transistor amplifier is typically composed of one or more transistors in addition to capacitors, resistors, and possibly inductors to provide the amplifier with the desired gain (amplification), frequency response, and power output. Amplifiers composed of individual transistors, resistors, and capacitors have been replaced for many applications by integrated circuits (ICs), which are only slightly larger in size than some transistors. An IC may contain thousands of tiny transistors, diodes, resistors, and capacitors on a small chip of silicon less than 1 cm square. In addition to amplifying signals, ICs have been designed to serve as memory or logic units in computers or other applications. These ICs can be programmed to perform arithmetic operations, as in a calculator, or to perform control operations, as in many new appliances.

# CHAPTER
# *29*

## *Light*

**29-1**
INTRODUCTION

The nature of light has been a mystery for centuries. It is still the subject of intense study and refinement of current theory.

Isaac Newton thought light was made up of streams of particles (which at the time were called corpuscles). The primary evidence supporting this theory was that light travels in straight lines. Obviously, sight lines are used daily in common tasks as well as industry.

Not everyone, however, was convinced that light consisted of streams of particles. Others, following the work of Christian Huygens, felt that the nature of light could be more accurately explained by a wave theory. Huygens believed light to be a line of waves generated from a light source.

The particle theory persisted for a long time because of the great prestige of Isaac Newton. It was finally abandoned in the early 1800s when diffraction and interference of light were discovered—properties that simply could not be explained by the particle theory.

The wave theory was then refined through the years and from it was developed the electromagnetic theory of the nature of light. This theory suggests that light is a form of radiation, like radio waves or X rays.

Beginning in the late 1800s experiments were developed which confirmed the electromagnetic theory and showed light to be a small portion of the electromagnetic spectrum. Such waves are all similar in nature, travel at the same speed in a vacuum ($3.00 \times 10^8$ m/s or 186,000 mi/h), but differ in their frequencies and wavelengths.

We then define *light* as *radiant energy that can be seen by the human eye*. Note that other radiations may be observed by indirect means such as photographic film.

The electromagnetic theory, though, is still not a complete explanation of the nature of light. Early in this century, the *photoelectric effect* was discovered: the emission (or giving off) of electrons by a substance when struck by electromagnetic radiation. The electromagnetic wave theory fails to fully explain the nature of these emissions of electrons.

Later work by Max Planck and Albert Einstein developed the idea that light was energy radiated in the form of wave packets of energy which were called *photons*. Further work showed that in some circumstances light acted like a stream of particles. This theory of Planck is called the *quantum theory*.

Light therefore appears to have at least a dual character, having properties of both waves and particles. This dual character may be shown by considering how energy may be transported from one point to another.

The first is to transport particles of matter which carry energy with them. Examples of this method are electron conduction in a wire, shooting a bullet, natural gas pipelines, and gasoline transport.

The second method is the propagation of a wave disturbance through the medium between two points. Sound and water waves are examples of this method of energy transport.

Light is unusual in that it appears to combine characteristics of each method. When light is traveling through a medium, it appears to behave like a wave, with the following characteristics: (1) reflection at the surface of a medium; (2) refraction when passing from one medium to another; (3) interference (cancellation) when two waves are properly superimposed; and (4) diffraction (bending) when the waves pass the corners of an obstacle. When light interacts with matter, such as when it is absorbed or emitted, it behaves as if it were a massless particle.

**29-2
THE SPEED
OF LIGHT**

One of the most important measured quantities in physics is the speed of light. At first, light was assumed to be instantaneous. Galileo probably first suggested that time was required for light to travel from place to place.

A Danish astronomer, Olaus Roemer, made the first estimate of the speed of light from his study of the time of eclipse of one of the moons of the planet Jupiter as viewed from different places of the orbit of the earth in 1675. Since then, laboratory methods for measuring the speed of light have been developed, most notably by Albert Michelson. His measurements were made by using rotating mirrors to reflect a beam of light to a mirror on a mountain 22 miles away and back. He received the Nobel Prize in physics in 1907 for his work and was the first American to receive the prize. He also developed an instrument called an interferometer, with which in 1920 he made the first accurate measurement of a star's diameter.

Modern laboratory methods are used to make the measurement in a vacuum, instead of air, and have accurately measured the speed of light *(c)* to be $2.997925 \times 10^8$ m/s. This number is usually rounded to $3.00 \times 10^8$ m/s or 186,000 mi/s.

As stated earlier, light is one form of a class of radiation called electromagnetic radiation, which also includes radio and television waves, infrared, gamma rays, and X rays. A chart of the entire electromagnetic spectrum is shown in the following table.

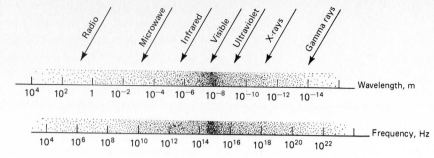

**THE ELECTROMAGNETIC SPECTRUM**

The distance traveled by any form of electromagnetic radiation can be found by combining this velocity with the equation relating distance traveled, velocity, and time: $x = vt$. Since the velocity is $c$, this equation becomes

$$\boxed{x = ct}$$

## EXAMPLE

Find the distance (in mi) traveled by an X ray in 0.100 s.

DATA:

$$c = 186{,}000 \text{ mi/s}$$
$$t = 0.100 \text{ s}$$
$$x = ?$$

BASIC EQUATION:

$$x = ct$$

WORKING EQUATION:  Same

SUBSTITUTION:

$$x = (186{,}000 \text{ mi/s})(0.100 \text{ s})$$
$$= 18{,}600 \text{ mi}$$

Very large distances, such as those between stars, cannot be conveniently expressed in common distance units. Astronomers therefore use the unit light year to measure such distances. A *light-year* is the distance traveled by light in one earth year. So, 1 light-year equals $9.5 \times 10^{15}$ m.

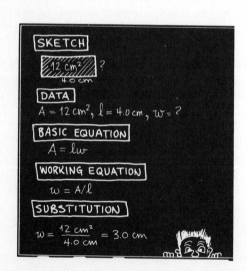

## PROBLEMS

1. Find the distance (in metres) traveled by a radio wave in 5.00 s.
2. Find the distance (in metres) traveled by a lightwave in 6.40 s.
3. A television signal is sent to a communications satellite that is $20{,}\overline{0}00$ mi above a relay station. How long does it take for the signal to reach the satellite?
4. How long does it take for a radio signal from earth to reach an astronaut on the moon? The distance from the earth to the moon is $2.40 \times 10^5$ mi.
5. The sun is $9.30 \times 10^7$ mi from the earth. How long does it take light to travel from the sun to the earth?
6. A radar wave that is bounced off an airplane returns to the radar receiver in $2.50 \times 10^{-5}$ s. How far (in km) is the airplane from the radar receiver?
7. How long does it take for a radio wave to travel $30\overline{0}0$ mi across the country?
8. How long does it take for a flash of light to travel $10\overline{0}$ m?
9. How long does it take for a police radar beam to travel to a truck and back if the truck is 115 m from the radar unit?
10. How far away is an ICBM if the radar wave returns from the scanning unit in $1.24 \times 10^{-7}$ s?

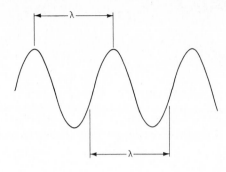

Light and the other forms of electromagnetic radiation are composed of oscillations in the electric and magnetic fields that exist in space. These oscillations are set up by rapid movement of charged particles such as electrons in radio antennas and electrons in a hot object such as a light-bulb filament. All waves are characterized by the distance that separates two points on the wave which are at the same point of vibration. This distance is called the *wavelength* and is denoted by the Greek lowercase letter lambda, $\lambda$. The wavelength of visible light ranges from about $4.00 \times 10^{-7}$ m to $7.60 \times 10^{-7}$ m. The wavelengths of other electromagnetic radiations can be found at the top of page 324.

Another characteristic of waves is the *frequency, f*. Frequency is the number of complete oscillations per second. The measurement unit of frequency (cycles/s) is named the hertz (Hz) after Heinrich Hertz, a leader in the study of electromagnetic theory. For example, 1 cycle per second = 1 Hz = 1/s. Since a "cycle" has no units, it does not appear in the Hz unit (see above: 1 Hz = 1/s).

A basic relationship exists for all waves, which relates the frequency, wavelength, and the velocity of the wave. The velocity is equal to the product of the frequency and the wavelength:

$$\boxed{c = f\lambda}$$

## EXAMPLE

Find the frequency of a light wave that has a wavelength of $5.00 \times 10^{-7}$ m.

DATA:

$$\lambda = 5.00 \times 10^{-7} \text{ m}$$
$$c = 3.00 \times 10^{8} \text{ m/s}$$
$$f = ?$$

BASIC EQUATION:

$$c = f\lambda$$

WORKING EQUATION:

$$f = \frac{c}{\lambda}$$

SUBSTITUTION:

$$f = \frac{3.00 \times 10^{8} \text{ m/s}}{5.00 \times 10^{-7} \text{ m}}$$
$$= 6.00 \times 10^{14} \text{ Hz (cycles/s)}$$

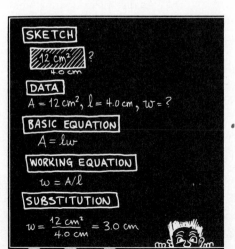

## PROBLEMS

1. $c = 3.00 \times 10^{8}$ m/s
   $\lambda = 4.00 \times 10^{-5}$ m
   $f = ?$

2. $c = 3.00 \times 10^{8}$ m/s
   $\lambda = 7.80 \times 10^{-10}$ m
   $f = ?$

3. $c = 3.00 \times 10^{8}$ m/s
   $f = 9.70 \times 10^{11}$ Hz
   $\lambda = ?$

4. $c = 3.00 \times 10^{8}$ m/s
   $f = 2.42 \times 10^{7}$ Hz
   $\lambda = ?$

5. Find the wavelength of a radio wave from an AM station broadcasting at a frequency of $14\overline{0}0$ kHz.

6. Find the wavelength of a radio wave from a FM station broadcasting at a frequency of 94.5 MHz.

As mentioned earlier, light sometimes behaves as if it were a massless particle. These particles are called photons and each carries a portion of the energy of the wave. This energy is given by

$$E = hf$$

where: $f$ = frequency
$h = 6.62 \times 10^{-34}$ Js   (Planck's constant)
$E$ = energy

## EXAMPLE

What is the energy of a source of electromagnetic radiation with a frequency of $1.00 \times 10^{12}$ Hz?

DATA:

$$h = 6.62 \times 10^{-34} \text{ Js}$$
$$f = 1.00 \times 10^{12} \text{ Hz} = 1.00 \times 10^{12}/\text{s}$$
$$E = ?$$

BASIC EQUATION:

$$E = hf$$

WORKING EQUATION:   Same

SUBSTITUTION:

$$E = (6.62 \times 10^{-34} \text{ Js})(1.00 \times 10^{12}/\text{s})$$
$$= 6.62 \times 10^{-22} \text{ J}$$

## PROBLEMS

1. What is the energy of a source of electromagnetic radiation with a frequency of $1.00 \times 10^{10}$ Hz?

2. What is the frequency of a source of electromagnetic radiation with energy of $3.96 \times 10^{-22}$ J?

3. What is the energy of a source of electromagnetic radiation with a frequency of $1.00 \times 10^{8}$ Hz?

4. What is the frequency of emission of a source with energy of $2.00 \times 10^{-24}$ J?

You will recall that light is produced along with other forms of radiation when substances are heated like our greatest source of natural light, the sun. Light may be produced in other ways, however, such as electrons bombarding gas molecules in neon lights or chemically, like fireflies. Our most common source of artificial light, however, is the incandescent lamp.

Incandescent lamps (like light bulbs) produce light by the heating of a material (the filament) and the giving off of a wide range of radiation in addition to visible light. Such objects, which produce light, are called *luminous*. On the other hand, objects like the moon, which are not producers of light but only reflect light from another source, are called *illuminated*.

When light strikes the surface of most objects, some light is reflected, some transmitted, and some absorbed.

The study of the measurement of light is called *photometry*. Two important measurable quantities in photometry are the luminous intensity or, *luminous flux, I*, of a light source, and the illumination of a surface, or, *illuminance, E*.

Luminous intensity measures the brightness of a light source. The unit for luminous flux *(I)* is the candle or *candela* (cd). The early use of certain candles for standards of intensity led to the name of the unit. We now use a platinum source at a certain temperature as the standard for comparison. Another unit, the *lumen,* ℓm, is often used for the measurement of the intensity of a source. The conversion factor between candles and lumens is

$$1 \text{ cd} = 4\pi \text{ } \ell m$$

Thus, a certain 40-W light bulb that is rated at 35 cd would have an intensity rating of 440 ℓm.

$$35 \text{ cd} \times \frac{4\pi \text{ } \ell m}{1 \text{ cd}} = 440 \text{ } \ell m$$

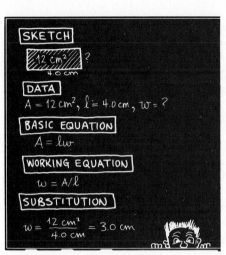

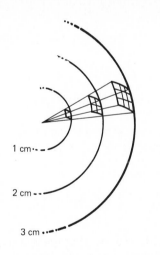

## PROBLEMS

1. $I = 48.0$ cd
   $I = $ _____ ℓm
2. $I = 197$ cd
   $I = $ _____ ℓm
3. $I = 543$ ℓm
   $I = $ _____ cd
4. $I = 432$ ℓm
   $I = $ _____ cd
5. $I = 75.0$ cd
   $I = $ _____ ℓm
6. $I = 650$ ℓm
   $I = $ _____ cd

The illumination on a surface may be varied by either changing the luminous intensity of the source (use a brighter bulb) or changing the position of the source (moving it closer to or farther from the surface to be illuminated). Of course, the illuminance is also less if the surface illuminated is slanted and not directly facing the light source.

*The amount of illumination on a surface varies inversely with the square of the distance from the source.* For example, if the distance of the illuminated surface from the source is doubled, the illumination is reduced to one-fourth of its former intensity. This can be illustrated by considering a point source of light at the center of concentric (having the same center) spheres. The solid angle is measured in units called *steradians.*

If the source radiates light uniformly in all directions, the light is uniformly distributed over a spherical surface centered at the source. Since the surface area of a sphere is $4\pi r^2$, the intensity of illumination, or illuminance, *E,* at the surface is given by

$$\boxed{E = \frac{I}{4\pi r^2}}$$

where: $I =$ intensity of the source (lumens)
$r =$ distance between the source and the illuminated surface

The unit of illuminance, *E,* is the *lux:*

$$1 \frac{\ell m}{m^2} = 1 \text{ lux}$$

## EXAMPLE 1

Find the intensity of illumination (illuminance) *E* on a surface located 2.00 m from a source with an intensity of 40$\overline{0}$ ℓm.

DATA:

$$I = 40\overline{0} \text{ } \ell m$$
$$r = 2.00 \text{ m}$$
$$E = ?$$

BASIC EQUATION:

$$E = \frac{I}{4\pi r^2}$$

WORKING EQUATION:  Same

SUBSTITUTION:

$$E = \frac{40\overline{0} \text{ lm}}{4\pi(2.00 \text{ m})^2}$$

$$= 7.96 \frac{\text{lm}}{\text{m}^2} \quad \text{or} \quad 7.96 \text{ lux}$$

The unit used for intensity of illumination is the lux ($\text{lm/m}^2$) in the metric system as shown above and the $\text{lm/ft}^2$ in the English system. Another unit often used is the foot-candle, which is equal to 1 $\text{lm/ft}^2$

$$1 \text{ ft-candle} = 1 \text{ lm/ft}^2$$

## EXAMPLE 2

Find the intensity of illumination 4.00 ft from a source with an intensity of $60\overline{0}$ lm.

DATA:

$$I = 60\overline{0} \text{ lm}$$
$$r = 4.00 \text{ ft}$$
$$E = ?$$

BASIC EQUATION:

$$E = \frac{I}{4\pi r^2}$$

WORKING EQUATION:  Same

SUBSTITUTION:

$$E = \frac{60\overline{0} \text{ lm}}{4\pi(4.00 \text{ ft})^2}$$

$$= 2.98 \frac{\text{lm}}{\text{ft}^2} \cdot \frac{1 \text{ ft-candle}}{1 \text{ lm/ft}^2}$$

$$= 2.98 \text{ ft-candles}$$

In photography, photoelectric cells are used in light meters or exposure meters which can be used to measure illuminance for the taking of photographs. The electricity produced is proportional to the illuminance and can be directly calibrated on the instrument scale. The units of measurement of such meters, however, are not standardized and the scale may be arbitrarily selected.

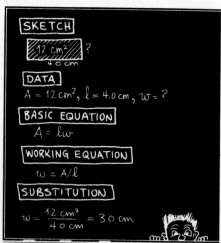

## PROBLEMS

1. $I = 90\overline{0}$ lm
   $r = 7.00$ ft
   $E = ?$

2. $I = 741$ lm
   $r = 4.00$ m
   $E = ?$

3. $I = 893$ lm
   $r = 4.50$ ft
   $E = ?$

4. $E = 4.32$ lux
   $r = 9.00$ m
   $I = ?$

5. $E = 10.5$ ft-candles
   $r = 6.00$ ft
   $I = ?$

6. Find the intensity of the light source necessary to produce an illuminance of 5.50 ft-candles a distance of 9.85 ft from the source.

7. Find the intensity of the light source necessary to produce an illuminance of 2.39 lux a distance of 4.50 m from the source.

8. Find the intensity of the light source necessary to produce an illuminance of 5.28 lux a distance of 6.50 m from the source.

9. If an observer triples her distance from a light source:
   (a) Does the illumination at that point increase or decrease?
   (b) In what proportion does the illumination increase or decrease?

10. If the illuminated surface is slanted at an angle of 35.0°, what part of the full-front illumination is lost?

# CHAPTER
# *30*

## *Reflection*

**30-1**
**MIRRORS**
**AND IMAGES**

Although we may not completely understand the nature of light, we know very well how light behaves. Every day we experience and unconsciously use our knowledge of what light does and depend on the ways it works. We are able to see objects because of the first characteristic we will study: reflection. *Reflection* is the name given to turning back a beam of light from a surface. Unlike sound, light does not require a medium (some kind of matter) to travel through and may be transmitted or passed through empty space. When light does strike a medium, the light may be reflected, absorbed, transmitted, or a combination of the three.

Mirrors, of course, show how light may be reflected. Any dark cloth shows how light may be absorbed. Window glass illustrates how light may be transmitted, or passed through a medium.

A medium may be classified according to how well light may be transmitted through it.

1. *Transparent* (almost all light passes through)—examples: window glass, clear water
2. *Translucent* (some but not all light passes through)—examples: murky water, light fog, skylight panels for farm buildings, stained glass
3. *Opaque* (almost all light reflected or absorbed)—examples: wood, metal, plaster

These classifications are relative because some light is reflected from the surface of any medium, whereas some passes into or through it.

In studying reflection we will look at what happens when light is turned back from a surface. The beam of a flashlight directed at a mirror shows several things about reflection. First, upon striking the surface of the glass, some of the light is reflected in all directions. This is called *scattering*. If there were no scattering, no light would reach our eye and we would be unable to observe the beam at all. However, only a very small part of the beam of light is scattered. Rough or uneven surfaces produce more scattering than do smooth ones. This scattering of light by uneven surfaces is called *diffusion*. Diffused lighting has many applications at home and in industry where bright glare is not desirable.

**330**

Nearly complete reflection (with very little scattering) is called *regular reflection*. Regular reflection occurs when parallel or nearly parallel rays of light (such as sunlight and spotlight beams) are still parallel after being reflected from a surface.

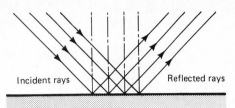

Incident rays          Reflected rays

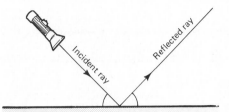

A flashlight beam on a mirror in a darkened room also shows something else about light striking a regular reflecting surface: The reflected rays of light leave the surface at the same angle at which the incident (coming-in) rays strike the surface.

Expressed another way, the angles measured from the normal or perpendicular to the reflecting surface are equal. These angles are called the following:

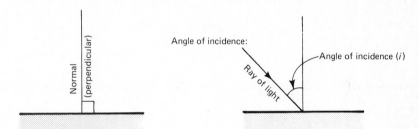

Angle of incidence:

and the angle of reflection:

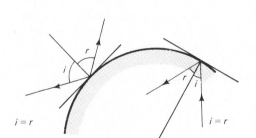

The same principle applies to curved surfaces. (See figure at left.) This behavior of light rays is defined by the following law:

---

*FIRST LAW OF REFLECTION:*

*The angle of incidence (i) is equal to the angle of reflection (r); that is,*

$$\angle i = \angle r$$

---

Further observation of the light beam readily shows a second law:

---

*SECOND LAW OF REFLECTION:*

The incident ray, the reflected ray, and the normal (perpendicular) all lie in the same plane.

---

These laws of reflection apply not only to light, but to all kinds of waves.

Mirrors of glass or any highly reflecting surface have countless practical applications, from rear-view mirrors in automobiles to watching for shoplifting in stores.

We look next at how images are formed by three widely used kinds of mirrors: plane, concave, and convex. *Plane* mirrors are, of course, flat. *Concave* mirrors are curved away from the observer (like the inside of a bowl), and *convex* mirrors are curved toward the observer (like Christmas tree ornaments).

For convenience we will use spherical (sections of spheres) mirrors, although parabolic mirrors (reflecting surface in the shape of a parabola) have wider practical use.

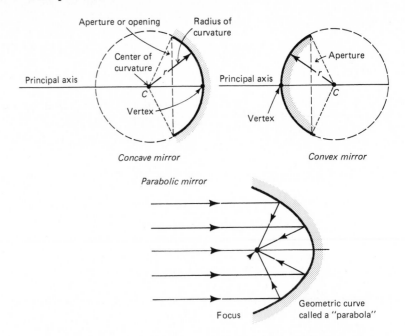

*Concave mirror*          *Convex mirror*

*Parabolic mirror*

Focus          Geometric curve called "parabola"

We consider next how images are formed by plane, concave, and convex mirrors. Images formed by mirrors may be *real images* (images actually formed by rays of light) or *virtual images* (images that only appear to the eye to be formed by rays of light).

Real images are always inverted (upside down) and may be larger, smaller, or the same size as the object. They can be shown on a screen.

Virtual images are always erect and may be larger, smaller, or the same size as the object. They cannot be shown on a screen.

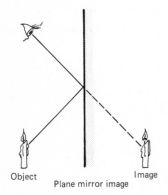

Object          Image
Plane mirror image

## 30-2
## IMAGES FORMED
## BY PLANE MIRRORS

Plane mirror images are always erect, virtual, and appear as far behind the mirror as the distance the object is in front of the mirror.

Note that plane mirrors also reverse right and left, so the right hand held in front of a plane mirror appears, in the mirror, to be a left hand.

We will use light-ray diagrams as a method to illustrate how our eyes see images in the various kinds of mirrors. We will do this by representing rays of light and lines of sight with straight lines as follows:

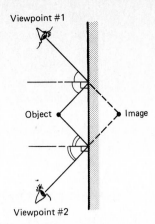

This method can be used to construct diagrams and locate the images formed. Simply view the object from two or more separate places and construct the light-ray lines.

**30-3**
**IMAGES FORMED**
**BY CONCAVE MIRRORS**

Find a tablespoon and look at your image in it. Now turn it over and look again. Of course, the images are very different, in that one is erect and the other inverted (upside down).

We can use ray diagrams to show just why this happens. As we shall see, the kind of image produced depends on the location of the object with respect to the mirror.

Becoming familiar with how light is reflected from spherical surfaces will enable us to construct these diagrams. *Note:* Rays parallel to the principal

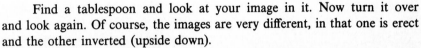

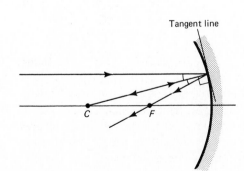

axis will always be reflected through a single point called the *focal point* or *focus, F*. Also, note that any ray through the center of curvature is reflected straight back because any such ray is perpendicular to the surface of the mirror at the point the ray strikes the mirror. For small apertures, the focal length is one-half the radius of curvature. That is, $f = R/2$.

If the object is placed at the focal point, no image will be formed because the rays of light will be reflected back parallel to the principal axis.

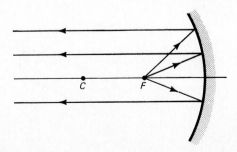

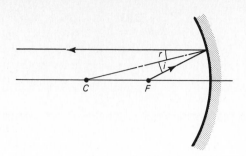

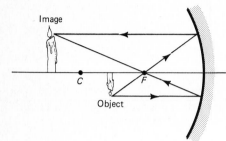

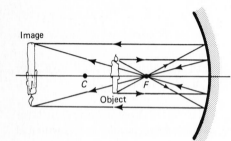

The location of the reflected ray in the above figure may be found by using the laws of reflection. The angle of incidence is equal to the angle of reflection. The normal or perpendicular is a line made between the center of curvature and the point the light ray strikes the mirror.

Now consider the more common case, where the object is beyond the center of curvature (for example, looking into a tablespoon). See figure at upper left. Note that again we use the fact that a ray parallel to the principal axis is reflected through the focal point and a ray through the focal point is reflected parallel to the principal axis. Then, where the two rays intersect, a point on the image is formed. (In this case it is the flame of the candle. Can you see that the candle base image lies on the principal axis because the object base is also on that line?) The same method would be used for the case where the candle base extends below the principal axis. See figure at left.

Now apply these principles to the following diagrams of other images formed by concave mirrors. Decide whether the image is real or virtual; erect or inverted; larger, smaller, or the same size; and where located.

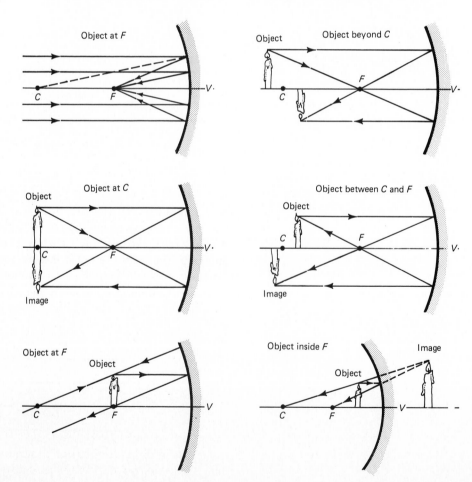

Note that the only time a virtual image is produced is when the object is between the focal point and the mirror. The construction is the same as the other cases except that the light rays are converging (coming together) and must be extended behind the mirror (where the image appears to be) forming the virtual image.

**30-4**
**IMAGES FORMED**
**BY CONVEX MIRRORS**

By looking into the back side of our tablespoon we see an erect, virtual, smaller image. Use the diagram shown to see how such an image is formed.

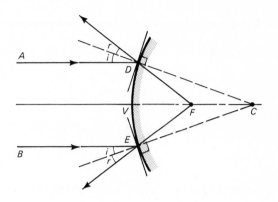

Note that even for convex mirrors, $i = r$, so:

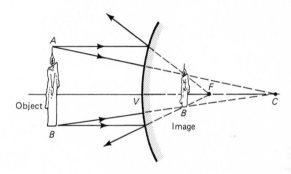

Curved surface mirrors are used in some telescopes, spotlights, and automobile headlights. But because spherical mirrors produce clear images over only a very small portion of their surfaces (small aperture), the surfaces used commercially are actually another geometric shape, that of a parabola or parabolic, as mentioned before. For apertures wider than about 10°, spherical mirrors produce fuzzy images because all parallel rays are not reflected through the focal point. This is called *spherical aberration*.

**30-5**
**THE MIRROR**
**FORMULA**

As we might expect from the foregoing cases, the focal point (distance of the focus from the lens), $f$; the distance the object is from the mirror, $s_o$; and the distance the image is from the mirror, $s_i$, are all related. This relationship can be expressed as the *mirror formula:*

$$\frac{1}{f} = \frac{1}{s_o} + \frac{1}{s_i}$$

where:  $f$ = focal length of mirror
$s_o$ = distance of object from mirror
$s_i$ = distance of image from mirror

Therefore, if two of the three distances $f$, $s_o$, and $s_i$ are known, the third can be calculated, and the three working equations become

$$f = \frac{s_o s_i}{s_o + s_i}$$

$$s_i = \frac{f s_o}{s_o - f}$$

$$s_o = \frac{f s_i}{s_i - f}$$

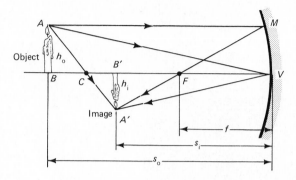

A second useful formula shows how the height of the object, $h_o$, and the height of the image, $h_i$, depend on the object distance, $s_o$, and the image distance, $s_i$:

$$\boxed{\frac{h_i}{h_o} = \frac{s_i}{s_o}}$$

where:  $h_i$ = image height
$h_o$ = object height
$s_i$ = image distance
$s_o$ = object distance

In using *both* of the preceding formulas for concave and convex mirrors we must remember that virtual images are *always* negative, and in the case of the convex mirror the focal distance is also negative.

## EXAMPLE

An object 10.0 cm in front on a convex mirror forms an image 5.00 cm behind the mirror. What is the focal length of the mirror?

SKETCH:

Object

Image

10.0 cm    5.00 cm

DATA:

$s_o = 10.0$ cm

$s_i = -5.00$ cm     *Note*: The image is virtual (appears behind the mirror) so $s_i$ is given a (−) sign to show this [won't $f$ also be (−)?]

$f = ?$

BASIC EQUATION:

$$\frac{1}{f} = \frac{1}{s_o} + \frac{1}{s_i}$$

WORKING EQUATION:

$$f = \frac{s_o s_i}{s_o + s_i}$$

SUBSTITUTION:

$$f = \frac{(10.0 \text{ cm})(-5.00 \text{ cm})}{10.0 \text{ cm} + (-5.00 \text{ cm})}$$

$$= -10.0 \text{ cm}$$

Remember that $f$ and $s_i$ may be negative only when forming virtual images and/or using convex mirrors.

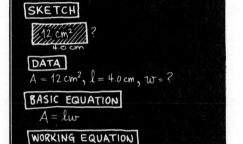

## PROBLEMS

1. Using $\frac{1}{f} = \frac{1}{s_o} + \frac{1}{s_i}$ and $s_o = 2.00$ cm and $s_i = 6.00$ cm, find $f$.

2. Using $\frac{1}{f} = \frac{1}{s_o} + \frac{1}{s_i}$ and $f = 15.0$ cm and $s_i = 3.00$ cm, find $s_o$.

3. Using $\frac{1}{f} = \frac{1}{s_o} + \frac{1}{s_i}$ and $s_i = 16.0$ cm and $f = 10.0$ cm, find $s_o$.

4. Using $\frac{1}{f} = \frac{1}{s_o} + \frac{1}{s_i}$ and $s_i = -10.0$ cm and $f = -5.00$ cm, find $s_o$.

5. Using $\frac{1}{f} = \frac{1}{s_o} + \frac{1}{s_i}$ and $s_o = 7.00$ cm and $s_i = 17.0$ cm, find $f$.

6. Using $\frac{h_i}{h_o} = \frac{s_i}{s_o}$ and $h_i = 2.00$ cm, $h_o = 4.50$ cm, and $s_i = 6.00$ cm, find $s_o$.

7. Using $\frac{h_i}{h_o} = \frac{s_i}{s_o}$ and $h_o = 12.0$ cm, $s_i = 13.0$ cm, and $s_o = 25.0$ cm, find $h_i$.

8. Using $\frac{h_i}{h_o} = \frac{s_i}{s_o}$ and $h_i = 3.50$ cm, $h_o = 2.50$ cm, and $s_i = 15.5$ cm, find $s_o$.

9. If an object is 2.50 m tall and is 8.60 m from a large mirror with an image formed 3.75 m from the mirror, what is the height of the image formed?

10. An object 30.0 cm tall is located 10.5 cm from a concave mirror with focal length 16.0 cm.
    (a) Where is the image located?
    (b) How high is it?

11. An object and its image in a concave mirror are the same height when the object is 20.0 cm from the mirror. What is the focal length of the mirror?

12. An object 12.6 cm in front of a convex mirror forms an image 6.00 cm behind the mirror. What is the focal length of the mirror?

# CHAPTER
# *31*

# *Refraction*

**31-1**
**THE LAW**
**OF REFRACTION**

Does light travel in a straight line? "Sure" would be most people's first answer. But does it always? Why does a stick appear to bend at the surface when placed in water?

The answers to our questions may be found in the study of another property of light—refraction. *Refraction* is the bending of light as it passes at an angle from one medium to another of different optical density. Optical density is a measurable property of a material that is related to the speed of light in that particular material. For example, water is optically denser than air and the speed of light in water is less than the speed of light in air. It is this change of speed when passing from one medium to another that produces refraction. A wave from the diagram illustrates how this occurs. Note that when passing from one medium to another perpendicular to the surface (called the interface) the wave is not bent, although the speed of the wave is slowed.

When the wave passes obliquely (at an angle) through, the entire wave front does not all strike the surface at the same time. The first part of the wave to strike the glass is slowed before the part striking later—thus the bending of the wave. (See figure at top of page 339.)

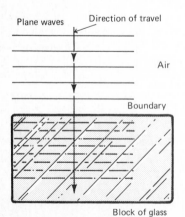

Plane waves    Direction of travel

Air

Boundary

Block of glass

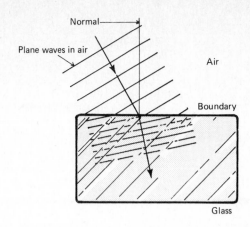

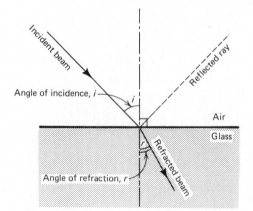

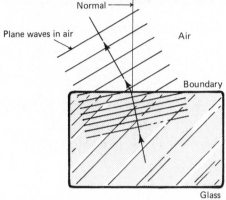

Draw your own diagram to show whether a fish in a pond, as viewed from the bank, is actually nearer the surface or the bottom of the pond than it appears to be.

As we might expect, since the speed of light increases when it leaves a denser medium to enter the air, bending also occurs. In this case, however, instead of the light bending toward the perpendicular or normal, the light is bent away from the normal when passing from the denser medium to the less dense. (See figure at left.)

This illustrates the following law:

---

*LAW OF REFRACTION*

When a beam of light passes, at an angle, from a medium of less optical density to a denser medium, the light is bent *toward* the normal. When a beam of light passes, at an angle, from a medium of more optical density to one less dense, the light is bent *away from* the normal.

---

Note that the angles of incidence and refraction are measured from the *normal*.

A Dutch mathematician, Willebord Snell, determined that a formula could be used to identify a property of substances that pass light called the index of refraction. This index is a constant for a particular material and is independent of the angle the light strikes.

It may be expressed in two ways:

1. As Snell's law, the index of refraction, *n*, equals the sine of the angle of incidence divided by the sine of the angle of refraction:

$$n = \frac{\sin i}{\sin r}$$

2. As the ratio of the speed of light in a vacuum (nearly the same as in air) to the speed of light in the particular substance:

$$n = \frac{\text{speed of light in vacuum}}{\text{speed of light in substance}}$$

## EXAMPLE 1

The angle of incidence of light passing from air to water is $\overline{6}0°$. The angle of refraction is 41°. What is the index of refraction of the water?

DATA:

$$i = \overline{60}°$$
$$r = 41°$$
$$n = ?$$

BASIC EQUATION:

$$n = \frac{\sin i}{\sin r}$$

WORKING EQUATION:   Same

SUBSTITUTION:

$$n = \frac{\sin 60°}{\sin 41°}$$
$$= \frac{0.866}{0.656}$$
$$= 1.32$$

## EXAMPLE 2

The index of refraction of water is 1.33. What is the speed of light in water?

DATA:

$$n = 1.33$$
$$c = 3.00 \times 10^8 \text{ m/s}$$
$$v_{water} = ?$$

BASIC EQUATION:

$$n = \frac{\text{speed of light in vacuum}}{\text{speed of light in substance}}$$

WORKING EQUATION:

$$\text{speed of light in water} = \frac{\text{speed of light in vacuum}}{n}$$

SUBSTITUTION:

$$\text{speed of light in water} = \frac{3.00 \times 10^8 \text{ m/s}}{1.33}$$
$$= 2.26 \times 10^8 \text{ m/s}$$

The index of refraction for each of several common substances is given in the following table.

**INDEX OF REFRACTION**

| Substance | Index of refraction |
|---|---|
| Air, dry (STP) | 1.00029 |
| Alcohol, ethyl | 1.360 |
| Benzene | 1.501 |
| Carbon dioxide (STP) | 1.00045 |
| Carbon disulfide | 1.625 |
| Carbon tetrachloride | 1.459 |
| Diamond | 2.417 |
| Glass, crown flint | 1.575 |
| Lucite | 1.50 |
| Quartz, fused | 1.45845 |
| Water, distilled | 1.333 |
| Water vapor (STP) | 1.00025 |

If the angle of refraction is 90° or greater, a beam of light does not leave the medium but is reflected back inside the medium. This is called *total internal reflection*. Total internal reflection occurs when the angle of incidence is greater than the *critical angle*. The critical angle is the smallest angle that produces an angle of refraction of 90° or more. It may be expressed by the formula

$$\sin i_c = \frac{1}{n}$$

where: $i_c$ = critical angle of incidence
$n$ = index of refraction of denser medium

## EXAMPLE

What is the critical angle of incidence for a liquid that has an index refraction of 1.33?

DATA:

$$n = 1.33$$
$$i_c = ?$$

BASIC EQUATION:

$$\sin i_c = \frac{1}{n}$$

WORKING EQUATION:   Same

SUBSTITUTION:

$$\sin i_c = \frac{1}{1.33} = 0.752$$

Therefore,   $i_c = 49°$

Where there is total internal reflection, no light enters the air; it is totally reflected within the glass. The property of having a very small critical angle gives a diamond its brilliance by multiple internal reflections before the light passes out through the top.

An example of the practical application of this principle is fiber optics. Light may be transferred inside flexible fibers which are transparent but keep nearly all the light inside by means of it being reflected along the inside surface of the fiber.

Many, many technical applications use the principles of refraction from testing to find out the nature of liquids to microscopes to eyeglasses.

Attention will now be directed to the use of refraction in applications using lenses. Lenses may be converging or diverging. *Converging lenses* bend the light passing through them to some point beyond the lens. Converging lenses are always thicker in the middle than on the edges.

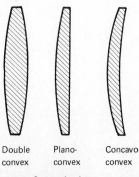

Double    Plano-    Concavo-
convex    convex    convex

Converging lenses

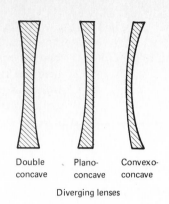

Double concave    Plano-concave    Convexo-concave

Diverging lenses

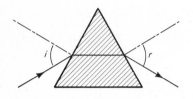

*Diverging lenses* bend the light passing through them so as to spread the light. Diverging lenses are thicker on the edges than at the center.

Understanding how light is bent in lenses may be made easier by looking at the bending of light passing through a prism.

Recall and apply the law of refraction, noting that light is bent toward the normal when passing into the glass and away from the normal when passing from the glass back to the air.

**31-4**
**IMAGES FORMED BY CONVERGING LENSES**

As with mirrors, we may use light-ray diagrams to help us to understand how light can be bent with lenses. Every lens has a focal length—that distance from the lens center to the point where parallel beams directed through the lens come together if converging or *appear* to come together if diverging.

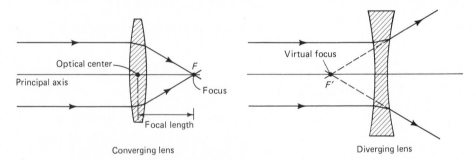

Converging lens                    Diverging lens

The location of the focus depends upon the curvature of the lens and the index of refraction of the glass or material of which the lens is made.

Rays of light passing through the optical center of the lens are refracted so little that we may consider them as going straight through as shown at left.

Now apply these principles to the following diagrams of other images formed by converging lenses, depending on the object location, and decide whether the image is real, virtual, or no image formed at all; erect or inverted; larger, smaller, or the same size; and where located.

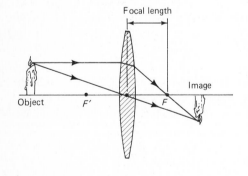

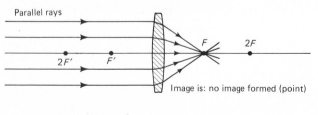

Image is: no image formed (point)

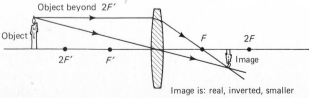

Image is: real, inverted, smaller

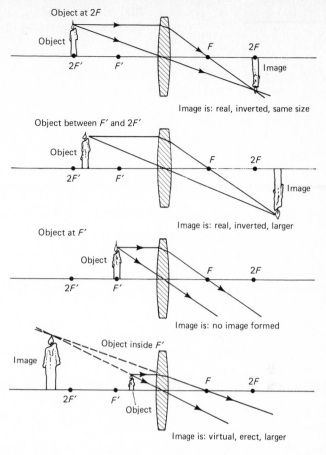

Object at 2F

Object

2F'    F'    F    2F

Image

Image is: real, inverted, same size

Object between F' and 2F'

Object

2F'    F'    F    2F

Image

Image is: real, inverted, larger

Object at F'

Object

2F'    F'    F    2F

Image is: no image formed

Object inside F'

Image

2F'    F'    F    2F

Object

Image is: virtual, erect, larger

**31-5**
**IMAGES FORMED**
**BY DIVERGING LENSES**

Virtual images are the only images produced by diverging lenses.

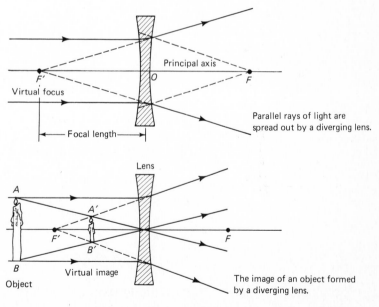

Principal axis

F'    O    F

Virtual focus

Focal length

Parallel rays of light are spread out by a diverging lens.

Lens

A

A'

F'    F

B'

B    Virtual image

Object

The image of an object formed by a diverging lens.

The same formulas that apply to mirrors also apply to lenses.

$$\frac{1}{f} = \frac{1}{s_o} + \frac{1}{s_i}$$

where:   $f$ = focal length
$s_o$ = object distance from lens center
$s_i$ = image distance from lens center

Therefore, if two of the three distances $f$, $s_o$, and $s_i$ are known, the third can be calculated, and the three working equations become

$$f = \frac{s_o s_i}{s_o + s_i}$$

$$s_i = \frac{f s_o}{s_o - f}$$

$$s_o = \frac{f s_i}{s_i - f}$$

Also,

$$\boxed{\frac{h_i}{h_o} = \frac{s_i}{s_o}}$$

where:   $h_i$ = height of image
$h_o$ = height of object
$s_i$ = image distance from lens center
$s_o$ = object distance from lens center

So if three of the four quantities $h_o$, $h_i$, $s_o$, and $s_i$ are known, the fourth can be calculated.

## EXAMPLE

An object 3.00 cm tall is placed 24.0 cm from a converging lens. A real image is formed 8.00 cm from the lens.
(a) What is the focal length of the lens?
(b) What is the size of the image?

(a) DATA:

$$s_o = 24.0 \text{ cm}$$

$$s_i = 8.00 \text{ cm}$$

$$f = ?$$

BASIC EQUATION:

$$\frac{1}{f} = \frac{1}{s_o} + \frac{1}{s_i}$$

WORKING EQUATION:

$$f = \frac{s_o s_i}{s_o + s_i}$$

SUBSTITUTION:

$$f = \frac{(24.0 \text{ cm})(8.00 \text{ cm})}{24.0 \text{ cm} + 8.00 \text{ cm}}$$

$$= 6.00 \text{ cm}$$

(b) DATA:

$$s_o = 24.0 \text{ cm}$$

$$s_i = 8.00 \text{ cm}$$

$$h_o = 3.00 \text{ cm}$$

$$h_i = ?$$

BASIC EQUATION:

$$\frac{h_i}{h_o} = \frac{s_i}{s_o}$$

WORKING EQUATION:

$$h_i = \frac{s_i h_o}{s_o}$$

SUBSTITUTION:

$$h_i = \frac{(8.00 \text{ cm})(3.00 \text{ cm})}{24.0 \text{ cm}}$$

$$= 1.00 \text{ cm}$$

Remember, when the image is virtual, $s_i$ is negative (−), and for diverging lenses both $s_i$ and $f$ are negative.

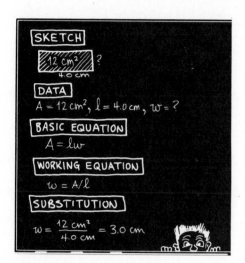

## PROBLEMS

1. What is the index of refraction of a medium for which the angle of incidence of a light beam is $3\bar{0}°$ and angle of refraction is $25°$?

2. If the index of refraction of a medium is 2.40 and the angle of incidence is $15°$, what is the angle of refraction?

3. If the index of refraction of a liquid is 1.50, what is the speed of light in that liquid?

4. The angle of incidence of light passing from air to a liquid is $38°$. The angle of refraction is $24°$. What is the index of refraction of the liquid?

5. If the critical angle of a liquid is $42°$, what is the index of refraction for that liquid?

6. If the index of refraction of a substance is 2.4, what is its critical angle of incidence?

7. A converging lens has a focal length of 15.0 cm. If it is placed 48.0 cm from an object, how far from the lens will the image be formed?

8. An object 2.50 cm tall is placed 20.0 cm from a converging lens. A real image is formed 9.00 cm from the lens.
   (a) What is the focal length of the lens?
   (b) What is the size of the image?

9. The focal length of a lens is 5.00 cm. How far from the lens must the object be to produce an image 1.50 cm from the lens?

10. If the distance from the lens in your eye to the retina is 19.0 mm, what is the focal length of the lens when reading a sign 40.0 cm from the lens?

11. If an object 5.00 cm tall is placed 15.0 cm from a converging lens and a real image is formed 7.50 cm from the lens:
    (a) What is the focal length of the lens?
    (b) What is the size of the image?

12. An object 4.50 cm tall is placed 18.0 cm from a converging lens with a focal length of 26.0 cm.
    (a) What is the location of the image?
    (b) What is its size?

# Review of Algebra

**A-1**
**SIGNED NUMBERS**

Signed numbers have many and wide applications to the study of physics. The rules for working with signed numbers follow.

*Adding signed numbers*    *To add two positive numbers,* add their absolute values.* A positive sign may or may not be placed before the result. It is usually omitted.

### EXAMPLE 1

Add each of the following.

(a)   +4            (b) $(+3) + (+5) = +8$   or   8
     +7
     $\overline{+11}$   or   11

*To add two negative numbers,* add their absolute values and place a negative sign before the result.

### EXAMPLE 2

Add each of the following.

(a) $-2$      (b) $(-6) + (-7) = -13$
    $-5$      (c) $(-8) + (-4) = -12$
    $\overline{-7}$

*To add a negative number and a positive number,* find the difference of their absolute values. The sign of the number having the larger absolute value is placed before the result.

### EXAMPLE 3

Add each of the following.

(a) +4      (b) $-2$      (c) $-8$      (d) +9
    $-6$          +8            +3            $-4$
    $\overline{-2}$      $\overline{+6}$      $\overline{-5}$      $\overline{+5}$

* The absolute value of a number is its nonnegative value. For example, the absolute value of $-6$ is 6; the absolute value of $+10$ is 10; and the absolute value of 0 is 0.

(e) $(+7)+(-2)=+5$    (f) $(-9)+(+6)=-3$
(g) $(-3)+(+10)=+7$    (h) $(+4)+(-12)=-8$

*To add three or more signed numbers:*

1. Add the positive numbers.
2. Add the negative numbers.
3. Add the sums from steps 1 and 2 according to the rules for addition of signed numbers.

## EXAMPLE 4

Add: $(-2)+4+(-6)+10+(-7)$.

$$
\begin{array}{llll}
\textit{Step } 1: & +4 & \textit{Step } 2: \; -2 & \textit{Step } 3: \; -15 \\
& +10 & -6 & +14 \\
& \underline{+14} & \underline{-7} & \underline{+14} \\
& & -15 & -1
\end{array}
$$

Therefore, $(-2)+4+(-6)+10+(-7)=-1$.

*Subtracting signed numbers*    To subtract two signed numbers, change the sign of the *number being subtracted* and *add* according to the rules for addition.

## EXAMPLE 5

Subtract each of the following:

(a) Subtract:  $\begin{array}{r} +3 \\ +7 \\ \hline -4 \end{array}$  $\leftrightarrow$  Add:  $\begin{array}{r} +3 \\ -7 \\ \hline -4 \end{array}$    To subtract, change the sign of the number being subtracted, $+7$, and add.

(b) Subtract:  $\begin{array}{r} -9 \\ -6 \\ \hline -3 \end{array}$  $\leftrightarrow$  Add:  $\begin{array}{r} -9 \\ +6 \\ \hline -3 \end{array}$    To subtract, change the sign of the number being subtracted, $-6$, and add.

(c) Subtract:  $\begin{array}{r} +8 \\ -4 \\ \hline +12 \end{array}$  $\leftrightarrow$  Add:  $\begin{array}{r} +8 \\ +4 \\ \hline +12 \end{array}$

(d) Subtract:  $\begin{array}{r} -6 \\ +8 \\ \hline -14 \end{array}$  $\leftrightarrow$  Add:  $\begin{array}{r} -6 \\ -8 \\ \hline -14 \end{array}$

(e) $(+6)-(+8)=(+6)+(-8)=-2$    To subtract, change the sign of the number being subtracted, $+8$, and add.

(f) $(-3)-(-5)=(-3)+(+5)=+2$
(g) $(+10)-(-3)=(+10)+(+3)=+13$
(h) $(-5)-(+2)=(-5)+(-2)=-7$

When more than two signed numbers are involved in subtraction, change the sign of *each* number being subtracted and add the resulting signed numbers.

## EXAMPLE 6

Subtract:  $(-2)-(+4)-(-1)-(-3)-(+5)$
         $=(-2)+(-4)+(+1)+(+3)+(-5)$.

$$
\begin{array}{llll}
\textit{Step } 1: +1 & \textit{Step } 2: \; -2 & \textit{Step } 3: \; +4 \\
+3 & -4 & -11 \\
\underline{+4} & \underline{-5} & \underline{-11} \\
& -11 & -7
\end{array}
$$

Therefore, $(-2)-(+4)-(-1)-(-3)-(+5)=-7$.

When combinations of addition and subtraction of signed numbers occur in the same problem, change *only* the sign of each number being subtracted. Then add the resulting signed numbers.

## EXAMPLE 7

Find the result:

$$(-2)+(-4)-(-3)-(+6)+(+1)-(+2)+(-7)-(-5)$$
$$=(-2)+(-4)+(+3)+(-6)+(+1)+(-2)+(-7)+(+5)$$

*Step* 1: $\begin{array}{r} +3 \\ +1 \\ +5 \\ \hline +9 \end{array}$     *Step* 2: $\begin{array}{r} -2 \\ -4 \\ -6 \\ -2 \\ -7 \\ \hline -21 \end{array}$     *Step* 3: $\begin{array}{r} +9 \\ -21 \\ \hline -12 \end{array}$

Therefore, $(-2)+(-4)-(-3)-(+6)+(+1)-(+2)+(-7)-(-5)=-12$.

*Multiplying signed numbers*    To multiply two signed numbers:

1. If the signs of the numbers are both positive or both negative, find the product of their absolute values. This product is always positive.
2. If the signs of the numbers are unlike, find the product of their absolute values and place a negative sign before the result.

## EXAMPLE 8

Multiply each of the following.

(a) $\begin{array}{r} +3 \\ +4 \\ \hline +12 \end{array}$    (b) $\begin{array}{r} -5 \\ -8 \\ \hline +40 \end{array}$    (c) $\begin{array}{r} -6 \\ +7 \\ \hline -42 \end{array}$    (d) $\begin{array}{r} +2 \\ -3 \\ \hline -6 \end{array}$

(e) $(+3)(+5)=+15$     (f) $(-7)(-8)=+56$

(g) $(-1)(+6)=-6$     (h) $(+4)(-2)=-8$

To multiply more than two signed numbers, first multiply the absolute values of the numbers. If there is an odd number of negative factors, place a negative sign before the result. If there is an even number of factors, the product is positive. *Note:* An *even* number is divisible by 2.

## EXAMPLE 9

Multiply each of the following.

(a) $(+5)(-6)(+2)(-1)=+60$
(b) $(-3)(-3)(+4)(-5)=-180$

*Dividing signed numbers*    The rules for dividing signed numbers are similar to those for multiplying signed numbers.
    To divide two signed numbers:

1. If the signs of the numbers are both positive or both negative, divide their absolute values. This quotient is always positive.
2. If the two numbers have different signs, divide their absolute values and place a negative sign before the quotient.

## EXAMPLE 10

Divide each of the following.

(a) $\dfrac{+10}{+2}=+5$    (b) $\dfrac{-18}{-3}=+6$    (c) $\dfrac{+20}{-4}=-5$    (d) $\dfrac{-24}{+2}=-12$

*Perform the indicated operations.*

1. $(-5) + (-6)$
2. $(+1) + (-10)$
3. $(-3) + (+8)$
4. $(+5) + (+7)$
5. $(-5) + (+3)$
6. $0 + (-3)$
7. $(-7) - (-3)$
8. $(+2) - (-9)$
9. $(-4) - (+2)$
10. $(+4) - (+7)$
11. $0 - (+3)$
12. $0 - (-2)$
13. $(-9)(-2)$
14. $(+4)(+6)$
15. $(-7)(+3)$
16. $(+5)(-8)$
17. $(+6)(0)$
18. $(0)(-4)$
19. $\dfrac{+36}{+12}$
20. $\dfrac{-9}{-3}$
21. $\dfrac{+16}{-2}$
22. $\dfrac{-15}{+3}$
23. $\dfrac{0}{+6}$
24. $\dfrac{4}{0}$
25. $(+2) + (-1) + (+10)$
26. $(-7) + (+2) + (+9) + (-8)$
27. $(-9) + (-3) + (+3) + (-8) + (+4)$
28. $(+8) + (-2) + (-6) + (+7) + (-6) + (+9)$
29. $(-4) - (+5) - (-4)$
30. $(+3) - (-5) - (-6) - (+5)$
31. $(-7) - (-4) - (+6) - (+4) - (-5)$
32. $(-8) - (+7) - (+3) - (-7) - (-8) - (-2)$
33. $(+5) + (-2) - (+7)$
34. $(-3) - (-8) - (+3) + (-9)$
35. $(-2) - (+1) - (-10) + (+12) + (-9)$
36. $(-1) - (-11) + (+2) - (-10) + (+8)$
37. $(+3)(-5)(+3)$
38. $(-1)(+2)(+2)(-1)$
39. $(+2)(-4)(-6)(-3)(+2)$
40. $(-1)(+3)(-2)(-4)(+5)(-1)$

**A-2**
**POWERS**
**OF 10**

The ability to work quickly and accurately with powers of 10 is important in scientific and technical fields.

> *When multiplying two powers of 10, add the exponents. That is,*
> $$10^a \times 10^b = 10^{a+b}$$

### EXAMPLE 1

Multiply.

(a) $(10^6)(10^3) = 10^{6+3} = 10^9$
(b) $(10^4)(10^2) = 10^{4+2} = 10^6$
(c) $(10^1)(10^{-3}) = 10^{1+(-3)} = 10^{-2}$
(d) $(10^{-2})(10^{-5}) = 10^{[-2+(-5)]} = 10^{-7}$

> *When dividing two powers of 10, subtract the exponents as follows:*
> $$10^a \div 10^b = 10^{a-b}$$

EXAMPLE 2

Divide.

(a) $\dfrac{10^7}{10^4} = 10^{7-4} = 10^3$

(b) $\dfrac{10^3}{10^5} = 10^{3-5} = 10^{-2}$

(c) $\dfrac{10^{-2}}{10^{+3}} = 10^{(-2)-(+3)} = 10^{-5}$

(d) $\dfrac{10^4}{10^{-2}} = 10^{4-(-2)} = 10^6$

> *To raise a power of 10 to a power, multiply the exponents as follows:*
> $$(10^a)^b = 10^{ab}$$

EXAMPLE 3

Find each power.

(a) $(10^2)^3 = 10^{(2)(3)} = 10^6$
(b) $(10^{-3})^2 = 10^{(-3)(2)} = 10^{-6}$
(c) $(10^4)^{-5} = 10^{(4)(-5)} = 10^{-20}$
(d) $(10^{-3})^{-4} = 10^{(-3)(-4)} = 10^{12}$

Next, we will show that $10^0 = 1$. To do this, we need to use the substitution principle, which states that

$$\text{if } a = b \text{ and } a = c, \quad \text{then } b = c$$

First,

$$\dfrac{10^n}{10^n} = 10^{n-n} \quad \text{To divide powers, subtract the exponents.}$$

$$= 10^0$$

Second,

$$\dfrac{10^n}{10^n} = 1 \quad \text{Any number other than zero divided by itself equals 1.}$$

That is, since

$$\dfrac{10^n}{10^n} = 10^0 \quad \text{and} \quad \dfrac{10^n}{10^n} = 1$$

then $10^0 = 1$.

We also will use the fact that $1/10^a = 10^{-a}$. To show this,

$$\dfrac{1}{10^a} = \dfrac{10^0}{10^a} \qquad (1 = 10^0)$$

$$= 10^{0-a} \quad \text{To divide powers, subtract the exponents.}$$

$$= 10^{-a}$$

We also need to show that $1/10^{-a} = 10^a$.

$$\dfrac{1}{10^{-a}} = \dfrac{10^0}{10^{-a}}$$

$$= 10^{0-(-a)}$$

$$= 10^a$$

In summary,

$$10^0 = 1; \qquad \frac{1}{10^a} = 10^{-a}; \qquad \frac{1}{10^{-a}} = 10^a$$

## PROBLEMS

*Do as indicated. Express the results using positive exponents.*

**1.** $(10^5)(10^3)$      **2.** $10^6 \div 10^2$      **3.** $(10^2)^4$

**4.** $(10^{-2})(10^{-3})$      **5.** $\dfrac{10^3}{10^5}$      **6.** $(10^{-3})^3$

**7.** $10^5 \div 10^{-2}$      **8.** $(10^{-2})^{-3}$      **9.** $(10^4)(10^{-1})$

**10.** $\dfrac{10^0}{10^{-4}}$      **11.** $(10^0)(10^{-4})$      **12.** $\dfrac{10^{-4}}{10^{-3}}$

**13.** $(10^0)^{-2}$      **14.** $10^{-3}$      **15.** $\dfrac{1}{10^{-5}}$

**16.** $\dfrac{(10^4)(10^{-2})}{(10^6)(10^3)}$      **17.** $\dfrac{(10^{-2})(10^{-3})}{(10^3)^2}$      **18.** $\dfrac{(10^2)^4}{(10^{-3})^2}$

**19.** $\left(\dfrac{1}{10^3}\right)^2$      **20.** $\left(\dfrac{10^2}{10^{-3}}\right)^2$      **21.** $\left(\dfrac{10 \cdot 10^2}{10^{-1}}\right)^2$

**22.** $\left(\dfrac{1}{10^{-3}}\right)^{-2}$      **23.** $\dfrac{(10^4)(10^{-2})}{10^{-8}}$      **24.** $\dfrac{(10^4)(10^6)}{(10^0)(10^{-2})(10^3)}$

**A-3**
**REVIEW**
**OF SOLVING**
**LINEAR EQUATIONS**

An equation is a mathematical sentence stating that two quantities are equal. To solve an equation means to find the number or numbers that can replace the variable in the equation to make the equation a true statement. The value we find that makes the equation a true statement is called the *root* of the equation. When the root of an equation is found, we say we have *solved* the equation.

---

*If $a = b$, then $a + c = b + c$ or $a - c = b - c$. (If two quantities are equal, then adding or subtracting the same quantity to both of them maintains the equality.)*

---

To solve an equation using this rule, think first of undoing what has been done to the variable.

### EXAMPLE 1

Solve $x - 5 = -9$ for $x$.

$$x - 5 = -9$$
$$x - 5 + 5 = -9 + 5 \qquad \text{Undo the subtraction by adding 5 to both sides.}$$
$$x = -4$$

### EXAMPLE 2

Solve $x + 4 = 29$ for $x$.

$$x + 4 = 29$$
$$x + 4 - 4 = 29 - 4 \qquad \text{Undo the addition by subtracting 4 from both sides.}$$
$$x = 25$$

---

*If $a = b$, then $ac = bc$ or $a/c = b/c$ with $c \neq 0$. (If two quantities are equal, then multiplying or dividing both sides of the equation by the same number will maintain the equality.)*

---

EXAMPLE 3

Solve $3x = 18$ for $x$.

$$3x = 18$$

$$\frac{3x}{3} = \frac{18}{3} \qquad \text{Undo the multiplication by dividing both sides by 3.}$$

$$x = 6$$

EXAMPLE 4

Solve $x/4 = 9$ for $x$.

$$\frac{x}{4} = 9$$

$$4\left(\frac{x}{4}\right) = 4 \cdot 9 \qquad \text{Undo the division by multiplying both sides by 4.}$$

$$x = 36$$

EXAMPLE 5

Solve $3x + 5 = 17$.

In this example more than one operation is indicated on the variable. There is an addition of 5 and a multiplication by 3. *In general, to solve such an equation, undo additions and subtractions first; then undo multiplications and divisions.*

$$3x + 5 = 17$$

$$3x + 5 - 5 = 17 - 5 \qquad \text{Subtract 5 from both sides.}$$

$$3x = 12$$

$$\frac{3x}{3} = \frac{12}{3} \qquad \text{Divide both sides by 3.}$$

$$x = 4$$

EXAMPLE 6

Solve $2x - 7 = 10$ for $x$.

$$2x - 7 = 10$$

$$2x - 7 + 7 = 10 + 7 \qquad \text{Add 7 to both sides.}$$

$$2x = 17$$

$$\frac{2x}{2} = \frac{17}{2} \qquad \text{Divide both sides by 2.}$$

$$x = \frac{17}{2} = 8.5$$

EXAMPLE 7

Solve $x/5 - 10 = 22$ for $x$.

$$\frac{x}{5} - 10 = 22$$

$$\frac{x}{5} - 10 + 10 = 22 + 10 \qquad \text{Add 10 to both sides.}$$

$$\frac{x}{5} = 32$$

$$5\left(\frac{x}{5}\right) = 5(32) \qquad \text{Multiply both sides by 5.}$$

$$x = 160$$

*Solve each equation for* x *or* y.

**1.** $3x = 4$        **2.** $\dfrac{y}{2} = 10$

**3.** $x - 5 = 12$        **4.** $x + 1 = 9$

**5.** $2x + 10 = 10$        **6.** $4x = 28$

**7.** $2x - 2 = 33$        **8.** $4 = \dfrac{x}{10}$

**9.** $172 - 43x = 43$        **10.** $9x + 7 = 4$

**11.** $6y - 24 = 0$        **12.** $3y + 15 = 75$

**13.** $15 = \dfrac{105}{y}$        **14.** $6x = x - 15$

**15.** $2 = \dfrac{50}{2y}$        **16.** $9y = 67.5$

**17.** $8x - 4 = 36$        **18.** $10 = \dfrac{136}{4x}$

**19.** $2x + 22 = 75$        **20.** $9x + 10 = x - 26$

To solve an equation with variables on both sides:

1. Add or subtract either variable term from both sides of the equation.
2. Add or subtract from both sides of the equation the constant term that now appears on the same side of the equation with the variable. Then solve.

## EXAMPLE 8

Solve $3x + 6 = 7x - 2$ for $x$.

$$3x + 6 = 7x - 2$$
$$3x + 6 - 3x = 7x - 2 - 3x \qquad \text{Subtract } 3x \text{ from both sides.}$$
$$6 = 4x - 2$$
$$6 + 2 = 4x - 2 + 2 \qquad \text{Add 2 to both sides.}$$
$$8 = 4x$$
$$\frac{8}{4} = \frac{4x}{4} \qquad \text{Divide both sides by 4.}$$
$$2 = x$$

## EXAMPLE 9

Solve $4x - 2 = -5x + 10$ for $x$.

$$4x - 2 = -5x + 10$$
$$4x - 2 + 5x = -5x + 10 + 5x \qquad \text{Add } 5x \text{ to both sides.}$$
$$9x - 2 = 10$$
$$9x - 2 + 2 = 10 + 2 \qquad \text{Add 2 to both sides.}$$
$$9x = 12$$
$$\frac{9x}{9} = \frac{12}{9} \qquad \text{Divide both sides by 9.}$$
$$x = \frac{4}{3}$$

There are equations written with portions of the equation included in parentheses. To solve these equations, first remove parentheses and then proceed as before. The rules for removing parentheses are the following:

1. If the parentheses are preceded by a plus (+) sign, they may be removed without changing any signs.
   Examples:  $2 + (3 - 5) = 2 + 3 - 5$
   $3 + (x + 4) = 3 + x + 4$

2. If the parentheses are preceded by a minus (−) sign, the parentheses may be removed if *all* the signs of the numbers (or letters) within the parentheses are changed.
   Examples:  $2 - (3 - 5) = 2 - 3 + 5$
   $5 - (x - 7) = 5 - x + 7$

3. If the parentheses are preceded by a number, the parentheses may be removed if each of the terms inside the parentheses is multiplied by that (signed) number.
   Examples:  $2(x + 4) = 2x + 8$
   $-3(x - 5) = -3x + 15$
   $2 - 4(3 - 5) = 2 - 12 + 20$

## EXAMPLE 10

Solve $3(x - 4) = 15$ for $x$.

$$3(x - 4) = 15$$
$$3x - 12 = 15 \qquad \text{Remove parentheses.}$$
$$3x - 12 + 12 = 15 + 12 \qquad \text{Add 12 to both sides.}$$
$$3x = 27$$
$$\frac{3x}{3} = \frac{27}{3} \qquad \text{Divide both sides by 3.}$$
$$x = 9$$

## EXAMPLE 11

Solve $2x - (3x + 15) = 4x - 1$ for $x$.

$$2x - (3x + 15) = 4x - 1$$
$$2x - 3x - 15 = 4x - 1 \qquad \text{Remove parentheses.}$$
$$-x - 15 = 4x - 1 \qquad \text{Combine like terms.}$$
$$-x - 15 + x = 4x - 1 + x \qquad \text{Add } x \text{ to both sides.}$$
$$-15 = 5x - 1$$
$$-15 + 1 = 5x - 1 + 1 \qquad \text{Add 1 to both sides.}$$
$$-14 = 5x$$
$$\frac{-14}{5} = \frac{5x}{5} \qquad \text{Divide both sides by 5.}$$
$$-2.8 = x$$

## PROBLEMS

*Solve each equation.*

1. $4x + 9 = 7x - 18$
2. $2x - 4 = 3x + 7$
3. $-2x + 5 = 3x - 10$
4. $5x + 3 = 2x - 18$
5. $3x + 5 = 5x - 11$
6. $-5x + 12 = 12x - 5$
7. $13x + 2 = 20x - 5$
8. $5x + 3 = -9x - 38$
9. $-4x + 2 = -10x - 20$
10. $9x + 3 = 6x + 8$
11. $3x + (2x - 7) = 8$
12. $11 - (x + 12) = 100$
13. $7x - (13 - 2x) = 5$
14. $20(7x - 2) = 180$
15. $-3x + 5(x - 6) = 12$
16. $3(x + 117) = 201$
17. $5(2x - 1) = 8(x + 3)$
18. $3(x + 4) = 8 - 3(x - 2)$
19. $-2(3x - 2) = 3x - 2(5x + 1)$
20. $\dfrac{x}{5} - 2\left(\dfrac{2x}{5} + 1\right) = 28$

A quadratic equation in one variable is one in which the largest exponent of the variable is 2. The most general quadratic equation in variable $x$ is written as

$$ax^2 + bx + c = 0$$

## EXAMPLE 1

Solve $x^2 = 16$ for $x$.

To solve a quadratic equation of this type, all we need to do is take the square root of both sides of the equation.

$$x^2 = 16 \qquad \text{Take the square root of both sides.}$$
$$x = \pm 4$$

In general, equations of the form $ax^2 = b$, where $a \neq 0$, can be solved as follows:

$$ax^2 = b$$
$$x^2 = \frac{b}{a} \qquad \text{Divide both sides by } a.$$
$$x = \pm\sqrt{\frac{b}{a}} \qquad \text{Take the square root of both sides.}$$

## EXAMPLE 2

Solve $2x^2 - 18 = 0$ for $x$.

$$2x^2 - 18 = 0$$
$$2x^2 = 18 \qquad \text{Add 18 to both sides.}$$
$$x^2 = 9 \qquad \text{Divide both sides by 2.}$$

So $\qquad\qquad x = \pm 3. \qquad \text{Take the square root of both sides.}$

## EXAMPLE 3

Solve $5y^2 = 100$ for $y$.

$$5y^2 = 100$$
$$y^2 = 20 \qquad \text{Divide both sides by 5.}$$
$$y = \pm\sqrt{20} \qquad \text{Take the square root of both sides.}$$
$$y = \pm 4.47$$

## PROBLEMS

*Solve each equation.*

1. $x^2 = 36$
2. $y^2 = 100$
3. $2x^2 = 98$
4. $5x^2 = 0.05$
5. $3x^2 - 27 = 0$
6. $2y^2 - 15 = 17$
7. $10x^2 + 4.9 = 11.3$
8. $2(32)(48 - 15) = v^2 - 27^2$
9. $2(107) = 9.8t^2$
10. $65 = \pi r^2$
11. $2050 = \pi r^2$
12. $24^2 = a^2 + 16^2$

The solutions of the general quadratic equation

$$ax^2 + bx + c = c \qquad (\text{where } a \neq 0)$$

is given by the formula (called the *quadratic formula*):

355

$$x = \frac{-b \pm \sqrt{b^2 - 4ac}}{2a}$$

where:  $a =$ coefficient of the $x^2$ term
$b =$ coefficient of the $x$ term
$c =$ constant term

The symbol ($\pm$) is used to combine two expressions or equations into one. For example, $a \pm 2$ means $a + 2$ or $a - 2$. Similarly,

$$x = \frac{-b \pm \sqrt{b^2 - 4ac}}{2a}$$

means

$$x = \frac{-b + \sqrt{b^2 - 4ac}}{2a} \quad \text{or} \quad x = \frac{-b - \sqrt{b^2 - 4ac}}{2a}$$

## EXAMPLE 4

In the equation, $4x^2 - 3x - 7 = 0$, identify $a$, $b$, and $c$.

$$a = 4, \quad b = -3, \quad \text{and} \quad c = -7$$

## EXAMPLE 5

Solve $x^2 + 2x - 8 = 0$ using the quadratic formula.
First, $a = 1$, $b = 2$, and $c = -8$. Then

$$x = \frac{-b \pm \sqrt{b^2 - 4ac}}{2a}$$

So

$$x = \frac{-2 \pm \sqrt{(2)^2 - 4(1)(-8)}}{2(1)}$$

$$= \frac{-2 \pm \sqrt{4 - (-32)}}{2}$$

$$= \frac{-2 \pm \sqrt{36}}{2}$$

$$= \frac{-2 \pm 6}{2}$$

$$= \frac{-2 + 6}{2} \quad \text{or} \quad \frac{-2 - 6}{2}$$

$$= \frac{4}{2} \quad \text{or} \quad \frac{-8}{2}$$

$$= 2 \quad \text{or} \quad -4$$

The solutions are 2 and $-4$.

If the number under the radical sign is not a perfect square, find the square root of the number by using a calculator and proceed as before.

## EXAMPLE 6

Solve $4x^2 - 7x = 32$ using the quadratic formula.
Before identifying $a$, $b$, and $c$, the equation must be set equal to zero. That is,

$$4x^2 - 7x - 32 = 0$$

First, $a = 4$, $b = -7$, and $c = -32$. Then

$$x = \frac{-b \pm \sqrt{b^2 - 4ac}}{2a}$$

$$= \frac{-(-7) \pm \sqrt{(-7)^2 - 4(4)(-32)}}{2(4)}$$

$$= \frac{7 \pm \sqrt{49 - (-512)}}{8}$$

$$= \frac{7 \pm \sqrt{561}}{8}$$

$$= \frac{7 \pm 23.7}{8} \qquad (\sqrt{561} = 23.7)$$

$$= \frac{7 + 23.7}{8} \quad \text{or} \quad \frac{7 - 23.7}{8}$$

$$= 3.84 \quad \text{or} \quad -2.09$$

The approximate solutions are 3.84 and $-2.09$.

## PROBLEMS

*Find the values of* a, b, *and* c *in each quadratic equation.*

**1.** $3x^2 + x - 5 = 0$                     **2.** $-2x^2 + 7x + 4 = 0$

**3.** $6x^2 + 8x + 2 = 0$                   **4.** $5x^2 - 2x - 15 = 0$

**5.** $9x^2 + 6x = 4$                        **6.** $6x^2 = x + 9$

**7.** $5x^2 + 6x = 0$                         **8.** $7x^2 - 45 = 0$

**9.** $9x^2 = 64$                              **10.** $16x^2 = 49$

*Solve each quadratic equation using the quadratic formula.*

**11.** $x^2 - 10x + 21 = 0$                 **12.** $2x^2 + 13x + 15 = 0$

**13.** $6x^2 + 7x = 20$                     **14.** $15x^2 = 4x + 4$

**15.** $6x^2 - 2x = 19$                     **16.** $4x^2 = 28x - 49$

**17.** $18x^2 - 15x = 26$                 **18.** $48x^2 + 9 = 50x$

**19.** $16.5x^2 + 8.3x - 14.7 = 0$       **20.** $125x^2 - 167x + 36 = 0$

# APPENDIX

# B

## Scientific Hand Calculators—Brief Instructions on Use

**B-1**
INTRODUCTION

There are several kinds and brands of calculators. Some are very simple to use and of pocket size. Others are as small but do more difficult calculations. Still others are large and do the most difficult calculations. Some can be programmed. However, most calculators operate in one of two ways: by algebraic logic or reverse Polish notation. We will demonstrate various operations on calculators that use algebraic logic, which follows the steps commonly used in mathematics. If your calculator uses reverse Polish notation, please consult your instruction manual.

To demonstrate how to use a calculator, we will (1) use a flow chart, (2) show what buttons are pushed and the order in which they are pushed, and (3) show the display at each step. We assume that you know how to add, subtract, multiply, and divide on the calculator.

We have chosen to illustrate the most common types of calculators. Yours may differ in the number of digits displayed, or in the order of buttons pushed if yours has a function (F) button. If so, consult your manual.

**B-2**
SCIENTIFIC NOTATION

Numbers expressed in scientific notation can be entered into many calculators. The results may then also be given in scientific notation.

### EXAMPLE 1

Multiply $(6.5 \times 10^8)(1.4 \times 10^{-15})$ and write the result in scientific notation. See figure at top of page 359.

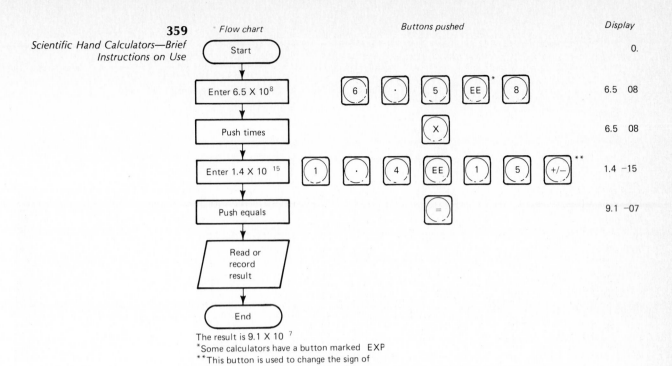

| Flow chart | Buttons pushed | Display |
|---|---|---|
| Start | | 0. |
| Enter 6.5 X $10^8$ | 6  ·  5  EE*  8 | 6.5   08 |
| Push times | X | 6.5   08 |
| Enter 1.4 X $10^{15}$ | 1  ·  4  EE  1  5  +/−** | 1.4   −15 |
| Push equals | = | 9.1   −07 |
| Read or record result | | |
| End | | |

The result is 9.1 X $10^7$
*Some calculators have a button marked EXP
**This button is used to change the sign of
the *last number* that has been entered.

## EXAMPLE 2

Divide $\dfrac{3.24 \times 10^{-5}}{7.2 \times 10^{-12}}$ and write the result in scientific notation.

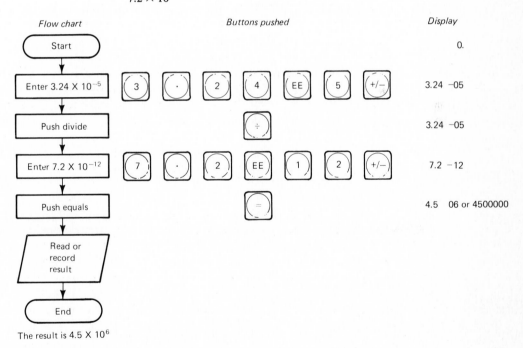

| Flow chart | Buttons pushed | Display |
|---|---|---|
| Start | | 0. |
| Enter 3.24 X $10^{-5}$ | 3  ·  2  4  EE  5  +/− | 3.24   −05 |
| Push divide | ÷ | 3.24   −05 |
| Enter 7.2 X $10^{-12}$ | 7  ·  2  EE  1  2  +/− | 7.2   −12 |
| Push equals | = | 4.5   06 or 4500000 |
| Read or record result | | |
| End | | |

The result is 4.5 X $10^6$

EXAMPLE 3

Find the value of $\dfrac{(-6.3 \times 10^4)(-5.07 \times 10^{-9})(8.11 \times 10^{-6})}{(5.63 \times 10^{12})(-1.84 \times 10^7)}$ and write the result rounded to three significant digits in scientific notation.

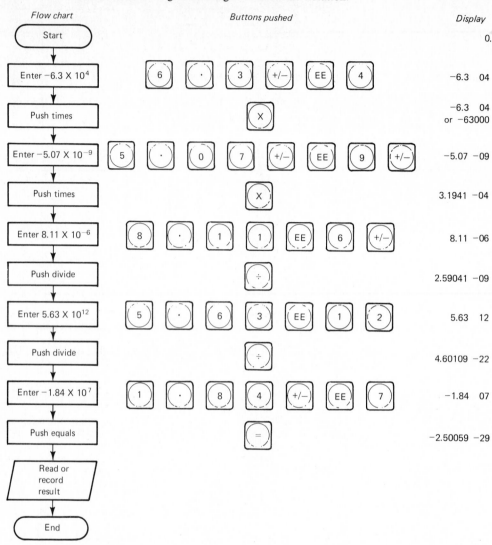

| Flow chart | Buttons pushed | Display |
|---|---|---|
| Start | | 0. |
| Enter $-6.3 \times 10^4$ | 6 . 3 +/− EE 4 | −6.3  04 |
| Push times | X | −6.3  04 or −63000 |
| Enter $-5.07 \times 10^{-9}$ | 5 . 0 7 +/− EE 9 +/− | −5.07 −09 |
| Push times | X | 3.1941 −04 |
| Enter $8.11 \times 10^{-6}$ | 8 . 1 1 EE 6 +/− | 8.11 −06 |
| Push divide | ÷ | 2.59041 −09 |
| Enter $5.63 \times 10^{12}$ | 5 . 6 3 EE 1 2 | 5.63  12 |
| Push divide | ÷ | 4.60109 −22 |
| Enter $-1.84 \times 10^7$ | 1 . 8 4 +/− EE 7 | −1.84  07 |
| Push equals | = | −2.50059 −29 |
| Read or record result | | |
| End | | |

The result rounded to three significant digits is $-2.50 \times 10^{-29}$

## B-3 SQUARES AND SQUARE ROOTS

EXAMPLE 1

Find the value of $(46.8)^2$.

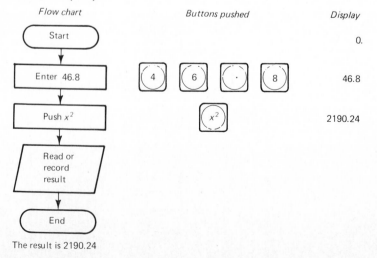

| Flow chart | Buttons pushed | Display |
|---|---|---|
| Start | | 0. |
| Enter 46.8 | 4 6 . 8 | 46.8 |
| Push $x^2$ | $x^2$ | 2190.24 |
| Read or record result | | |
| End | | |

The result is 2190.24

## EXAMPLE 2

Find the value of $(6.3 \times 10^{-18})^2$.

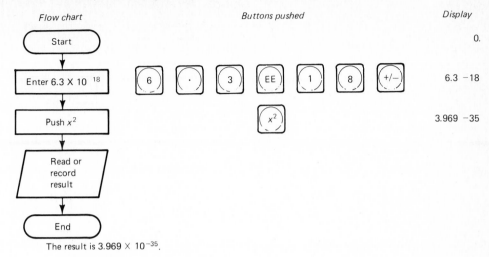

The result is $3.969 \times 10^{-35}$.

## EXAMPLE 3

Find the value of $\sqrt{158.65}$ and round to four significant digits.

The result rounded to four significant digits is 12.60.

## EXAMPLE 4

Find the value of $\sqrt{6.95 \times 10^{-15}}$ and round to three significant digits.

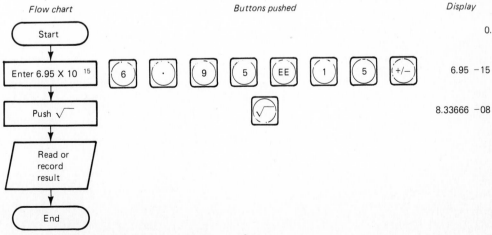

The result rounded to three significant digits is $8.34 \times 10^{8}$.

EXAMPLE 5

Find the value of $\sqrt{15.7^2 + 27.6^2}$ and round to three significant digits.

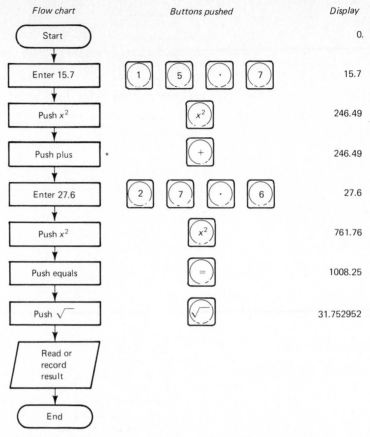

| Flow chart | Buttons pushed | Display |
|---|---|---|
| Start | | 0. |
| Enter 15.7 | 1 5 . 7 | 15.7 |
| Push $x^2$ | $x^2$ | 246.49 |
| Push plus * | + | 246.49 |
| Enter 27.6 | 2 7 . 6 | 27.6 |
| Push $x^2$ | $x^2$ | 761.76 |
| Push equals | = | 1008.25 |
| Push $\sqrt{\phantom{x}}$ | $\sqrt{\phantom{x}}$ | 31.752952 |
| Read or record result | | |
| End | | |

*Some calculators require the use of a memory to do this calculation.
The result is 31.8 rounded to three significant digits.

**B-4**
**USING A CALCULATOR
WITH MEMORY**

For some combinations of operations, a calculator with a memory is very helpful. A memory stores numbers for future use. Entries can then be added to or subtracted from what is already in the memory. *Note:* Consult the instruction manual if your calculator has more than one memory.

EXAMPLE 1

Find the value of $6 \times \$1.44 + 15 \times \$8.36 + 9 \times \$0.98 - 5 \times \$15.95$.
See figure on page 363.

| Flow chart | Buttons pushed | Display |
|---|---|---|
| Start | | 0. |
| Enter 6 | 6 | 6 |
| Push times | X | 6 |
| Enter 1.44 | 1 . 4 4 | 1.44 |
| Push equals | = | 8.64 |
| Push store or add to memory | STO or M⁺ | 8.64 |
| Enter 15 | 1 5 | 15 |
| Push times | X | 15 |
| Enter 8.36 | 8 . 3 6 | 8.36 |
| Push equals | = | 125.4 |
| Push add to memory | SUM or M⁺ | 125.4 |
| Enter 9 | 9 | 9 |
| Push times | X | 9 |
| Enter 0.98 | . 9 8 | 0.98 |
| Push equals | = | 8.82 |
| Push add to memory | SUM or M⁺ | 8.82 |
| Enter 5 | 5 | 5 |
| Push times | X | 5 |
| Enter 15.95 | 1 5 . 9 5 | 15.95 |
| Push equals | = | 79.75 |
| Push subtract from memory | +/− → SUM or M⁻ | 79.75 |
| Push memory recall | RCL or MR | 63.11 |
| Read or record result | | |
| End | | |

The result is $63.11.

EXAMPLE 2

Find the value of $(15.63)^2 + (18.74)^2 - 2(15.63)(18.74)(0.7547)$ and round the result to four significant digits.

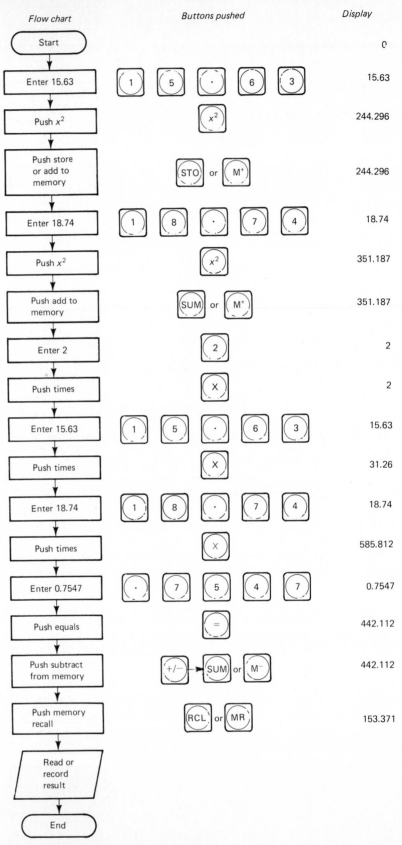

| Flow chart | Buttons pushed | Display |
|---|---|---|
| Start | | 0 |
| Enter 15.63 | 1 5 · 6 3 | 15.63 |
| Push $x^2$ | $x^2$ | 244.296 |
| Push store or add to memory | STO or M+ | 244.296 |
| Enter 18.74 | 1 8 · 7 4 | 18.74 |
| Push $x^2$ | $x^2$ | 351.187 |
| Push add to memory | SUM or M+ | 351.187 |
| Enter 2 | 2 | 2 |
| Push times | X | 2 |
| Enter 15.63 | 1 5 · 6 3 | 15.63 |
| Push times | X | 31.26 |
| Enter 18.74 | 1 8 · 7 4 | 18.74 |
| Push times | X | 585.812 |
| Enter 0.7547 | · 7 5 4 7 | 0.7547 |
| Push equals | = | 442.112 |
| Push subtract from memory | +/− → SUM or M− | 442.112 |
| Push memory recall | RCL or MR | 153.371 |
| Read or record result | | |
| End | | |

The result rounded to four significant digits is 153.4.

Most hand calculators have buttons to evaluate the sine, cosine, and tangent functions.

### EXAMPLE 1

Find sin 26° rounded to four significant digits.

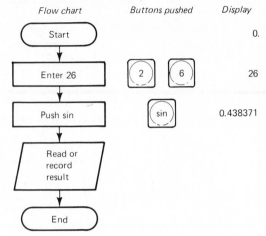

That is, sin 26° = 0.4384 rounded to four significant digits.

### EXAMPLE 2

Find cos 36.75° rounded to four significant digits.

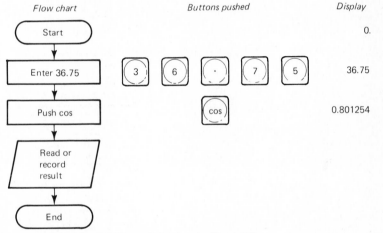

That is, cos 36.75° = 0.8013 rounded to four significant digits.

### EXAMPLE 3

Find tan 70.6° rounded to four significant digits.

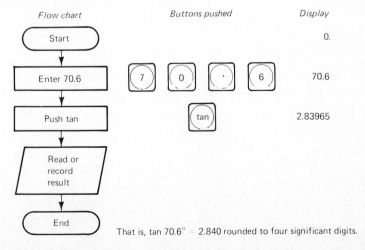

That is, tan 70.6° = 2.840 rounded to four significant digits.

Since we deal almost exclusively with right triangles, we will discuss how to find the angle of a right triangle when the value of the trigonometric ratio is known. That is, we will be finding an angle $A$ when $0° \leq A \leq 90°$.

### EXAMPLE 4

Given $\sin A = 0.4321$, find angle $A$ to the nearest tenth of a degree.

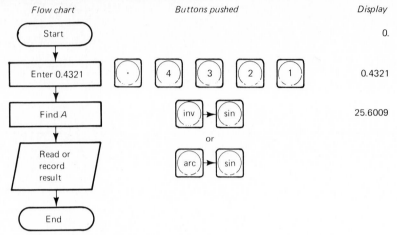

Thus, $A = 25.6°$ to the nearest tenth degree.

### EXAMPLE 5

Given $\cos B = 0.6046$, find angle $B$ to the nearest tenth of a degree.

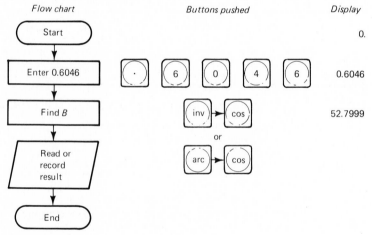

Thus, $B = 52.8°$ to the nearest tenth degree.

### EXAMPLE 6

Given $\tan A = 2.584$, find angle $A$ to the nearest tenth of a degree.

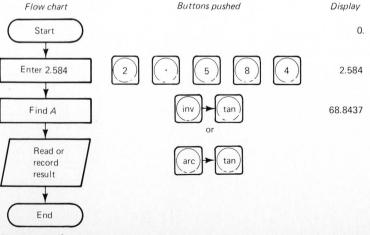

Thus, $A = 68.8°$ to the nearest tenth degree.

A trigonometric function often occurs in an expression that must be evaluated.

EXAMPLE 1

Given $a = (\tan 54°)(25.6 \text{ m})$, find $a$ rounded to three significant digits.

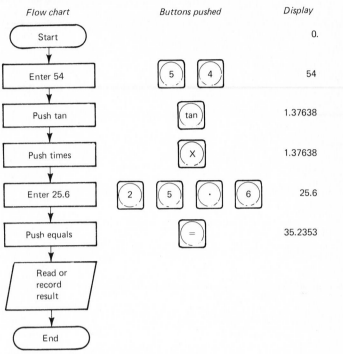

Thus, $a = 35.2$ m rounded to three significant digits.

EXAMPLE 2

Given $b = \dfrac{452 \text{ m}}{\cos 37.5°}$, find $b$ rounded to three significant digits.

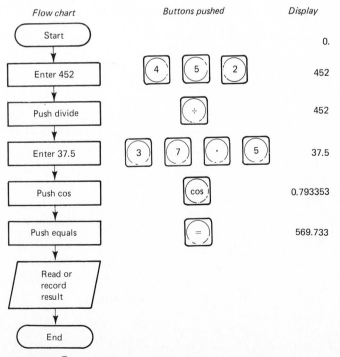

Thus, $b = 57\overline{0}$ m rounded to three significant digits.

**B-7**
USING
THE $y^x$ BUTTON

To raise a number to a power, use the $y^x$ button as follows.

EXAMPLE 1
___

Find the value of $4^5$.

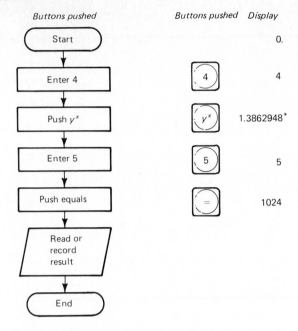

*Not all calculators display this number. Some display 4.
That is, $4^5 = 1024$.

EXAMPLE 2
___

Find the value of $1.5^{-4}$ rounded to two significant digits.

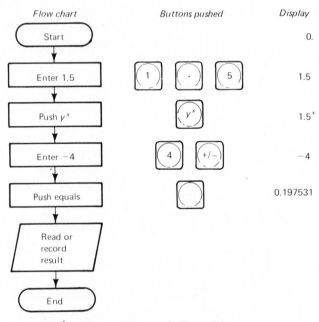

That is, $1.5^{-4} = 0.20$ rounded to two significant digits.
*Some calculators display 0.4054648.

## PROBLEMS

*Do as indicated and round each result to three significant digits.*

**1.** $(6.43 \times 10^8)(5.16 \times 10^{10})$  

**2.** $(4.16 \times 10^{-5})(3.45 \times 10^{-7})$

**3.** $(1.456 \times 10^{12})(-4.69 \times 10^{-18})$  

**4.** $(-5.93 \times 10^9)(7.055 \times 10^{-12})$

5. $(7.45 \times 10^8) \div (8.92 \times 10^{18})$

6. $(1.38 \times 10^{-6}) \div (4.324 \times 10^6)$

7. $\dfrac{-6.19 \times 10^{12}}{7.755 \times 10^{-8}}$

8. $\dfrac{1.685 \times 10^{10}}{1.42 \times 10^{24}}$

9. $\dfrac{(5.26 \times 10^{-8})(8.45 \times 10^6)}{(-6.142 \times 10^9)(1.056 \times 10^{-12})}$

10. $\dfrac{(-2.35 \times 10^{-9})(1.25 \times 10^{11})(4.65 \times 10^{17})}{(8.75 \times 10^{23})(-5.95 \times 10^{-6})}$

11. $(68.4)^2$

12. $(3180)^2$

13. $\sqrt{46,500}$

14. $\sqrt{0.000634}$

15. $(1.45 \times 10^5)^2$

16. $(1.095 \times 10^{-18})^2$

17. $\sqrt{4.63 \times 10^{18}}$

18. $\sqrt{9.49 \times 10^{-15}}$

19. $\sqrt{(4.68)^2 + (9.63)^2}$

20. $\sqrt{(18.4)^2 - (6.5)^2}$

21. $3 \times \$1.95 + 5 \times \$0.49 + 2 \times \$0.85 + 7 \times \$0.19 + 14 \times \$0.09 + 6 \times \$3.95$

22. $4 \times \$16.50 + 2 \times \$9.95 + 3 \times \$4.49 + 6 \times \$5.59 + \$1.29 - 2 \times \$3.98$

23. $\dfrac{32.4}{46.5} - \dfrac{18.5}{6.32}$

24. $(18.6)(5.92) - (4.63)(28.5)$

25. $(39.6)^2 - (7.54)^2$

26. $\sqrt{19.7} - \sqrt{86.4}$

27. $\dfrac{91.4 - 48.6}{91.4 - 15.9}$

28. $\dfrac{14.7 + 9.6}{45.7 + 68.2}$

29. $(69.2)(46.8)^2 + (19.7)^2$

30. $(58.7)^2 - (69.2)(0.965)$

31. $(257)^2 + (352)^2 - 2(257)(352)(-0.2588)$

32. $\sqrt{(0.356)^2 + (0.427)^2 - 2(0.365)(0.427)(0.7071)}$

33. $\sin 13°$

34. $\cos 22°$

35. $\tan 52.3°$

36. $\tan 31.25°$

37. $\cos 59.36°$

38. $\sin 84.55°$

39. $\sin 48°$

40. $\cos 48°$

41. $\tan 75°$

42. $\sin 8°$

43. $\sin 8.7°$

44. $\cos 35°$

*Find each angle rounded to the nearest tenth of a degree.*

45. $\sin A = 0.6527$

46. $\cos B = 0.2577$

47. $\tan A = 0.4568$

48. $\sin B = 0.4658$

49. $\cos A = 0.5563$

50. $\tan B = 1.496$

51. $\sin B = 0.1465$

52. $\cos A = 0.4968$

53. $\tan B = 1.987$

54. $\sin A = 0.2965$

55. $\cos B = 0.3974$

56. $\tan A = 0.8885$

*Find each angle to the nearest tenth of a degree between 0° and 90° and each side to three significant digits.*

57. $b = (\sin 58.2°)(296 \text{ m})$

58. $a = (\cos 25.2°)(54.5 \text{ m})$

59. $c = \dfrac{37.5 \text{ m}}{\cos 65.2°}$

60. $b = \dfrac{59.7 \text{ m}}{\tan 41.2°}$

61. $\tan A = \dfrac{512 \text{ km}}{376 \text{ km}}$

62. $\cos B = \dfrac{75.2 \text{ m}}{89.5 \text{ m}}$

63. $a = (\cos 19.5°)(15.7 \text{ cm})$

64. $c = \dfrac{236 \text{ km}}{\sin 65.2°}$

65. $b = \dfrac{36.7 \text{ m}}{\tan 59.2°}$

66. $a = (\tan 5.7°)(135 \text{ m})$

*Find the value of each power and round each result to three significant digits.*

67. $12^4$

68. $1.8^3$

69. $0.46^5$

70. $9^{-3}$

71. $14^{-5}$

72. $0.65^{-4}$

# APPENDIX
# C

# *Tables*

**TABLE 1**
English Weights and Measures

Units of length

Standard unit—inch (in. or ″)
12 inches = 1 foot (ft or ′)
3 feet = 1 yard (yd)
$5\frac{1}{2}$ yards or $16\frac{1}{2}$ feet = 1 rod (rd)
5280 feet = 1 mile (mi)

Units of weight

Standard unit—pound (lb)
16 ounces (oz) = 1 pound
2000 pounds = 1 ton (T)

Volume measure

*Liquid*

16 ounces (fl oz) = 1 pint (pt)
2 pints = 1 quart (qt)
4 quarts = 1 gallon (gal)

*Dry*

2 pints (pt) = 1 quart (qt)
8 quarts = 1 peck (pk)
4 pecks = 1 bushel (bu)

**TABLE 2**
Metric System Prefixes

| Multiple or submultiple[a] decimal form | Power of 10 | Prefix[b] | Prefix symbol | Pronun-ciation | Meaning |
|---|---|---|---|---|---|
| 1,000,000,000,000 | $10^{12}$ | tera | T | tĕr′ă | one trillion times |
| 1,000,000,000 | $10^{9}$ | giga | G | jĭg′ă | one billion times |
| 1,000,000 | $10^{6}$ | mega | M | mĕg′ă | one million times |
| 1,000 | $10^{3}$ | kilo | k | kĭl′ō | one thousand times |
| 100 | $10^{2}$ | hecto | h | hĕk′tō | one hundred times |
| 10 | $10^{1}$ | deka | da | dĕk′ă | ten times |
| 0.1 | $10^{-1}$ | deci | d | dĕs′ĭ | one tenth of |
| 0.01 | $10^{-2}$ | centi | c | sĕnt′ĭ | one hundredth of |
| 0.001 | $10^{-3}$ | milli | m | mĭl′ĭ | one thousandth of |
| 0.000001 | $10^{-6}$ | micro | μ | mĭ′krō | one millionth of |
| 0.000000001 | $10^{-9}$ | nano | n | năn′ō | one billionth of |
| 0.000000000001 | $10^{-12}$ | pico | p | pē′kō | one trillionth of |

[a] Factor by which the unit is multiplied.
[b] The same prefixes are used with all SI metric units.

As an example, the prefixes are used below with the metric standard unit of length, metre (m).

1 *tera*metre (Tm) = 1,000,000,000,000 m        1 m = 0.000000000001 Tm

1 *giga*metre (Gm) = 1,000,000,000 m        1 m = 0.000000001 Gm

1 *mega*metre (Mm) = 1,000,000 m        1 m = 0.000001 Mm

1 *kilo*metre (km) = 1,000 m        1 m = 0.001 km

1 *hecto*metre (hm) = 100 m        1 m = 0.01 hm

1 *deka*metre (dam) = 10 m        1 m = 0.1 dam

1 *deci*metre (dm) = 0.1 m        1 m = 10 dm

1 *centi*metre (cm) = 0.01 m        1 m = 100 cm

1 *milli*metre (mm) = 0.001 m        1 m = 1,000 mm

1 *micro*metre (μm) = 0.000001 m        1 m = 1,000,000 μm

1 *nano*metre (nm) = 0.000000001 m        1 m = 1,000,000,000 nm

1 *pico*metre (pm) = 0.000000000001 m        1 m = 1,000,000,000,000 pm

**TABLE 3**
Conversion Table for Length

|  | cm | m | km | in. | ft | mile |
|---|---|---|---|---|---|---|
| 1 centimetre = | 1 | $10^{-2}$ | $10^{-5}$ | 0.394 | $3.28 \times 10^{-2}$ | $6.21 \times 10^{-6}$ |
| 1 metre = | 100 | 1 | $10^{-3}$ | 39.4 | 3.28 | $6.21 \times 10^{-4}$ |
| 1 kilometre = | $10^5$ | 1000 | 1 | $3.94 \times 10^4$ | 3280 | 0.621 |
| 1 inch = | 2.54 | $2.54 \times 10^{-2}$ | $2.54 \times 10^{-5}$ | 1 | $8.33 \times 10^{-2}$ | $1.58 \times 10^{-5}$ |
| 1 foot = | 30.5 | 0.305 | $3.05 \times 10^{-4}$ | 12 | 1 | $1.89 \times 10^{-4}$ |
| 1 mile = | $1.61 \times 10^5$ | 1610 | 1.61 | $6.34 \times 10^4$ | 5280 | 1 |

**CONVERSION FACTORS**    Conversion factors may be made directly from the tables. For example, 1 cm = 0.394 in. For further instructions on conversion factors, see page 7 of the text.

**TABLE 4**
Conversion Table for Area

| Metric | English |
|---|---|
| 1 m² = 10,000 cm² | 1 ft² = 144 in² |
| = 1,000,000 mm² | 1 yd² = 9 ft² |
| 1 cm² = 100 mm² | 1 rd² = 30.25 yd² |
| = 0.0001 m² | 1 acre = 160 rd² |
| 1 km² = 1,000,000 m² | = 4840 yd² |
| | = 43,560 ft² |
| | 1 mi² = 640 acres |

|  | m² | cm² | ft² | in² |
|---|---|---|---|---|
| 1 square metre = | 1 | $10^4$ | 10.8 | 1550 |
| 1 square centimetre = | $10^{-4}$ | 1 | $1.08 \times 10^{-3}$ | 0.155 |
| 1 square foot = | $9.29 \times 10^{-2}$ | 929 | 1 | 144 |
| 1 square inch = | $6.45 \times 10^{-4}$ | 6.45 | $6.94 \times 10^{-3}$ | 1 |
| 1 circular mil = $5.07 \times 10^{-6}$ cm² = $7.85 \times 10^{-7}$ in² | | | | |

**TABLE 5**
Conversion Table for Volume

| Metric | English |
|---|---|
| 1 m³ = $10^6$ cm³ | 1 ft³ = 1728 in³ |
| 1 cm³ = $10^{-6}$ m³ | 1 yd³ = 27 ft³ |
| = $10^3$ mm³ | |

|  | m³ | cm³ | L | ft³ | in³ |
|---|---|---|---|---|---|
| 1 m³ = | 1 | $10^6$ | 1000 | 35.3 | $6.10 \times 10^4$ |
| 1 cm³ = | $10^{-6}$ | 1 | $1.00 \times 10^{-3}$ | $3.53 \times 10^{-5}$ | $6.10 \times 10^{-2}$ |
| 1 litre = | $1.00 \times 10^{-3}$ | 1000 | 1 | $3.53 \times 10^{-2}$ | 61.0 |
| 1 ft³ = | $2.83 \times 10^{-2}$ | $2.83 \times 10^4$ | 28.3 | 1 | 1728 |
| 1 in³ = | $1.64 \times 10^{-5}$ | 16.4 | $1.64 \times 10^{-2}$ | $5.79 \times 10^{-4}$ | 1 |
| 1 U.S. fluid gallon = 4 U.S. fluid quarts = 8 U.S. pints = 128 U.S. fluid ounces = 231 in³ = 0.134 ft³ | | | | | |
| 1 L = 1000 cm³ = 1.06 qt    1 fl oz = 29.5 cm³ | | | | | |
| 1 ft³ = 7.47 gal = 28.3 L | | | | | |

EXAMPLE

1 m³ = 35.3 ft³.

**TABLE 6**
Conversion Table for Mass

|  | g | kg | slug | oz | lb | ton |
|---|---|---|---|---|---|---|
| 1 gram = | 1 | 0.001 | $6.85 \times 10^{-5}$ | $3.53 \times 10^{-2}$ | $2.21 \times 10^{-3}$ | $1.10 \times 10^{-6}$ |
| 1 kilogram = | 1000 | 1 | $6.85 \times 10^{-2}$ | 35.3 | 2.21 | $1.10 \times 10^{-3}$ |
| 1 slug = | $1.46 \times 10^{4}$ | 14.6 | 1 | 515 | 32.2 | $1.61 \times 10^{-2}$ |
| 1 ounce = | 28.4 | $2.84 \times 10^{-2}$ | $1.94 \times 10^{-3}$ | 1 | $6.25 \times 10^{-2}$ | $3.13 \times 10^{-5}$ |
| 1 pound = | 454 | 0.454 | $3.11 \times 10^{-2}$ | 16 | 1 | $5.00 \times 10^{-4}$ |
| 1 ton = | $9.07 \times 10^{5}$ | 907 | 62.2 | $3.2 \times 10^{4}$ | 2000 | 1 |
| 1 metric ton = 1000 kg = 2205 lb | | | | | | |
| 1 g = 15.4 grains | | | | | | |

Quantities in the shaded areas are not mass units. When we write, for example, 1 kg " = " 2.21 lb, this means that a kilogram is a mass that weighs 2.21 pounds under standard conditions of gravity ($g = 9.80$ m/s² = 32.2 ft/s²).

**TABLE 7**
Conversion Table for Density

|  | slug/ft³ | kg/m³ | g/cm³ | lb/ft³ | lb/in³ |
|---|---|---|---|---|---|
| 1 slug per ft³ = | 1 | 515.4 | 0.515 | 32.2 | $1.86 \times 10^{-2}$ |
| 1 kilogram per m³ = | $1.94 \times 10^{-3}$ | 1 | 0.001 | $6.24 \times 10^{-2}$ | $3.61 \times 10^{-5}$ |
| 1 gram per cm³ = | 1.94 | 1000 | 1 | 62.4 | $3.61 \times 10^{-2}$ |
| 1 pound per ft³ = | $3.11 \times 10^{-2}$ | 16.0 | $1.60 \times 10^{-2}$ | 1 | $5.79 \times 10^{-4}$ |
| 1 pound per in³ = | 53.7 | $2.77 \times 10^{4}$ | 27.7 | 1728 | 1 |

Quantities in the shaded areas are weight densities and, as such, are dimensionally different from mass densities.

Note that $D_w = D_w g$, where:

$$D_w = \text{weight density}$$
$$D_m = \text{mass density}$$
$$g = 9.80 \text{ m/s}^2 = 32.2 \text{ ft/s}^2$$

**TABLE 8**
Conversion Table for Time

|  | yr | day | h | min | s |
|---|---|---|---|---|---|
| 1 year = | 1 | 365 | $8.77 \times 10^{3}$ | $5.26 \times 10^{5}$ | $3.16 \times 10^{7}$ |
| 1 day = | $2.74 \times 10^{-3}$ | 1 | 24 | 1440 | $8.64 \times 10^{4}$ |
| 1 hour = | $1.14 \times 10^{-4}$ | $4.17 \times 10^{-2}$ | 1 | 60 | 3600 |
| 1 minute = | $1.90 \times 10^{-6}$ | $6.94 \times 10^{-4}$ | $1.67 \times 10^{-2}$ | 1 | 60 |
| 1 second = | $3.17 \times 10^{-8}$ | $1.16 \times 10^{-5}$ | $2.78 \times 10^{-4}$ | $1.67 \times 10^{-2}$ | 1 |

**TABLE 9**
Conversion Table for Speed

| | ft/s | km/h | m/s | mi/h | cm/s | knot |
|---|---|---|---|---|---|---|
| 1 foot per second = | 1 | 1.10 | 0.305 | 0.682 | 30.5 | 0.593 |
| 1 kilometre per hour = | 0.911 | 1 | 0.278 | 0.621 | 27.8 | 0.540 |
| 1 metre per second = | 3.28 | 3.60 | 1 | 2.24 | 100 | 1.94 |
| 1 mile per hour = | 1.47 | 1.61 | 0.447 | 1 | 44.7 | 0.869 |
| 1 centimetre per second = | $3.28 \times 10^{-2}$ | $3.60 \times 10^{-2}$ | 0.01 | $2.24 \times 10^{-2}$ | 1 | $1.94 \times 10^{-2}$ |
| 1 knot = | 1.69 | 1.85 | 0.514 | 1.15 | 51.4 | 1 |
| 1 knot = 1 naut mi/h | 1 mi/min = 88.0 ft/s = 60.0 mi/h | | | 55.0 mi/h = 88.6 km/h | | |

**TABLE 10**
Conversion Table for
Force

| | N | lb |
|---|---|---|
| 1 newton = | 1 | 0.225 |
| 1 pound = | 4.45 | 1 |

**TABLE 11**
Conversion Table for Pressure

| | atm | inch of water | mm-Hg | N/m² (Pa) | lb/in² | lb/ft² |
|---|---|---|---|---|---|---|
| 1 atmosphere = | 1 | 407 | $76\bar{0}$ | $1.01 \times 10^5$ | 14.7 | 2120 |
| 1 inch of water[a] at 4°C = | $2.46 \times 10^{-3}$ | 1 | 1.87 | 249 | $3.61 \times 10^{-2}$ | 5.20 |
| 1 millimetre of mercury[a] at 0°C = | $1.32 \times 10^{-3}$ | 0.535 | 1 | 133 | $1.93 \times 10^{-2}$ | 2.79 |
| 1 newton per metre² = (pascal) | $9.87 \times 10^{-6}$ | $4.02 \times 10^{-3}$ | $7.50 \times 10^{-3}$ | 1 | $1.45 \times 10^{-4}$ | $2.09 \times 10^{-2}$ |
| 1 pound per in² = | $6.81 \times 10^{-2}$ | 27.7 | 51.7 | $6.90 \times 10^3$ | 1 | 144 |
| 1 pound per ft² = | $4.73 \times 10^{-4}$ | 0.192 | 0.359 | 47.9 | $6.94 \times 10^{-3}$ | 1 |

[a] Where the acceleration of gravity has the standard value, $9.80 \text{ m/s}^2 = 32.2 \text{ ft/s}^2$.

**TABLE 12**
Conversion Table for Energy, Work, Heat

| | Btu | ft lb | hp-h | J | cal | kWh |
|---|---|---|---|---|---|---|
| 1 British thermal unit = | 1 | 778 | $3.93 \times 10^{-4}$ | 1060 | 252 | $2.93 \times 10^{-4}$ |
| 1 foot pound = | $1.29 \times 10^{-3}$ | 1 | $5.05 \times 10^{-7}$ | 1.36 | 0.324 | $3.77 \times 10^{-7}$ |
| 1 horsepower-hour = | 2550 | $1.98 \times 10^6$ | 1 | $2.69 \times 10^6$ | $6.41 \times 10^5$ | 0.746 |
| 1 joule = | $9.48 \times 10^{-4}$ | 0.738 | $3.73 \times 10^{-7}$ | 1 | 0.239 | $2.78 \times 10^{-7}$ |
| 1 calorie = | $3.97 \times 10^{-3}$ | 3.09 | $1.56 \times 10^{-6}$ | 4.19 | 1 | $1.16 \times 10^{-6}$ |
| 1 kilowatt-hour = | 3410 | $2.66 \times 10^6$ | 1.34 | $3.60 \times 10^6$ | $8.60 \times 10^5$ | 1 |

**TABLE 13**

Conversion Table for Power

|  | Btu/h | ft lb/s | hp | cal/s | kW | W |
|---|---|---|---|---|---|---|
| 1 British thermal unit per hour = | 1 | 0.216 | $3.93 \times 10^{-4}$ | $7.00 \times 10^{-2}$ | $2.93 \times 10^{-4}$ | 0.293 |
| 1 foot pound per second = | 4.63 | 1 | $1.82 \times 10^{-3}$ | 0.324 | $1.36 \times 10^{-3}$ | 1.36 |
| 1 horsepower = | 2550 | 550 | 1 | 178 | 0.746 | 746 |
| 1 calorie per second = | 14.3 | 3.09 | $5.61 \times 10^{-3}$ | 1 | $4.19 \times 10^{-3}$ | 4.19 |
| 1 kilowatt = | 3410 | 738 | 1.34 | 239 | 1 | 1000 |
| 1 watt = | 3.41 | 0.738 | $1.34 \times 10^{-3}$ | 0.239 | 0.001 | 1 |

**TABLE 14**

Heat Constants

|  | Melting point (°C) | Boiling point (°C) | Specific heat (cal/g°C or kcal/kg°C or Btu/lb°F) | Specific heat (J/kg°C) | Heat of fusion (cal/g or kcal/kg) | Heat of fusion (J/kg) | Heat of vaporization (cal/g or kcal/kg) | Heat of vaporization (J/kg) |
|---|---|---|---|---|---|---|---|---|
| Alcohol, ethyl | −117 | 78.5 | 0.58 | 2400 | 24.9 | $1.04 \times 10^5$ | 204 | $8.54 \times 10^5$ |
| Aluminum | 660 | 2057 | 0.22 | 920 | 76.8 | $3.21 \times 10^5$ | | |
| Brass | 840 | | 0.092 | 390 | | | | |
| Copper | 1083 | 2330 | 0.092 | 390 | 49.0 | $2.05 \times 10^5$ | | |
| Glass | | | 0.21 | 880 | | | | |
| Ice | 0 | | 0.51 | 2100 | $8\overline{0}$ | $3.35 \times 10^5$ | | |
| Iron (steel) | 1540 | 3000 | 0.115 | 481 | 7.89 | $3.30 \times 10^4$ | | |
| Lead | 327 | 1620 | 0.031 | 130 | 5.86 | $2.45 \times 10^4$ | | |
| Mercury | −38.9 | 357 | 0.033 | 140 | 2.82 | $1.18 \times 10^4$ | 65.0 | $2.72 \times 10^5$ |
| Silver | 961 | 1950 | 0.056 | 230 | 26.0 | $1.09 \times 10^5$ | | |
| Steam | | | 0.48 | $200\overline{0}$ | | | | |
| Water (liquid) | 0 | $10\overline{0}$ | 1.00 | 4190 | | | $54\overline{0}$ | $2.26 \times 10^6$ |
| Zinc | 419 | 907 | 0.092 | 390 | 23.0 | $9.63 \times 10^4$ | | |

**TABLE 15**

Conversion Table for Charge

1 electronic charge = $1.60 \times 10^{-19}$ coulomb
1 ampere-hour = 3600 C

**TABLE 16**
Copper Wire Table

| Gauge no. | Diameter (mils) | Diameter (mm) | Cross section | | Ohms per 1000 ft | | Weight per 1000 ft (lb) |
|---|---|---|---|---|---|---|---|
| | | | cir mils | in² | 25°C (77°F) | 65°C (149°F) | |
| 0000 | 460.0 | | 212,000 | 0.166 | 0.0500 | 0.0577 | 641.0 |
| 000 | 410.0 | | 168,000 | 0.132 | 0.0630 | 0.0727 | 508.0 |
| 00 | 365.0 | | 133,000 | 0.105 | 0.0795 | 0.0917 | 403.0 |
| 0 | 325.0 | | 106,000 | 0.0829 | 0.100 | 0.116 | 319.0 |
| 1 | 289.0 | 7.35 | 83,700 | 0.0657 | 0.126 | 0.146 | 253.0 |
| 2 | 258.0 | 6.54 | 66,400 | 0.0521 | 0.159 | 0.184 | 201.0 |
| 3 | 229.0 | 5.83 | 52,600 | 0.0413 | 0.201 | 0.232 | 159.0 |
| 4 | 204.0 | 5.19 | 41,700 | 0.0328 | 0.253 | 0.292 | 126.0 |
| 5 | 182.0 | 4.62 | 33,100 | 0.0260 | 0.319 | 0.369 | 100.0 |
| 6 | 162.0 | 4.12 | 26,300 | 0.0206 | 0.403 | 0.465 | 79.5 |
| 7 | 144.0 | 3.67 | 20,800 | 0.0164 | 0.508 | 0.586 | 63.0 |
| 8 | 128.0 | 3.26 | 16,500 | 0.0130 | 0.641 | 0.739 | 50.0 |
| 9 | 114.0 | 2.91 | 13,100 | 0.0103 | 0.808 | 0.932 | 39.6 |
| 10 | 102.0 | 2.59 | 10,400 | 0.00815 | 1.02 | 1.18 | 31.4 |
| 11 | 91.0 | 2.31 | 8,230 | 0.00647 | 1.28 | 1.48 | 24.9 |
| 12 | 81.0 | 2.05 | 6,530 | 0.00513 | 1.62 | 1.87 | 19.8 |
| 13 | 72.0 | 1.83 | 5,180 | 0.00407 | 2.04 | 2.36 | 15.7 |
| 14 | 64.0 | 1.63 | 4,110 | 0.00323 | 2.58 | 2.97 | 12.4 |
| 15 | 57.0 | 1.45 | 3,260 | 0.00256 | 3.25 | 3.75 | 9.86 |
| 16 | 51.0 | 1.29 | 2,580 | 0.00203 | 4.09 | 4.73 | 7.82 |
| 17 | 45.0 | 1.15 | 2,050 | 0.00161 | 5.16 | 5.96 | 6.20 |
| 18 | 40.0 | 1.02 | 1,620 | 0.00128 | 6.51 | 7.51 | 4.92 |
| 19 | 36.0 | 0.91 | 1,290 | 0.00101 | 8.21 | 9.48 | 3.90 |
| 20 | 32.0 | 0.81 | 1,020 | 0.000802 | 10.4 | 11.9 | 3.09 |
| 21 | 28.5 | 0.72 | 810 | 0.000636 | 13.1 | 15.1 | 2.45 |
| 22 | 25.3 | 0.64 | 642 | 0.000505 | 16.5 | 19.0 | 1.94 |
| 23 | 22.6 | 0.57 | 509 | 0.000400 | 20.8 | 24.0 | 1.54 |
| 24 | 20.1 | 0.51 | 404 | 0.000317 | 26.2 | 30.2 | 1.22 |
| 25 | 17.9 | 0.46 | 320 | 0.000252 | 33.0 | 38.1 | 0.970 |
| 26 | 15.9 | 0.41 | 254 | 0.000200 | 41.6 | 48.0 | 0.769 |
| 27 | 14.2 | 0.36 | 202 | 0.000158 | 52.5 | 60.6 | 0.610 |
| 28 | 12.6 | 0.32 | 160 | 0.000126 | 66.2 | 76.4 | 0.484 |
| 29 | 11.3 | 0.29 | 127 | 0.0000995 | 83.4 | 96.3 | 0.384 |
| 30 | 10.0 | 0.26 | 101 | 0.0000789 | 105 | 121 | 0.304 |
| 31 | 8.9 | 0.23 | 79.7 | 0.0000626 | 133 | 153 | 0.241 |
| 32 | 8.0 | 0.20 | 63.2 | 0.0000496 | 167 | 193 | 0.191 |
| 33 | 7.1 | 0.18 | 50.1 | 0.0000394 | 211 | 243 | 0.152 |
| 34 | 6.3 | 0.16 | 39.8 | 0.0000312 | 266 | 307 | 0.120 |
| 35 | 5.6 | 0.14 | 31.5 | 0.0000248 | 335 | 387 | 0.0954 |
| 36 | 5.0 | 0.13 | 25.0 | 0.0000196 | 423 | 488 | 0.0757 |
| 37 | 4.5 | 0.11 | 19.8 | 0.0000156 | 533 | 616 | 0.0600 |
| 38 | 4.0 | 0.10 | 15.7 | 0.0000123 | 673 | 776 | 0.0476 |
| 39 | 3.5 | 0.09 | 12.5 | 0.0000098 | 848 | 979 | 0.0377 |
| 40 | 3.1 | 0.08 | 9.9 | 0.0000078 | 1070 | 1230 | 0.0200 |

**TABLE 17**
Conversion Table for Plane Angles

| | ° | ′ | ″ | rad | rev |
|---|---|---|---|---|---|
| 1 degree = | 1 | 60 | 3600 | $1.75 \times 10^{-2}$ | $2.78 \times 10^{-3}$ |
| 1 minute = | $1.67 \times 10^{-2}$ | 1 | 60 | $2.91 \times 10^{-4}$ | $4.63 \times 10^{-5}$ |
| 1 second = | $2.78 \times 10^{-4}$ | $1.67 \times 10^{-2}$ | 1 | $4.85 \times 10^{-6}$ | $7.72 \times 10^{-7}$ |
| 1 radian = | 57.3 | 3440 | $2.06 \times 10^{5}$ | 1 | 0.159 |
| 1 revolution = | 360 | $2.16 \times 10^{4}$ | $1.30 \times 10^{6}$ | 6.28 or $2\pi$ | 1 |

**TABLE 18**

Formulas from Geometry

---

Plane figures

---

In the following, *a, b, c, d,* and *h* are lengths of sides and altitudes, respectively.

|  |  | *Perimeter* | *Area* |
|---|---|---|---|
| Rectangle | 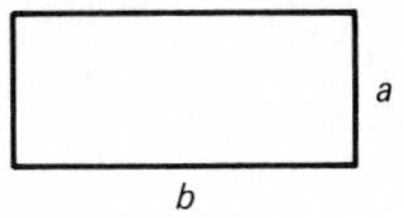 | $P = 2(a + b)$ | $A = ab$ |
| Square | 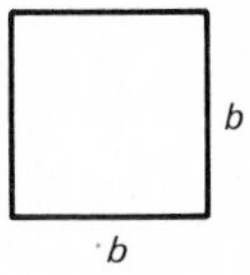 | $P = 4b$ | $A = b^2$ |
| Parallelogram | 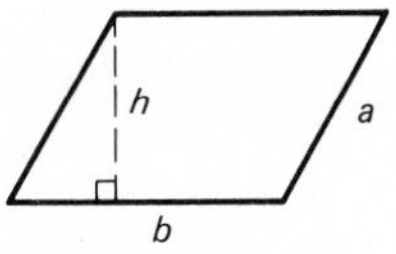 | $P = 2(a + b)$ | $A = bh$ |
| Rhombus | 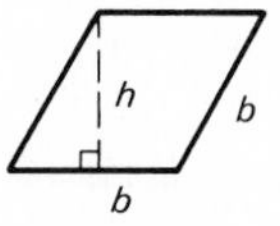 | $P = 4b$ | $A = bh$ |
| Trapezoid | 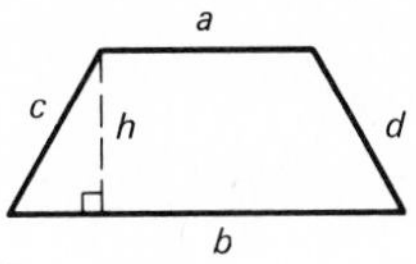 | $P = a + b + c + d$ | $A = \left(\dfrac{a + b}{2}\right)h$ |
| Triangle | 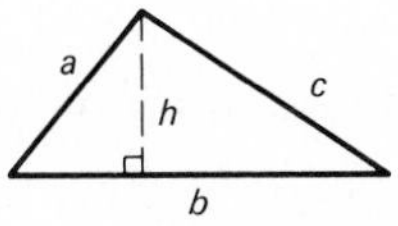 | $P = a + b + c$ | $A = \dfrac{1}{2}bh$ |

The sum of the measures of the angles of a triangle $= 180°$.

In a right triangle:

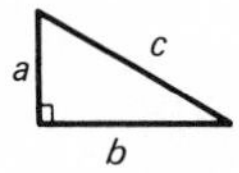

$$c^2 = a^2 + b^2 \quad \text{or} \quad c = \sqrt{a^2 + b^2}$$

|  |  | *Circumference* | *Area* |
|---|---|---|---|
| Circle |  | $C = \pi d$ <br> $C = 2\pi r$ | $A = \pi r^2 \quad d = 2r$ <br> $A = \dfrac{\pi d^2}{4}$ |

The sum of the measures of the central angles of a circle $= 360°$.

In the following, *B*, *r*, and *h* are the area of base, length of radius, and height, respectively.

*Volume*   *Lateral Surface Area*

Prism      $V = Bh$

Cylinder   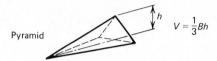   $V = \pi r^2 h$

$V = \dfrac{\pi d^2 h}{4}$   $A = 2\pi rh$

Pyramid      $V = \dfrac{1}{3}Bh$

Cone      $V = \dfrac{1}{3}\pi r^2 h$   $A = \pi rs$, *s* is the slant height

Sphere   $V = \dfrac{4}{3}\pi r^3$   $A = 4\pi r^2$

$V = \dfrac{\pi}{6}d^3$

## TABLE 19
## Formulas from Right-Triangle Trigonometry

$\sin A = \dfrac{\text{side opposite angle } A}{\text{hypotenuse}}$

$\cos A = \dfrac{\text{side adjacent to angle } A}{\text{hypotenuse}}$

$\tan A = \dfrac{\text{side opposite angle } A}{\text{side adjacent to angle } A}$

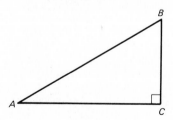

**TABLE 20**
Trigonometric Ratios

| Angle (°) | Sine | Cosine | Tangent | Angle (°) | Sine | Cosine | Tangent |
|---|---|---|---|---|---|---|---|
| 0 | 0.000 | 1.000 | 0.000 | 45 | 0.707 | 0.707 | 1.000 |
| 1 | 0.017 | 0.999 | 0.017 | 46 | 0.719 | 0.695 | 1.036 |
| 2 | 0.035 | 0.999 | 0.035 | 47 | 0.731 | 0.682 | 1.072 |
| 3 | 0.052 | 0.999 | 0.052 | 48 | 0.743 | 0.669 | 1.111 |
| 4 | 0.070 | 0.998 | 0.070 | 49 | 0.755 | 0.656 | 1.150 |
| 5 | 0.087 | 0.996 | 0.087 | 50 | 0.766 | 0.643 | 1.192 |
| 6 | 0.105 | 0.995 | 0.105 | 51 | 0.777 | 0.629 | 1.235 |
| 7 | 0.122 | 0.993 | 0.123 | 52 | 0.788 | 0.616 | 1.280 |
| 8 | 0.139 | 0.990 | 0.141 | 53 | 0.799 | 0.602 | 1.327 |
| 9 | 0.156 | 0.988 | 0.158 | 54 | 0.809 | 0.588 | 1.376 |
| 10 | 0.174 | 0.985 | 0.176 | 55 | 0.819 | 0.574 | 1.428 |
| 11 | 0.191 | 0.982 | 0.194 | 56 | 0.829 | 0.559 | 1.483 |
| 12 | 0.208 | 0.978 | 0.213 | 57 | 0.839 | 0.545 | 1.540 |
| 13 | 0.225 | 0.974 | 0.231 | 58 | 0.848 | 0.530 | 1.600 |
| 14 | 0.242 | 0.970 | 0.249 | 59 | 0.857 | 0.515 | 1.664 |
| 15 | 0.259 | 0.966 | 0.268 | 60 | 0.866 | 0.500 | 1.732 |
| 16 | 0.276 | 0.961 | 0.287 | 61 | 0.875 | 0.485 | 1.804 |
| 17 | 0.292 | 0.956 | 0.306 | 62 | 0.883 | 0.469 | 1.881 |
| 18 | 0.309 | 0.951 | 0.325 | 63 | 0.891 | 0.454 | 1.963 |
| 19 | 0.326 | 0.946 | 0.344 | 64 | 0.899 | 0.438 | 2.050 |
| 20 | 0.342 | 0.940 | 0.364 | 65 | 0.906 | 0.423 | 2.145 |
| 21 | 0.358 | 0.934 | 0.384 | 66 | 0.914 | 0.407 | 2.246 |
| 22 | 0.375 | 0.927 | 0.404 | 67 | 0.921 | 0.391 | 2.356 |
| 23 | 0.391 | 0.921 | 0.424 | 68 | 0.927 | 0.375 | 2.475 |
| 24 | 0.407 | 0.914 | 0.445 | 69 | 0.934 | 0.358 | 2.605 |
| 25 | 0.423 | 0.906 | 0.466 | 70 | 0.940 | 0.342 | 2.747 |
| 26 | 0.438 | 0.899 | 0.488 | 71 | 0.946 | 0.326 | 2.904 |
| 27 | 0.454 | 0.891 | 0.510 | 72 | 0.951 | 0.309 | 3.078 |
| 28 | 0.469 | 0.883 | 0.532 | 73 | 0.956 | 0.292 | 3.271 |
| 29 | 0.485 | 0.875 | 0.554 | 74 | 0.961 | 0.276 | 3.487 |
| 30 | 0.500 | 0.866 | 0.577 | 75 | 0.966 | 0.259 | 3.732 |
| 31 | 0.515 | 0.857 | 0.601 | 76 | 0.970 | 0.242 | 4.011 |
| 32 | 0.530 | 0.848 | 0.625 | 77 | 0.974 | 0.225 | 4.331 |
| 33 | 0.545 | 0.839 | 0.649 | 78 | 0.978 | 0.208 | 4.705 |
| 34 | 0.559 | 0.829 | 0.675 | 79 | 0.982 | 0.191 | 5.145 |
| 35 | 0.574 | 0.819 | 0.700 | 80 | 0.985 | 0.174 | 5.671 |
| 36 | 0.588 | 0.809 | 0.727 | 81 | 0.988 | 0.156 | 6.314 |
| 37 | 0.602 | 0.799 | 0.754 | 82 | 0.990 | 0.139 | 7.115 |
| 38 | 0.616 | 0.788 | 0.781 | 83 | 0.993 | 0.122 | 8.144 |
| 39 | 0.629 | 0.777 | 0.810 | 84 | 0.995 | 0.105 | 9.514 |
| 40 | 0.643 | 0.766 | 0.839 | 85 | 0.996 | 0.087 | 11.43 |
| 41 | 0.656 | 0.755 | 0.869 | 86 | 0.998 | 0.070 | 14.30 |
| 42 | 0.669 | 0.743 | 0.900 | 87 | 0.999 | 0.052 | 19.08 |
| 43 | 0.682 | 0.731 | 0.933 | 88 | 0.999 | 0.035 | 28.64 |
| 44 | 0.695 | 0.719 | 0.966 | 89 | 0.999 | 0.017 | 57.29 |
| 45 | 0.707 | 0.707 | 1.000 | 90 | 1.000 | 0.000 | — |

**TABLE 21**
The Greek Alphabet

| Capital | Lowercase | Name |
|---------|-----------|------|
| A | $\alpha$ | alpha |
| B | $\beta$ | beta |
| Γ | $\gamma$ | gamma |
| Δ | $\delta$ | delta |
| E | $\epsilon$ | epsilon |
| Z | $\zeta$ | zeta |
| H | $\eta$ | eta |
| Θ | $\theta$ | theta |
| I | $\iota$ | iota |
| K | $\kappa$ | kappa |
| Λ | $\lambda$ | lambda |
| M | $\mu$ | mu |
| N | $\nu$ | nu |
| Ξ | $\xi$ | xi |
| O | $o$ | omicron |
| Π | $\pi$ | pi |
| P | $\rho$ | rho |
| Σ | $\sigma$ | sigma |
| T | $\tau$ | tau |
| Y | $\upsilon$ | upsilon |
| Φ | $\phi$ | phi |
| X | $\chi$ | chi |
| Ψ | $\psi$ | psi |
| Ω | $\omega$ | omega |

# Answers to Odd-Numbered Problems

# CHAPTER 1

Page 5

**1.** $3.26 \times 10^2$ **3.** $2.65 \times 10^3$ **5.** $8.264 \times 10^2$ **7.** $4.13 \times 10^{-3}$ **9.** $6.43 \times 10^0$
**11.** $6.5 \times 10^{-5}$ **13.** $5.4 \times 10^5$ **15.** $7.5 \times 10^{-6}$ **17.** $5 \times 10^{-8}$ **19.** $7.32 \times 10^{17}$

Page 6

**1.** 86,200 **3.** 0.000631 **5.** 0.768 **7.** 777,000,000 **9.** 69.3 **11.** 96,100 **13.** 1.4
**15.** 0.0000084 **17.** 700,000,000,000 **19.** 0.00000072 **21.** 4,500,000,000,000
**23.** 0.000000000055

Pages 7–8

**1.** 3 yd; 9 ft; 274.32 cm **3.** 412.16 km **5.** (a) 19.48 ft (b) 2.795 in. (c) 3.048 cm **7.** Smaller **9.** 9

Pages 8–9

**1.** 720 s **3.** 28,800 s **5.** 26,280 h **7.** $2 \times 10^9$ h **9.** 0.024 $\mu$s **11.** 0.075 ms **13.** 18,000 $\mu$s
**15.** 15,000 ns **17.** 375 min

Page 10

**1.** 15,575 N **3.** 890 N **5.** 449.44 lb **7.** 7.5 lb **9.** 35.96 oz **11.** 418.3 N

Pages 12–13

**1.** 40 cm² **3.** 39 in² **5.** 22 in² **7.** 45 ft² **9.** 20.9 m² **11.** 1938 ft² **13.** 0.69 ft² **15.** 7.9 in²
**17.** 9.68 cm² **19.** 25.8 cm²

Pages 15–16

**1.** 72 in³ **3.** 40 cm³ **5.** 36 cm² **7.** 76 cm² **9.** 40 mL **11.** 513 ft³ **13.** 1.77 in³ **15.** 1442 cm³
**17.** 13,824 in³ **19.** 254,900 cm³ **21.** 622.71 cm³

Page 18

**1.** 3 **3.** 4 **5.** 2 **7.** 3 **9.** 5 **11.** 3 **13.** 5 **15.** 4 **17.** 3 **19.** 2 **21.** 1 **23.** 4 **25.** 2
**27.** 3 **29.** 4

**1.** 1 V **3.** 1 m **5.** 0.0001 in. **7.** 10 km **9.** 0.01 m **11.** 0.00001 in. **13.** 0.01 m **15.** 1 kg
**17.** 0.0001 in. **19.** 1000 Ω **21.** 1 A **23.** 0.01 m **25.** 100 kg **27.** 0.000001 A or $1 \times 10^{-6}$ A
**29.** 10,000 V or $1 \times 10^4$ V **31.** (a) 15.7 in. (b) 0.018 in. **33.** (a) 16.01 cm (b) 0.734 cm **35.** (a) 0.0350 A (b) 0.00040 A
**37.** (a) 27,0̄00 L (b) 4.75 L **39.** (a) All have one significant digit (b) 50 Ω **41.** (a) 0.05 in. (b) 16.4 in. **43.** (a) 0.65 m (b) 27.5 m
**45.** (a) 0.00005 g (b) 0.75 g **47.** (a) 3 V (b) 45,000 V **49.** (a) 20 Ω (b) 40̄0,000 Ω

**1.** 14,100 ft **3.** 83.3 cm **5.** 7̄0,000 V **7.** 803 m or 80,300 cm **9.** 18 A **11.** 500 kg **13.** 41.0 g
**15.** 3200 km **17.** 900,000 V **19.** 0.40 m or 4̄0 cm **21.** 4900 m² **23.** 1,4̄00,000 km² or $1.40 \times 10^6$ km² **25.** 738 m²
**27.** 5560 cm³ **29.** $2.91 \times 10^7$ in³ **31.** 3̄0 ft **33.** 3.06 cm **35.** 75 km/h **37.** 1̄000 V/A **39.** 1100 ft lb/s
**41.** 370 V/A **43.** 43.2 m **45.** 530 V²/Ω **47.** 4530 kg m/s² **49.** 10,300 m³

## CHAPTER 2

**1.** $s = vt$ **3.** $m = \dfrac{w}{g}$ **5.** $g = \dfrac{PE}{mh}$ **7.** $m = \dfrac{2\,(KE)}{v^2}$ **9.** $t = \dfrac{w}{P}$ **11.** $s = \dfrac{W}{F}$ **13.** $I = \dfrac{V - E}{-r}$ or $I = \dfrac{E - V}{r}$

**15.** $v_i = 2v_{\text{avg}} - v_f$ **17.** $s = \dfrac{v^2 - v_i^2 + 2as_i}{2a}$ **19.** $R = \dfrac{QJ}{I^2 t}$ **21.** $r = \sqrt{\dfrac{A}{\pi}}$

**1.** 162 ft² **3.** 7.50 cm **5.** 6.0 cm **7.** 10.9 yd **9.** 33.2 km **11.** 6.11 m **13.** 154 m² **15.** 21.6 in.
**17.** 121.6 m **19.** 122.4 ft²

**1.** 25,900 cm³ **3.** 284 cm³ **5.** 102.1 cm **7.** 10,100 ft³ **9.** 864 ft³ **11.** 12.0 cm² **13.** 1.58 cm²
**15.** 36.0 m **17.** 9.39 m **19.** 137 m **21.** 65.5 ft **23.** 60 panels **25.** 22.9 m³ **27.** 4.44 yd³

## CHAPTER 3

**1.** 3.0 **3.** 1.4 **5.** 3.6 **7.** **9.** **11.**

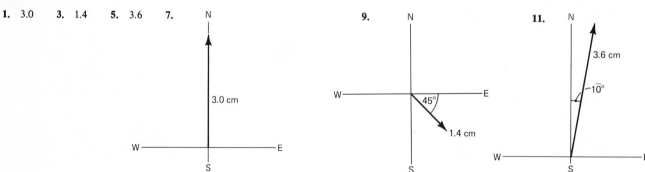

**13.** 2.0 **15.** 2.8 **17.** 6.3 **19.** **21.** **23.**

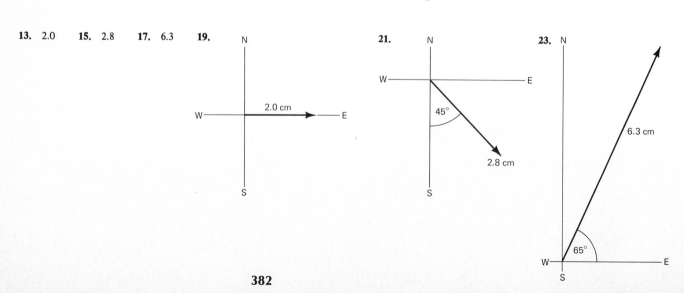

**25.** $1\frac{1}{4}$    **27.** $2\frac{5}{8}$    **29.** $\frac{15}{16}$    **31.**

N

$1\frac{1}{4}$ in.

W— ←————————— —E

S

**33.**

N

W ——————————— E
          45°

$2\frac{5}{8}$ in.

S

**35.**

$\frac{15}{16}$ in.

N
    25°

W ——————————— E

S

---

Page 43

**1.** 61 km at 55° north of east    **3.** 1300 mi at 1° west of south    **5.** 36 km at 5° east of north    **7.** 38 km at 25° north of west
**9.** 1500 mi at 71° north of east    **11.** 120 km at 72° south of east    **13.** 47 mi at 49° north of east

Page 45

**1.** $8\bar{0}$    **3.** $9\bar{0}$    **5.** $5\bar{0}$    **7.** 21.6    **9.** $12\bar{0}$    **11.** 25 m    **13.** 25    **15.** 79.2    **17.** 72.2
**19.** $8\bar{0}$ km/h, east    **21.** 125 mi/h, south    **23.** 61.1 km/h at $3\bar{0}°$ south of east

Page 48

**1.** $1\bar{0}$ ft/s²    **3.** $1\bar{0}$ ft/s²    **5.** 3.2 m/s²    **7.** 6.25 m/s²    **9.** 0.206 m/s²    **11.** 67.5 mi/h    **13.** 720 km/h    **15.** 16.7 s

Pages 51–52

**1.** 5.05 ft/s    **3.** 127.8 ft    **5.** 2.1 ft/s²    **7.** 9.00 mi/h    **9.** 550 ft    **11.** 1.1 m/s²    **13.** 43.2 m/s
**15.** −1.6 m/s²    **17.** (a) 23.5 m/s (b) 28.2 m    **19.** (a) 3190 m (b) 25.5 s (c) 51.0 s

## CHAPTER 4

Pages 58–59

**1.** 30.0    **3.** 744    **5.** 252    **7.** 1.71    **9.** 11.7    **11.** 0.518    **13.** 5250 N    **15.** 1320 lb    **17.** $40\bar{0}$ ft/s²
**19.** (a) 14.0 ft/s² (b) 10.6 ft/s²

Page 62

**1.** 380 N    **3.** 1100 N    **5.** 0.080

Page 64

**1.** 3.0; right    **3.** 15.0; left    **5.** 4; left    **7.** 4.00 ft/s²    **9.** 0.509 m/s²

Page 66

**1.** 322    **3.** 1.73    **5.** 0.652    **7.** 77.2    **9.** 14,700 N    **11.** 2450 N

Pages 71–72

**1.** 80.0    **3.** 765    **5.** $9.5 \times 10^8$    **7.** $6.89 \times 10^8$    **9.** (a) 12,600 slug ft/s (b) 173 km/h (c) $58\bar{0}0$ lb; 2580 lb
**11.** (a) 55,200 kg m/s (b) 47.2 m/s    **13.** $2.89 \times 10^{-3}$ s    **15.** 34.4 kg m/s

## CHAPTER 5

Page 75

**1.** *a*    **3.** *c*    **5.** *a*    **7.** *B*    **9.** *B*

Pages 78–79

**1.** 0.9455 **3.** 1.804 **5.** 0.9799 **7.** 0.6477 **9.** 0.3065 **11.** 0.4617 **13.** 16°
**15.** 43° **17.** 48° **19.** 36.6° **21.** 46.5° **23.** 30.0° **25.** 22.28° **27.** 16.75°
**29.** 35.50° **31.** $B = 65.0°$; $a = 8.45$ m; $b = 18.1$ m **33.** $A = 47.7°$; $B = 42.3°$; $a = 12.4$ km
**35.** $A = 24.4°$; $B = 65.6°$; $c = 24.2$ mi **37.** $B = 70°$; $b = 24$ m; $c = 25$ m **39.** $A = 49.35°$; $a = 17.98$ cm; $b = 15.44$ cm
**41.** (a) 10.0° (b) 2.12 cm (c) 8.24 cm **43.** $C = 2.7$ in.; $D = 2.3$ in.


Pages 80–81

**1.** $b = 8.49$ cm **3.** $c = 21.6$ mi **5.** $a = 10.2$ ft **7.** $c = 24.8$ cm **9.** $a = 8.60$ m **11.** $B = 8.00$ cm; $C = 16.1$ cm


Pages 84–85

**1.**
x comp: −4
y comp: −4

**3.**
x comp: +5
y comp: +4

**5.**
x comp: −4
y comp: −8

**7.**
x comp: −7
y comp: +4

**9.**
x comp: +3
y comp: +9

**11.**
x comp: +9
y comp: +5

**13.**
x comp: −10
y comp: +10

**15.**
x comp: +4
y comp: +12

**17.**
x comp: +8
y comp: −4

**19.**
x comp: −6.5
y comp: 0


Pages 92–94

**1.**
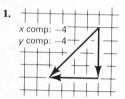
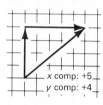
v = 20 m at 25°

**3.**
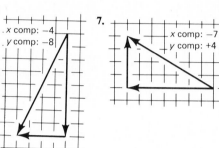
u = 25 m at 245°

**5.**
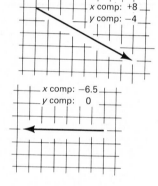
t = 15 m at 105°

**7.**
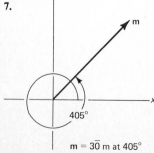
m = 30 m at 405°

| | x component | y component |
|---|---|---|
| **9.** | 9.96 m | 8.97 m |
| **11.** | −18.2 km | −45.1 km |
| **13.** | 97.4 km/h | −14.4 km/h |
| **15.** | 38.2 m | 7.09 m |
| **17.** | 6.17 km | −7.35 km |
| **19.** | −5.88 m/s | 28.9 m/s |

21. 10.0 m at 36.9°    23. 26.2 mi at 315.0°    25. 9.70 m/s at 98.3°    27. 53.3 m at 291.5°    29. 22.3 mi at 48.8°
31. 10.6 m/s at 155.7°    33. 37$\overline{0}$ km/h at 90.0°    35. 239 km/h at 190.8°    37. 127 mi/h at 239.5°    39. 226 km/h at 162.3°

## CHAPTER 6

Page 97

1. 10$\overline{0}$ lb    3. 319 N    5. 26$\overline{0}$ N    7. 69$\overline{00}$ N    9. 57$\overline{0}$ N    11. Yes

Pages 101–103

1. $F_1 = 70.7$ N; $F_2 = 70.7$ N    3. $F_1 = 823$ N; $F_2 = 475$ N    5. $F_1 = 577$ lb; $F_2 = 289$ lb    7. $F_1 = 433$ lb; $F_2 = 50\overline{0}$ lb
9. $T_1 = 1440$ lb; $T_2 = 1440$ lb    11. $C = 30\overline{00}$ lb; $T = 26\overline{00}$ lb    13. 5540 N    15. $T = 2320$ lb; $C = 3670$ lb

## CHAPTER 7

Pages 106–107

1. 34.3 ft lb    3. 0.455 N    5. 21,600 ft lb    7. 4410 J    9. 0.300 N    11. 24,100 J

Pages 110–111

1. 18.9 ft lb/s    3. 5.00 W    5. 12.4 ft lb/s    7. (a) 1.68 hp (b) 219 N    9. 59.4 s    11. 1.49 kW    13. 6.17 kW
15. (a) 6.9 kW (b) 9.2 hp (c) 11 kW

Page 114

1. 8080 ft lb    3. 217 J    5. (a) 80.7 ft/s (b) $3.09 \times 10^6$ ft lb    7. 4.48 ft/s    9. (a) 13.2 ft lb (b) 21.3 ft lb
11. 2650 kW    13. $1.59 \times 10^8$ J or 159 MJ

Page 116

1. 27.8 ft/s    3. 72.7 m/s

## CHAPTER 8

Pages 122–123

1. 14.8    3. 36.3    5. 52.4    7. 2.39    9. 48.8    11. 1.55    13. 2.27    15. 41.1    17. 4.00
19. 2.50    21. 4.00    23. 0.500

Page 124

1. 14.0    3. 271    5. 524    7. 48.8    9. 20.4    11. 438 lb    13. 429 N    15. 2010 N    17. 3.34

Pages 127–128

1. 1    3. 3    5. 6    7. 2    9.    11.    13.    15. 2.00    17. 30.0 ft    19. 82.0 m
21. 10$\overline{0}$ lb; 12$\overline{0}$ ft

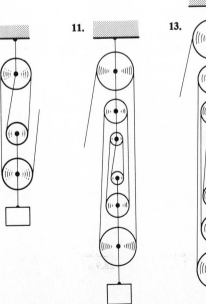

Page 130

**1.** 12.8 **3.** 21.2 **5.** 36.3 **7.** 4.62 **9.** 5.61 **11.** 4.00 **13.** No **15.** 3.84 **17.** 1.33 m

Pages 132–133

**1.** 2.30 **3.** 14.2 **5.** 2.28 **7.** 29.3 **9.** 35.2 **11.** 4.84 lb **13.** 2.23 cm **15.** 37.7 **17.** 188 lb
**19.** 33,500 N

Pages 133–134

**1.** 15.0 **3.** 40.0 **5.** 75.0 lb **7.** 125 N **9.** $1.80 \times 10^5$ N

## CHAPTER 9

Pages 139–140

**1.** (a) 40.8 rad (b) 2340° **3.** (a) 12.5 rev (b) $45\overline{0}0°$ **5.** 154 rpm **7.** 660 rpm **9.** 26.5 rev/s **11.** 140 rev/s
**13.** 7.78 rad/s **15.** 5730 rpm **17.** (a) 6.67 rev/s (b) $40\overline{0}$ rpm (c) 41.9 rad/s **19.** 8.40 rad **21.** 0.131 m
**23.** (a) 230 rad/s (b) 920 rad (c) 460 m/s **25.** (a) 0.105 rad/s (b) $1.75 \times 10^{-3}$ rad/s (c) $1.45 \times 10^{-4}$ rad/s **27.** $1.68 \times 10^3$ km/h

Page 141

**1.** 96.0 ft lb **3.** $1.49 \times 10^3$ lb **5.** 187 N **7.** $1.60 \times 10^3$ N **9.** 0.571 ft **11.** 159 lb

Page 143

**1.** $4.35 \times 10^3$ N **3.** 79.4 slugs **5.** 5.53 m/s **7.** 1.92 m **9.** $3.60 \times 10^3$ lb **11.** 48.0 ft **13.** 5430 N

Page 146

**1.** 18.7 hp **3.** 7280 ft lb/s **5.** $1.86 \times 10^3$ N m **7.** 343 hp **9.** 4.95/s **11.** $1.72 \times 10^5$ ft lb/s
**13.** (a) 299 hp (b) 599 hp **15.** 2.65 N m **17.** 74.4 kJ/min

## CHAPTER 10

Pages 150–151

**1.** 73.5 **3.** 200 **5.** 858 **7.** 13 **9.** 167 **11.** 44.9 rpm **13.** 75 teeth **15.** 75 teeth **17.** 42 teeth
**19.** 130 teeth **21.** 69 teeth

Pages 154–157

**1.** Counterclockwise **3.** Clockwise **5.** Clockwise **7.** Clockwise **9.** Clockwise **11.** $116\overline{0}$ rpm **13.** 576 rpm
**15.** 1480 rpm **17.** 40 teeth **19.** 20 teeth

Pages 158–159

**1.** 1930 **3.** $36\overline{0}0$ **5.** 147 **7.** 71.4 rpm **9.** 23.2 in **11.** Clockwise **13.** Counterclockwise **15.** Counterclockwise

## CHAPTER 11

Page 162

**1.** $10\overline{0}$ lb **3.** $50\overline{0}$ N **5.** $20\overline{0}$ lb **7.** $40\overline{0}$ N

Page 165

**1.** $F_1 = 103$ lb; $F_2 = 61.9$ lb **3.** $F_1 = 2.76 \times 10^4$ N; $F_2 = 1.07 \times 10^4$ N

Pages 166–167

**1.** $F_1 = 22.6$ **3.** $F_w = 42.0$ **5.** $F_1 = 1.52 \times 10^4$ N; $F_2 = 9.85 \times 10^3$ N **7.** $F_1 = 2.37 \times 10^4$ lb; $F_2 = 1.83 \times 10^4$ lb
**9.** $F_1 = 22.5$ lb; $F_2 = 20.5$ lb **11.** $F_1 = 1.03 \times 10^4$ N; $F_2 = 9.17 \times 10^3$ N **13.** $F_1 = 197$ N; $F_2 = 118$ N
**15.** $F_1 = 244$ lb; $F_2 = 96.3$ lb

# CHAPTER 12

Page 177

**1.** 49.9 in.   **3.** 48.9 N   **5.** 20$\overline{0}$ N/m   **7.** (a) 2.50 × 10⁻² cm (b) 1.56 × 10⁶ N   **9.** 3.58 × 10⁵ lb   **11.** 6.67 N
**13.** 1.09 × 10⁸ Pa

Pages 184–185

**1.** 2.80 × 10³ kg/m³   **3.** 1750 lb   **5.** 5.42 × 10⁻³ m³   **7.** 1.20 × 10³ lb/ft³   **9.** 2.70 × 10³ kg/m³   **11.** 680 kg/m³
**13.** 55.2 gal   **15.** 58.8 lb/ft³   **17.** 1.49 m³   **19.** 2.82 × 10³ kg/m³   **21.** (a) 1.00 × 10³ L (b) 1.47 × 10³ L (c) 73.5 L

*CHAPTER* **13**

Page 191

**1.** 2.5 m × 0.80 m:70.5 kPa; 2.5 m × 0.45 m:125 kPa; 0.80 m × 0.45 m:392 kPa   **3.** 21.7 lb/in²   **5.** 16.5 lb/ft³
**7.** 1.96 × 10³ lb   **9.** 1.97 in.   **11.** 245 kPa   **13.** 9.38 kPa   **15.** 1.08 × 10⁸ N

Pages 193–194

**1.** 60.0 lb   **3.** 12.5   **5.** 6.67   **7.** 33.3 N   **9.** 36$\overline{0}$0 lb   **11.** 18.0 kN

Page 196

**1.** 116.4 lb/in²   **3.** 84.7 lb/in²   **5.** 30.3 lb/in²

Page 198

**1.** 13.0 lb   **3.** 3.9 N   **5.** 3.20 N   **7.** 3.06 × 10⁴ ft³; 795 tons

Page 202

**1.** 340 L/min   **3.** (a) 9.59 cm (b) 0.340 m/s   **5.** (a) 34.0 ft/min (b) 136 ft/min

*CHAPTER* **14**

Pages 206–207

**1.** 21°C   **3.** 121°C   **5.** 257°F   **7.** −12.2°C   **9.** 41.0°F   **11.** 61$\overline{0}$°R   **13.** 14$\overline{0}$°R   **15.** 223 K   **17.** 5730°C
**19.** 899°C

Page 208

**1.** 23 cal   **3.** 1.17 × 10⁶ ft lb   **5.** 5.14 Btu

Page 212

**1.** 0.021 ft² °F h/Btu   **3.** 0.45 ft² °F h/Btu   **5.** 1.9 ft² °F h/Btu   **7.** 8.1 × 10⁶ Btu   **9.** 2.5 cal

Pages 213–214

**1.** 173   **3.** 38$\overline{0}$0   **5.** 51,0$\overline{0}$0   **7.** 278   **9.** 7.2   **11.** 64   **13.** 39,100 Btu   **15.** 2.79 × 10⁴ kcal   **17.** 1.89 × 10⁵ J

Pages 215—216

**1.** 428°F   **3.** 0.051 cal/g °C   **5.** 94°F   **7.** 8$\overline{0}$°C   **9.** 0.104 cal/g °C

*CHAPTER* **15**

Page 220

**1.** 0.30 ft   **3.** 0.10 m   **5.** 200.10 m   **7.** 0.54 ft   **9.** 0.751 in.

Page 221

**1.** 0.33 in²   **3.** 89.0 cm²   **5.** 60.1 cm³

Page 222

**1.** 653 L **3.** 299 ft³ **5.** 3754 ft³ **7.** 0.58 cm³ **9.** $215

## CHAPTER 16

Page 225–226

**1.** 288 K **3.** 44°C **5.** 532°R **7.** $9\overline{0}$°F **9.** 222 cm³ **11.** −16°F **13.** $38\overline{0}$ m³ **15.** 1380 L

Page 228

**1.** 252 cm³ **3.** 75.0 kPa **5.** 2.08 kg/m³ **7.** 0.180 lb/ft³ **9.** 90.4 kPa **11.** 3.33 kg/m³ **13.** 801 kPa
**15.** 19.0 ft³

Pages 229–230

**1.** 1280 in³ **3.** 506 m³ **5.** −39.3°C **7.** 143°C **9.** 2200 psi **11.** 399 kPa

## CHAPTER 17

Page 237

**1.** 1120 cal **3.** 10,700 Btu **5.** 25,600 cal **7.** 6070 Btu **9.** 11,840 Btu **11.** $6.70 \times 10^6$ J **13.** $5\overline{0}00$ kcal
**15.** 3.39 MJ **17.** 3.12 MJ

## CHAPTER 18

Page 241

**1.** $2.00 \times 10^{-3}$ s **3.** $80\overline{0}$ m/s **5.** $1.67 \times 10^{-15}$ s

## CHAPTER 19

Page 246

**1.** 325 m/s **3.** 317 m/s **5.** 5.00 s **7.** 2640 m

Page 248

**1.** 622 Hz **3.** 6200 Hz

## CHAPTER 21

Page 258

**1.** 1.07 Ω **3.** 0.0165 Ω/ft **5.** 1.04 Ω **7.** 0.684 Ω **9.** 0.0131 cm²

## CHAPTER 22

Page 260

**1.** 9.58 A **3.** $22\overline{0}$ V **5.** 153 Ω

Pages 264–265

**1.** 14 Ω **3.** 60.0 Ω **5.** 0.750 A **7.** 378 V **9.** 23.0 Ω

Pages 269–270

**1.** 3.44 Ω **3.** (a) 2.00 A (b) 1.25 A (c) 5.00 A **5.** 1.21 Ω **7.** (a) 1.67 A (b) 0.250 A **9.** (a) 30.3 V (b) 30.3 V

Pages 273–274

**1.** (a) R₂, R₃ (b) 3.00 Ω **3.** 8.89 A **5.** (a) 2.22 A (b) 6.67 A **7.** 21.2 Ω **9.** 44.6 V **11.** 5.42 A **13.** 10.0 Ω
**15.** 17.3 Ω **17.** 30.0 V **19.** 6.93 V **21.** ~~17.3 Ω~~ **23.** ~~60.0 V~~ **25.** ~~0.400 A~~

13.3Ω    60.0V    .400A

| | | V | I | R |
|---|---|---|---|---|
| **1.** | Batt. | 12.0 V | 9.00 A | 1.33 Ω |
| | $R_1$ | 12.0 V | 6.00 A | 2.00 Ω |
| | $R_2$ | 12.0 V | 3.00 A | 4.00 Ω |

| | | V | I | R |
|---|---|---|---|---|
| **3.** | Batt. | 36.0 V | 6.00 A | 6.00 Ω |
| | $R_1$ | 36.0 V | 2.00 A | 18.0 Ω |
| | $R_2$ | 36.0 V | 3.00 A | 12.0 Ω |
| | $R_3$ | 36.0 V | 1.00 A | 36.0 Ω |

| | | V | I | R |
|---|---|---|---|---|
| **5.** | Batt. | 50.0 V | 5.00 A | 10.0 Ω |
| | $R_1$ | 25.0 V | 2.00 A | 12.5 Ω |
| | $R_2$ | 25.0 V | 2.00 A | 12.5 Ω |
| | $R_3$ | 10.0 V | 3.00 A | 3.33 Ω |
| | $R_4$ | 40.0 V | 3.00 A | 13.3 Ω |

| | | V | I | R |
|---|---|---|---|---|
| **7.** | Batt. | 36.0 V | 9.00 A | 4.00 Ω |
| | $R_1$ | 12.0 V | 6.00 A | 2.00 Ω |
| | $R_2$ | 12.0 V | 3.00 A | 4.00 Ω |
| | $R_3$ | 24.0 V | 6.00 A | 4.00 Ω |
| | $R_4$ | 24.0 V | 3.00 A | 8.00 Ω |

| | | V | I | R |
|---|---|---|---|---|
| **9.** | Batt. | 80.0 V | 12.0 A | 6.67 Ω |
| | $R_1$ | 18.0 V | 4.00 A | 4.50 Ω |
| | $R_2$ | 18.0 V | 2.00 A | 9.00 Ω |
| | $R_3$ | 18.0 V | 6.00 A | 3.00 Ω |
| | $R_4$ | 48.0 V | 12.0 A | 4.00 Ω |
| | $R_5$ | 8.00 V | 4.00 A | 2.00 Ω |
| | $R_6$ | 8.00 V | 8.00 A | 1.00 Ω |
| | $R_7$ | 6.00 V | 12.0 A | 0.500 Ω |

| | | V | I | R |
|---|---|---|---|---|
| **11.** | Batt. | 65.0 V | 5.00 A | 13.0 Ω |
| | $R_1$ | 10.0 V | 0.500 A | 20.0 Ω |
| | $R_2$ | 10.0 V | 1.00 A | 10.0 Ω |
| | $R_3$ | 10.0 V | 2.50 A | 4.00 Ω |
| | $R_4$ | 10.0 V | 1.00 A | 10.0 Ω |
| | $R_5$ | 25.0 V | 5.00 A | 5.00 Ω |
| | $R_6$ | 30.0 V | 5.00 A | 6.00 Ω |

## CHAPTER 23

Pages 280–281

**1.** 1.49 V    **3.** 1.33 A    **5.** (a) 3.75 A (b) 9.00 V (c) 0.0200 Ω    **7.** 0.160 A    **9.** 0.120 A

Page 283

**1.** 957 W    **3.** 0.455 A    **5.** 6.82 A    **7.** $0.34    **9.** Yes    **11.** $0.017

## CHAPTER 24

Page 288

**1.** $3.00 \times 10^{-5}$ T    **3.** $4\bar{0}$ A

Page 289

**1.** $1.96 \times 10^{-2}$ T    **3.** 0.249 A

## CHAPTER 26

Page 298

**1.** 29.6 V    **3.** 113 V

Page 299

**1.** 1560 V    **3.** 117 V    **5.** 12.0 A

Page 301

**1.** 21.9 Ω    **3.** 3500 W or 3.5 kW    **5.** 1320 W or 1.32 kW    **7.** 588 W

Page 302

**1.** 12,600 kW    **3.** 130,000 kVA

## CHAPTER 27

Page 308

**1.** 10.0 turns    **3.** 58.5 V    **5.** 350 turns    **7.** 4.67 A    **9.** 0.0733 A    **11.** 2200 V    **13.** 0.255 A

## CHAPTER 28

Page 311

**1.** 1.89    **3.** 43.0    **5.** 251    **7.** 2.39    **9.** 0.153

Page 313

**1.** 221    **3.** 3300    **5.** 304    **7.** 19°    **9.** 48°    **11.** 0.226 A    **13.** 0.00455 A    **15.** 0.0263 A

Page 314

**1.** 3.98    **3.** 2.65    **5.** 0.00199

Page 316

**1.** 1280    **3.** 8.64    **5.** 33.3    **7.** 14°    **9.** 8°    **11.** 0.000781 A    **13.** 1.74 A    **15.** 0.450 A

Page 318

**1.** $Z = 141\ \Omega$; $I = 0.0355$ A    **3.** $Z = 636\ \Omega$; $I = 0.0236$ A

Page 320

**1.** $2.25 \times 10^4$/s    **3.** 37.5/s

## CHAPTER 29

Page 324

**1.** $1.5 \times 10^9$ m    **3.** 0.108 s    **5.** $50\overline{0}$ s    **7.** 16.1 ms    **9.** $7.67 \times 10^{-7}$ s

Page 325

**1.** $7.50 \times 10^{12}$ Hz    **3.** $3.09 \times 10^{-4}$ m    **5.** 214 m

Page 326

**1.** $6.62 \times 10^{-24}$ J    **3.** $6.62 \times 10^{-26}$ J

Page 327

**1.** 603 lm    **3.** 43.2 cd    **5.** 942 lm

Pages 328–329

**1.** 1.46 ft-cd    **3.** 3.51 ft-cd    **5.** $4.75 \times 10^3$ lm    **7.** 608 lm    **9.** (a) Decrease (b) $\frac{1}{9}$

## CHAPTER 30

Page 337

**1.** 1.50 cm    **3.** 26.7 cm    **5.** 4.95 cm    **7.** 6.24 cm    **9.** 1.09 m    **11.** 10.0 cm

## CHAPTER 31

Page 345

**1.** 1.2    **3.** $2.00 \times 10^8$ m/s    **5.** 1.5    **7.** 21.8 cm    **9.** −2.14 cm    **11.** (a) 5.00 cm (b) 2.5 cm

## APPENDIX A

Page 349

**1.** −11    **3.** 5    **5.** −2    **7.** −4    **9.** −6    **11.** −3    **13.** 18    **15.** −21    **17.** 0    **19.** 3    **21.** −8
**23.** 0    **25.** 11    **27.** −13    **29.** −5    **31.** −8    **33.** −4    **35.** 10    **37.** −45    **39.** −288

**1.** $10^8$   **3.** $10^8$   **5.** $\frac{1}{10^2}$   **7.** $10^7$   **9.** $10^3$   **11.** $\frac{1}{10^4}$   **13.** 1   **15.** $10^5$   **17.** $\frac{1}{10^{11}}$

**19.** $\frac{1}{10^6}$   **21.** $10^8$   **23.** $10^{10}$

**1.** $\frac{4}{3}$   **3.** 17   **5.** 0   **7.** 17.5   **9.** 3   **11.** 4   **13.** 7   **15.** 12.5   **17.** 5   **19.** 26.5

**1.** 9   **3.** 3   **5.** 8   **7.** 1   **9.** $-\frac{11}{3}$ or $-3\frac{2}{3}$   **11.** 3   **13.** 2   **15.** 21   **17.** $\frac{29}{2}$ or $14\frac{1}{2}$   **19.** $-6$

**1.** $\pm 6$   **3.** $\pm 7$   **5.** $\pm 3$   **7.** $\pm 0.8$   **9.** $\pm 4.67$   **11.** $\pm 25.5$

**1.** $a=3$; $b=1$; $c=-5$   **3.** $a=6$; $b=8$; $c=2$   **5.** $a=9$; $b=6$; $c=-4$   **7.** $a=5$; $b=6$; $c=0$   **9.** $a=9$; $b=0$;
$c=-64$   **11.** 3; 7   **13.** $\frac{4}{3}$; $-\frac{5}{2}$   **15.** 1.95; $-1.62$   **17.** 1.69; $-0.855$   **19.** 0.725; $-1.23$

## APPENDIX  B

**1.** $3.32 \times 10^{19}$   **3.** $-6.83 \times 10^{-6}$   **5.** $8.35 \times 10^{-11}$   **7.** $-7.98 \times 10^{19}$   **9.** $-68.5$   **11.** 4680   **13.** 216
**15.** $2.10 \times 10^{10}$   **17.** $2.15 \times 10^9$   **19.** 10.7   **21.** \$36.29   **23.** $-2.23$   **25.** 1510   **27.** 0.567   **29.** 152,000
**31.** 237,000   **33.** 0.225   **35.** 1.29   **37.** 0.510   **39.** 0.743   **41.** 3.73   **43.** 0.151
**45.** 40.7°   **47.** 24.6°   **49.** 56.2°   **51.** 8.4°   **53.** 63.3°   **55.** 66.6°   **57.** 252 m   **59.** 89.4 m
**61.** 53.7°   **63.** 14.8 cm   **65.** 21.9 m   **67.** 20,700   **69.** 0.0206   **71.** $1.86 \times 10^{-6}$

# Index

# A

# B

# C

# O

Ohm, unit of electrical resistance, 257
Ohmmeter, 274
Ohm's law, 259

# P

Parallel connection, of cells, 279
    of resistances, 265
Parallel forces, 160
Pascal's principle, 191
Phase relations in ac circuits, 312
Photon, 323
Potential difference, 255
Potential energy, 111
Pound, 9
Power:
    in ac circuits, 300
    defined, 107
    electric, 281
    rotary, 143
    units of, 108
Precision, 18
Pressure:
    absolute, 195
    air, 194
    gauge, 195
    liquid, 186
    transmission in liquids, 188
    units of, 186
Problem solving, 25
Pulleys and pulley systems, 124
Pythagorean theorem, 79

# R

Radian, 135
Rankine scale of temperature, 206
RCL circuits, 317
Reactance, 309
Reaction, 66
Rectification, 320
Reflection, 330
Refraction, 338
Resistance, electric, 257
Resistivity, electrical, 257
Resistors, 262
Resolution of vectors, 82

Resonance:
    electrical, 318
    sound, 248
Resultant vector, 82
Rotary motion, 135
$R$ value, 211

# S

Scalar, 38
Scientific notation, 4
Screw, 130
Series circuit, 262
Significant digits, 16
SI metric units, 2
Simple machines, 117
Sine, 75
Solids, 170
Sound, 244
Specific gravity, 198
Specific heat, defined, 212
Spectrum, electromagnetic, 324
Speed, defined, 43
    of light, 323
    of sound, 245
Static electricity, 250
Step-up transformer, 305
Storage cell, 277
Superposition, 241

# T

Tangent, 75
Temperature, 203
Tensile strength, 170
Tension, 98
Thermal conductivity, 211
Thermometers, 205
Time, measurement of, 8
Torque, defined, 140
Total internal reflection, 341
Transformers, 303
Trigonometry, 142

# U

Uniform circular motion, 142
Units, conversion of, 3